This book is devoted to the physics of electronic fluctuations (noise) in solids, and covers almost all important examples of the phenomenon. It is comprehensive, intelligible and well illustrated. Emphasis is given to main concepts, to physical mechanisms of fluctuations, and to conclusions on the nature and magnitude of noise.

In the Introduction, the fundamental concepts and methods of the theory of random processes are expounded. A detailed derivation of the fluctuation–dissipation relationships which, for equilibrium systems, relate the noise to the dissipative part of the response to external perturbation is presented. Where possible, both rigorous and less rigorous but instructive derivations are given. In the chapter on noise in nonequilibrium gases (for example, hot electrons in semiconductors) first a simple model is discussed which yields a qualitative understanding of the basic physics. This is then followed by a general and original approach which has been used by many authors to solve noise problems. Generation–recombination noise and the interplay between this noise and diffusion noise are briefly discussed. The chapter on noise in ballistic systems is, in fact, the first review of this modern development of fluctuation physics. The basic mechanisms of resistance (modulation) noise, used in the interpretation of experiments, are discussed. Noise in strongly disordered conductors is always abnormally high, and the main concepts and models of noise in disordered conductors are also covered. Qualitative and, for model systems, quantitative explanation of the high noise magnitude are presented. The book includes also the most comprehensive and complete review of flicker $(1/f)$ noise. Answers to many problems posed by this type of noise are given, and these answers are substantiated by numerous experiments. The final chapter is devoted to fluctuations in superconductive systems, namely, in Josephson junctions and type II superconductors, including the high-temperature ones. Throughout, the book is illustrated by experiments, many of which are fundamental and have become classics. The book also contains an extensive bibliography.

The book will be useful to graduate students and researchers in physics and electronic engineering, and especially to those carrying out research in the fields of noise phenomena and highly sensitive electronic devices, for example, radiation detectors, electronic devices for low-noise amplifiers, quantum magnetometers (SQUIDs), and frequency standards.

ELECTRONIC NOISE AND FLUCTUATIONS IN SOLIDS

ELECTRONIC NOISE AND FLUCTUATIONS IN SOLIDS

Sh. KOGAN

Institute of Radioengineering and Electronics, Russian Academy of Sciences
and
Los Alamos National Laboratory, USA

CAMBRIDGE UNIVERSITY PRESS
Cambridge, New York, Melbourne, Madrid, Cape Town, Singapore, São Paulo

Cambridge University Press
The Edinburgh Building, Cambridge CB2 8RU, UK

Published in the United States of America by Cambridge University Press, New York

www.cambridge.org
Information on this title: www.cambridge.org/9780521460347

First published 1996
This digitally printed version 2008

A catalogue record for this publication is available from the British Library

Library of Congress Cataloguing in Publication data

Kogan, Sh.
Electronic noise and fluctuations in solids / Sh. Kogan.
p.   cm.
Includes bibliographical references and index.
ISBN 0 521 46034 4 (hc)
1. Electronic noise. 2. Solids–Fluctuations. I. Title
QC176.8.E4K864   1996
621.382'24–dc20   95-38725   CIP

ISBN 978-0-521-46034-7 hardback
ISBN 978-0-521-07019-5 paperback

To the memory of my father, M.F. Kogan (1898–1948), who not only gave me life but also, foreseeing the Holocaust, saved our family from becoming its victims.

Contents

Preface

Science is built up of facts,
as a house is built up of stones;
but an accumulation of facts is no more a science
than a heap of stones is a house.

Henri Poincaré,
Science and Hypothesis, 1905, Chapter 9.

This book is an introduction to the physics of electronic noise and fluctuation phenomena in solids. It is written for physicists and electrical engineers interested in investigation of electric and magnetic noise and in development of sensitive solid-state devices. It can be used also in graduate schools of the Departments of Physics and Electrical Engineering.

Fluctuations or noise are spontaneous random (stochastic) variations of physical quantities in time or, more precisely, random deviations of these quantities from some mean values that are either constant or vary nonrandomly in time. Fluctuations are a manifestation of the thermal motion of matter and discreteness of its structure. The introduction of the concept of fluctuations and development of the physics of fluctuations is one of the greatest achievments of twentieth-century physics. The theory of Brownian motion developed by A. Einstein and by M. von Smoluchowski in the first decade of this century and its experimental proof by J.B. Perrin and T. Svedberg was a strong and, perhaps, the final argument in favor of the molecular–kinetic concept of heat. Hence, the physics of fluctuations is of great conceptual importance.

The applications of the physics of fluctuations stem from the fact that the ultimate accuracy of measurement of any physical quantity is limited just by fluctuations of this quantity, and the ultimate sensitivity of many devices is also limited by fluctuations. The spontaneous random fluctuations in electron devices are usually called *noise*, and both terms, fluctuations and noise, are used interchangeably. Not all sources of noise are inherent in the main kinetic process in a given device and irreducible. Some noise sources depend on the technology used in manufacturing the device and, in principle, by improving the technology, can be eliminated. The physics of fluctuations is, therefore, especially important for development of highly sensitive devices. Since the noise level in electron devices is sensitive to their quality, noise

measurements are used as a nondestructive test of quality and reliability of
these devices.

The physics of fluctuations is a part of physical kinetics which studies
the variations of physical quantities, occuring spontaneously or induced by
external fields. The investigation of fluctuation phenomena, which may be
called 'fluctuation spectroscopy', is a very informative method for the study
of kinetic processes in matter. Often, it is also a much more sensitive
method than the measurement of mean quantities. The role of the noise
spectrum analyzer, which also includes an amplifier, is similar to the role
of a microscope: it enables one to visualize the microscopic motion and
transitions of particles.

In equilibrium systems both kinetic properties, i.e., the properties of fluctuations, on one hand, and the response of the same system to small
external nonrandom perturbations, on the other hand, are related by very
general fluctuation–dissipation relations. Both kinds of kinetic properties,
fluctuations and response to deterministic perturbations, being controlled
by the same kinetic (relaxation) processes, are often related qualitatively
even in nonequilibrium systems. However, in nonequilibrium systems the
fluctuation–dissipation relations do not hold, and the information obtained
from measurements of the noise characteristics is independent of and comple-
mentary to the information obtained from the measurement of the response
to external perturbations. Moreover, some noise mechanisms, e.g., producing
fluctuations of resistance, can not be revealed in equilibrium noise. However,
even at small mean currents and voltages, when the deviation from equilibrium is in all other respects insignificant, the nonequilibrium excess noise
produced by resistance fluctuations is usually much higher than the equilibrium noise. Therefore, measurements of the spectral density of fluctuations
are used in many systems as a very sensitive method for studies of various
kinetic properties of solids.

Noise is a stochastic (random) process. The random character of the
fluctuations is a consequence of the fact that the thermal motion and quantum transition of particles (electrons and atoms) in solids are random. The
fluctuations are random even in the case of so called 'random telegraph
noise', which may be caused, for instance, by repeated random capture of
one electron into a single trap and emission of this electron from the trap
(Sec. 8.8).

Random (stochastic) motions are usually opposed to the deterministic
ones. However, in the last two decades, many dynamical systems, even with
few degrees of freedom, have been found to have extremely intricate, chaotic,
and unstable trajectories of motion which, nevertheless, may be viewed as

deterministic, i.e., they may be, in principle, exactly calculated beforehand for any given initial conditions by integrating the dynamical differential equations of motion. The mechanisms that lead to deterministic chaos are beyond the scope of this book, and this type of random processes is not discussed below. Only processes that can be called 'stochastic noise' are presented (we are using the terminology of the review by Gaspard & Wang (1993), in which the difference between deterministic chaos and stochastic noise is analyzed). Experimental studies of fluctuations are usually performed on systems that are either in equilibrium or are not too far from equilibrium, e.g., are not driven by a too strong external pumping that may result in dynamical chaotic behavior. In this book fluctuations in such systems only are discussed.

The reader of this book may ask a very delicate question: is there such a field of physics as the physics of fluctuation phenomena? Indeed, the physics of fluctuation phenomena in semiconductors, for instance, is quite different from the physics of these phenomena in superconductors. Is there something that is common for noise phenomena in different solids and justifies writing this book? Some arguments in favor of a positive answer may be given.

1. The fundamental concepts and methods of the theory of fluctuations are very general and can be applied to all physical systems. Such are the concepts of Markov processes, the Langevin approach, the theory of processes similar to shot noise, and the theory of noise enhancement in strongly disordered solids. In other words, many profound physical concepts are common for a great variety of fluctuation processes.
2. There are general relations, the fluctuation–dissipation relations, which can be applied to any physical system in equilibrium and to any kind of noise in these systems.
3. So called $1/f$ noise is found in almost all physical systems (Ch. 8). Despite the fact that no universal mechanism of this noise exists or even can exist, this kind of noise stems from some very *general* properties of disordered solids. Successful interpretation of the mechanism of this noise in one type of physical system helps to achieve such understanding in other, even completely different, physical systems.

A great number of books and reviews on fluctuations, or noise, have been published during the last several decades. However, the physics of fluctuation phenomena in solids is a rapidly developing field. During the last decade new phenomena and new features of known phenomena have been discovered and studied: noise in ballistic conductors, noise in mesoscopic conductors, noise in strongly disordered conductors, noise in high-T_c superconductors, magnetic $1/f$ noise in spin glasses and amorphous ferromagnets, random telegraph noise, i.e., resistance fluctuations generated by *individual* electron traps of atomic dimensions, or by hops of *individual* defects in metals.

The primary goal of this book is to present the *modern status of the field*. The main emphasis has been given to *the understanding of the physics of fluctuation phenomena*. Less attention has been given to mathematical methods of noise theory and to experimental methods of noise measurement. The author has tried to present simple qualitative explanations of all fluctuation phenomena discussed in this book, to use the simplest methods of theoretical description, and to avoid too lengthy derivations of equations and methods that are beyond the courses taught in graduate schools. The author hopes that the book will be useful for university students.

The main emphasis in this book is on the *theory* of the fluctuation phenomena and, perhaps, less on the *description and discussion of experiments*. This is a common shortcoming of many modern books on physics, even text-books on general physics for undergraduate students. Possibly, it is a result of the conciseness of the language of theoretical physics.

This book is far from complete. Many interesting fluctuation phenomena are not presented here. The main drawback is the absence of fluctuation phenomena in electron devices, photoconductors, photodetectors, and some other solids. The Barkhausen noise in ferromagnets is also not discussed. The critical fluctuations in the vicinity of second-order phase transitions are beyond the scope of this book. It was also impossible to cite all authors, even of very useful papers. I apologize to them.

The reader is assumed to have a general background in physics and mathematics. Some chapters require knowledge of quantum mechanics, including its Heisenberg picture, and statistical physics within the university courses. More special information is presented in the introductions to some chapters. The most used mathematics is Fourier transform (Fourier expansion) and matrix algebra.

The list of references and recommended books and reviews is given at the end of the book. Those which are of general importance, i.e., are not restricted to one narrow field, are presented at the beginning of the list of references. The equations are numbered within each section and subsection. Figures and tables are numbered in each chapter separately.

I wish to express gratitude to Dr. A.Ya. Shulman, Dr. V.D. Shadrin, and Dr. K.E. Nagaev of the Institute of Radioengineering and Electronics, Russian Academy of Sciences, Moscow (IRE RAS) for collaboration in this field, and to those scientists who carefully read some parts of the manuscript and made valuable comments: Prof. M.B. Weissman of the University of Illinois at Urbana-Champaign, Prof. K.K. Likharev of the State University of New York at Stony Brook, Prof. R. Katilius of the Semiconductor Physics Institute, Vilnius, Lithuania, Prof. B.K. Jones of the Lancaster University, United Kingdom, Prof. D.E. Khmelnitskii of the Cavendish Laboratory, Cambridge, UK and the L.D. Landau Institute for Theoretical Physics, Moscow, Dr. K.E. Nagaev of IRE RAS, and Prof. Shechao Feng of the University of California at Los Angeles. I would like to thank Prof. P. Mazetti for supplying me with reprints of papers by Italian authors on noise in amorphous ferromagnets. I acknowledge the significant role of Mr. R. Cooper, Editor at the Institute of Physics, Bristol, UK and Dr. S. Capelin of the Cambridge University Press who suggested writing a book on fluctuation phenomena. I am especially grateful to Mrs. Frances Nex of the Cambridge University Press who carefully edited this book. I would like to acknowledge the great help of Mr. V.L. Berman who made many of the figures.

This book was written partly in IRE RAS in Moscow, and in the Los Alamos National Laboratory (LANL), New Mexico, USA. Many people whose interests are far from the problems discussed in this book indirectly aided in writing it. I am thankful to Dr. D.L. Smith, Dr. D.E. Watkins, Dr. A. Redondo, Dr. T. Springer, Dr. I.H. Campbell, Dr. C. Heller, Dr. B. Laurich, Dr. P. Davids, and Dr. P. McDougall of the Materials Science & Technology Division of LANL for continuous aid. I highly appreciate friendly relations with A. and R. Browning, S. and M. Klein, S. and B. Krohn, J. and F. March, T.V. Grigor'eva and G.P. Berman, K. Milder, J. Samuels, and many other people in Los Alamos.

My family inspired and encouraged me to write this book, and I am very grateful for its support.

Abbreviations

2DEG	two-dimensional electron gas
BLE	Boltzmann–Langevin equation
CDW	charge density wave
CVC	current–voltage characteristic
FDR	fluctuation–dissipation relation
FDT	fluctuation–dissipation theorem
GMR	giant magnetoresistance
G–R	generation–recombination
IC	infinite cluster
JFET	junction field-effect transistor
MIM	metal–insulator–metal
MIS	metal–insulator–semiconductor
MOS	metal–oxide–semiconductor
MOSFET	metal–oxide–semiconductor field-effect transistor
NDR	negative differential resistance
QPC	quantum point contact
RKKY	Ruderman–Kittel–Kasuya–Yosida
RSJ	resistively shunted junction
RTN	random telegraph noise
S–c–S	superconductor–constriction–superconductor
SQUID	superconductive quantum interference device
TLF	two-level fluctuator
TLTS	two-level tunneling system

1
Introduction: Some basic concepts of the theory of random processes

Fluctuations in solids are typical random processes. By definition, a random process is a random function $x(t)$ of an independent variable t, which, in the theory of fluctuations, is the time. The theory of fluctuations is based on the highly developed mathematical theory of random processes. This chapter is an introduction to the basic concepts and methods of this theory which are used throughout this book. The list of recommended books and reviews is presented at the end of the book.

1.1 Probability density functions; moments; stationary processes

Let $P\{\ldots\}$ be the probability of an event indicated in the curly brackets. The distribution functions of the first, second, and higher order for a given random process $x(t)$ are defined as

$$W_1(x_1, t_1) = P\{x(t_1) \le x_1\}$$
$$W_2(x_1, t_1; x_2, t_2) = P\{x(t_1) \le x_1; x(t_2) \le x_2)\} \quad (1.1.1)$$
$$W_n(x_1, t_1; \ldots; x_n, t_n) = P\{x(t_1) \le x_1; \ldots; x(t_n) \le x_n\},$$

where x_i is the value of the random quantity at instant t_i.

For instance, $W_2(x_1, t_1; x_2, t_2)$ is the probability that the random quantity x at instant t_1 is less than a given value x_1, and at instant t_2 is less than x_2.

If $W_n(x_1, t_1; \ldots; x_n, t_n)$ are differentiable functions of the variables $x_1, \ldots, x_n$, one may introduce the corresponding probability density functions:

$$w_1(x_1, t_1) = \frac{\partial W_1(x_1, t_1)}{\partial x_1},$$
$$w_2(x_1, t_1; x_2, t_2) = \frac{\partial^2 W_2(x_1, t_1; x_2, t_2)}{\partial x_1 \partial x_2}. \quad (1.1.2)$$

The functions w_n are called n-dimensional probability density functions

1

of the random quantity $x(t)$. They may be defined in a different way. For instance, the two-dimensional probability density function,

$$w_2(x_1, t_1; x_2, t_2)dx_1dx_2 = P\{x_1 \leq x(t_1) < x_1 + dx_1; x_2 \leq x(t_2) < x_2 + dx_2\}.$$
$$(1.1.3)$$

The probability density functions must be consistent with each other, that is, each k-dimensional density function can be obtained from any n-dimensional function with $n > k$ by integration over 'extra' variables:

$$w_k(x_1, t_1; \ldots; x_k, t_k) = \int dx_{k+1} \ldots dx_n w_n(x_1, t_1; \ldots; x_n, t_n). \qquad (1.1.4)$$

Random processes are called stationary if all distributions $w_n(x_1, t_1; \ldots; x_n, t_n)$ remain invariable under any identical shift of all time points $t_1, \ldots, t_n$. It means that for a stationary process the probability density function $w_1(x_1, t_1)$ does not depend at all on the time t_1, the probability density function $w_2(x_1, t_1; x_2, t_2)$ depends only on the difference $t_1 - t_2$, that is, may be written as $w_2(x_1, x_2, t_1 - t_2)$, and so on.

The mean value of the random variable $x(t)$ is equal to (for the methods of averaging random quantities see below)

$$\langle x(t) \rangle = \int_{-\infty}^{+\infty} x(t)dW(x) = \int_{-\infty}^{+\infty} dx\, x w_1(x, t). \qquad (1.1.5)$$

The r-th order central moment is the average value of the random quantity $(\delta x(t))^r$, where $\delta x(t) = x(t) - \langle x \rangle$ is the deviation of the random quantity $x(t)$ from its mean value $\langle x \rangle$, i.e., the fluctuation:

$$\langle [\delta x(t)]^r \rangle = \int_{-\infty}^{+\infty} dx(\delta x)^r w_1(x, t). \qquad (1.1.6)$$

The second-order central moment is the variance, i.e., the mean value of the fluctuation squared.

If the probability density $w_n(x_1, t_1; \ldots; x_n, t_n)$ is known for different instants $t_1, \ldots, t_n$, one can calculate the correlation function:

$$\langle \delta x(t_1) \ldots \delta x(t_n) \rangle = \int dx_1\, dx_2 \ldots dx_n\, \delta x_1 \cdots \delta x_n\, w_n(x_1, t_1; \ldots; x_n, t_n). \quad (1.1.7)$$

The characteristic function of a random quantity is defined as the mean value of the function $\exp(iux)$, where x is the random variable and u is a real parameter:

$$\phi_x(u) \equiv \langle e^{iux} \rangle = \int_{-\infty}^{+\infty} e^{iux}dW(x). \qquad (1.1.8)$$

If the distribution function $W(x)$ is differentiable (see above), the characteristic function is simply the Fourier transform of the probability density function, Eq. (1.1.2):

$$\phi_x(u) = \int_{-\infty}^{+\infty} dx e^{iux} w(x). \tag{1.1.9}$$

The characteristic function has a very useful property: if the random variable X is a sum of N *independent* random variables, i.e.,

$$X = \sum_{i=1}^{N} x_i, \tag{1.1.10}$$

then the characteristic function of X is simply the product of the characteristic functions of the variables x_i :

$$\phi_X(u) = \phi_{x_1}(u) \cdot \phi_{x_2}(u) \cdots \phi_{x_N}(u). \tag{1.1.11}$$

Therefore, in many problems in which the random variable is given by Eq. (1.1.10) it is more convenient to calculate $\phi_X(u)$ first and then the density function $w(X)$ using the inverse Fourier transformation (Eq. (1.1.9)):

$$w(x) = \int_{-\infty}^{+\infty} \frac{du}{2\pi} \phi_x(u) e^{-iux}. \tag{1.1.12}$$

In the theory of fluctuations an important distribution is the normal (Gaussian) distribution. It arises when the random quantity $x(t)$ is a sum of many ($N \gg 1$) independent and identically distributed random quantities. An example is the noise generated in a macroscopic specimen by the random motion of a great number of independent but identical defects, or spins. Let $\xi_1, \ldots, \xi_N$ be independent and identically distributed random quantities and let $x = \xi_1 + \cdots + \xi_N$. If the summands $\xi_1, \ldots, \xi_N$ are small enough and their number N is great enough, that is, if as $N \to \infty$ the mean value of $x(t)$ is equal to $\langle x \rangle$, and the variance of $x(t)$ is equal to σ^2, the one-dimensional probability density function is

$$w_1(x) = \frac{1}{\sqrt{2\pi\sigma^2}} \exp\left[-\frac{(\delta x)^2}{2\sigma^2}\right], \tag{1.1.13}$$

where $\delta x = x - \langle x \rangle$ is the fluctuation. Such a distribution is called normal or Gaussian.

A random process is called Gaussian if *all* its probability density functions are normal for all $n = 1, \ldots$. The n-dimensional normal distribution reads:

$$w_n(x_1, \ldots, x_n) = \frac{1}{\sqrt{(2\pi)^n \det \hat{\lambda}}} \exp\left[-\frac{1}{2}\sum_{i=1}^{n}\sum_{j=1}^{n} \lambda_{ij}^{-1} \delta x_i \delta x_j\right]. \tag{1.1.14}$$

Here $\delta x_i = x_i - \langle x_i \rangle$. The matrix $\hat{\lambda}$ is called the covariance matrix. Its elements equal:

$$\lambda_{ij} = \langle \delta x_i \delta x_j \rangle = \lambda_{ji}. \tag{1.1.15}$$

In Eq. (1.1.14) $\det \hat{\lambda}$ is the determinant of the matrix $\hat{\lambda}$, λ_{ij}^{-1} are the elements of the matrix $\hat{\lambda}^{-1}$ which is the reciprocal of $\hat{\lambda}$.

In particular, the two-dimensional normal distribution equals:

$$w_2(x_1, x_2) = \frac{1}{2\pi\sigma_1\sigma_2\sqrt{1-\rho_{12}^2}} \exp\left\{-\frac{1}{2(1-\rho_{12}^2)}\left[\frac{(\delta x_1)^2}{\sigma_1^2} + \frac{(\delta x_2)^2}{\sigma_2^2}\right.\right.$$
$$\left.\left. - 2\rho_{12}\frac{\delta x_1}{\sigma_1}\frac{\delta x_2}{\sigma_2}\right]\right\}. \tag{1.1.16}$$

Here $\sigma_i^2 = \langle(\delta x_i)^2\rangle$ are the variances, $\rho_{12} = \langle \delta x_1 \delta x_2 \rangle / \sigma_1 \sigma_2$ is the correlation coefficient.

Equations (1.1.7) and (1.1.14) imply that for Gaussian random processes all nonzero n-th order ($n > 2$) moments can be expressed in terms of the second-order moments, i.e., the covariances λ_{ij} (pair correlations). In other words, the measurement of higher-order correlations can not give any new information apart from that which is contained in the pair correlation (for the consequences of this property of Gaussian random processes see Sec. 1.2).

Many random processes are similar to electron emission from the cathode of a thermionic tube. The instants of emission of individual electrons are random because the emission events are uncorrelated. Such random processes are called Poissonian. The hops of identical noninteracting defects in a solid are another example of such processes. Let n_1 be the mean rate of individual events, that is, the mean number of events per unit time. In the case of electron emission, n_1 is the mean number of electrons emitted per unit time. The mean number of such events during a definite time t_m is obviously $\bar{n} = n_1 t_m$. Of course, the actual number n of events that happen to occur during the time t_m is a random quantity which fluctuates around its mean value $\bar{n}$. The problem is to find the probability $P(n)$ of exactly n individual events occurring in the time $t_m = \bar{n}/n_1$.

Let us divide the time t_m into N equal intervals. N is taken so large or, equivalently, the intervals t_m/N are taken so small that occurence of more than one event in one interval is improbable. For each small interval two alternatives are possible: either one individual event or no event may occur. The corresponding probabilities are $\bar{n}/N$ and $1 - \bar{n}/N$, respectively. The probability of each definite sequence of n 'full' and $N - n$ 'empty' intervals (one event or no event occured, respectively) equals $(\bar{n}/N)^n(1 - \bar{n}/N)^{N-n}$.

The total number of different sequences (arrangements) of n 'full' and $N - n$ 'empty' intervals is the number of combinations $N!/n!(N - n)!$. Thus the probability of n events occurring equals

$$\frac{N!}{n!(N - n)!}\left(\frac{\bar{n}}{N}\right)^n\left(1 - \frac{\bar{n}}{N}\right)^{N-n}$$
$$= \frac{\bar{n}^n}{n!}\left(1 - \frac{\bar{n}}{N}\right)^N \cdot 1 \cdot \left(1 - \frac{1}{N}\right)\cdots\left(1 - \frac{n-1}{N}\right)\left(1 - \frac{\bar{n}}{N}\right)^{-n}.$$

The sought probability, which is called the Poisson distribution, follows from the last equation in the limit $N \to \infty$ at finite n. Since $\lim_{N\to\infty}(1 - \bar{n}/N)^N = \exp(-\bar{n})$,

$$P(n) = \frac{\bar{n}^n}{n!}e^{-\bar{n}}. \tag{1.1.17}$$

The statistical characteristics of a random process are nonrandom quantities, which can be, in principle, measured by a corresponding averaging of the random quantity (or quantities). There are two possible procedures of averaging.

1. Averaging over time, i.e., over a sufficiently long record of the random process. In this case, for example, the distribution function $W_1(x_1)$ of a stationary process is equal to that part of the total time t_m of measurement of this process during which the random quantity $x < x_1$. The mean value of the random quantity is then

$$\bar{x} = \lim_{t_m \to \infty} \frac{1}{t_m}\int_{-t_m/2}^{t_m/2} dt\, x(t). \tag{1.1.18}$$

2. Averaging over an ensemble of a large number N of identical systems, i.e., systems in which identical conditions for the random process and identical means of its measurement are provided. In this case the distribution function, $W_1(x_1, t_1)$, is equal to the fraction of the total number of ensemble systems in which at instant t_1 the random quantity $x < x_1$. The mean value in this case (denoted by angle brackets) equals:

$$\langle x \rangle = \lim_{N \to \infty} \frac{1}{N}\sum_{i=1}^{N} x_i(t). \tag{1.1.19}$$

Here $x_i(t)$ is the value of the random quantity at the instant t in the i-th system of the ensemble.

One may ask: does a definite mean value exist if one increases the time t_m of averaging or the number N of identical systems of the ensemble? This question is fundamental. It is often reduced to the following one: do both averaging procedures (see above) yield identical mean values? If the answers are positive, the random process is called ergodic. Ergodicity and nonergodicity are discussed in Sec. 1.4.

1.2 The correlation function

The correlation function is one of the most important characteristics of any random process. Let $x(t)$ be a quantity that randomly varies in time t and let $\langle x \rangle$ be its mean value. The variation in time of the fluctuation, i.e., the deviation of the random quantity from its mean value, $\delta x(t) = x(t) - \langle x \rangle$, is, of course, also random. The correlation function is a nonrandom characteristic of the kinetics of these random fluctuations: it shows how the fluctuations evolve in time *on average*.

Consider an ensemble of a large number N of systems in which identical conditions for the random process under consideration have been created. Let $\delta x_i(t_1)$ and $\delta x_i(t_2)$ be the fluctuations in the i-th system of the ensemble $(i = 1, \ldots, N)$ at instants t_1 and t_2, respectively. The correlation function is defined then by the following equation:

$$\psi_x(t_1, t_2) \equiv \langle \delta x(t_1) \delta x(t_2) \rangle = \langle x(t_1) x(t_2) \rangle - \langle x(t_1) \rangle \langle x(t_2) \rangle$$

$$= \lim_{N \to \infty} \frac{1}{N} \sum_{i=1}^{N} \delta x_i(t_1) \delta x_i(t_2). \tag{1.2.1}$$

Here and below the angle brackets denote averaging over an ensemble of identical systems.

In a stationary system the correlation function remains invariable if both instants, t_1 and t_2, are shifted identically. In such systems ψ_x depends only on the difference $t_1 - t_2$ (Sec. 1.1).

The correlation function may be determined also by averaging over a sufficiently long record of the random process $x(t)$ in one system, i.e., by averaging over a long enough time of measurement t_m:

$$\psi_x(t_1 - t_2) \equiv \overline{\delta x(t_1) \delta x(t_2)} = \lim_{t_m \to \infty} \frac{1}{t_m} \int_{-t_m/2}^{t_m/2} dt\, \delta x(t_1 + t) \delta x(t_2 + t). \tag{1.2.2}$$

Both correlation functions (1.2.1) and (1.2.2) coincide if the system is ergodic (see the discussion of ergodicity and nonergodicity of random systems in Sec. 1.4).

According to Eqs. (1.2.1) or (1.2.2), at $t_1 = t_2$ the correlation function is simply the variance, i.e., the average fluctuation squared, $\psi_x(t, t) = \langle (\delta x)^2 \rangle$. As $|t_1 - t_2| \to \infty$ the correlation function in stable systems falls off to zero. Let the fluctuation in the i-th system of the ensemble at instant t_1 be $\delta x_i(t_1)$. If the instant $t_2 > t_1$ but is close enough to t_1, the fluctuation in each system of the ensemble has no time to change its value appreciably, and the fluctuation $\delta x_i(t_2)$ most probably has the same sign as $\delta x_i(t_1)$. Therefore, the product $\delta x_i(t_1)\delta x_i(t_2) > 0$ in almost all systems of the ensemble. As the instants t_1 and t_2 become farther apart, the number of those systems in which the sign of $\delta x_i(t_2)$ is opposite to the sign of $\delta x_i(t_1)$ approaches the number of systems in which these signs coincide. Thus, in the sum of Eq. (1.2.1) besides positive terms negative ones also arise. At large enough $|t_1 - t_2|$ positive and negative summands are equally probable and $\psi_x(t_1 - t_2) \to 0$.

The decay of $\psi_x(t_1 - t_2)$ to zero as $|t_1 - t_2| \to \infty$ may be viewed as if the system 'forgets' its 'initial' fluctuation. In each physical system there is a characteristic time of 'forgetting' the initial fluctuation. This time is usually called the relaxation time. In real systems a number of kinetic processes (they may even be coupled) and, consequently, a number of relaxation times exist. Moreover, in many complex systems a continuous spectrum of such times even exists.

The correlation function can be written in terms of the two-dimensional probability density (Eq. 1.1.7):

$$\psi_x(t_1, t_2) = \int dx_1\, dx_2\, \delta x_1\, \delta x_2\, w_2(x_1, t_1; x_2, t_2). \tag{1.2.3}$$

Here $\delta x_1 = x_1 - \langle x(t_1)\rangle$, $\delta x_2 = x_2 - \langle x(t_2)\rangle$.

The two-dimensional probability density may be represented in the form of the product:

$$w_2(x_1, t_1; x_2, t_2) = w_1(x_1, t_1)P(x_2, t_2|x_1, t_1). \tag{1.2.4}$$

Here $w_1(x, t)$ is the one-dimensional probability density of the random quantity $x(t)$ at the instant t, $P(x_2, t_2|x_1, t_1)dx_2$ is the conditional probability of the quantity x at time t_2 being in the interval between x_2 and $x_2 + dx_2$, if in the previous instant t_1 the value of x was x_1. Let us denote by

$$\langle \delta x(t_2)|\delta x_1, t_1\rangle = \int dx_2 \delta x_2 P(\delta x_2, t_2|\delta x_1, t_1) \tag{1.2.5}$$

the mean value of fluctuation δx at the instant t_2 under the condition that at a previous instant t_1 the value of the random quantity was x_1. In these notations

$$\psi_x(t_1, t_2) = \int dx_1 \delta x_1 w_1(\delta x_1, t_1)\langle \delta x(t_2)|\delta x_1, t_1\rangle. \tag{1.2.6}$$

The quantity $\langle \delta x(t)|\delta x_0, t_0\rangle$ describes the average evolution (usually, a decay) of the fluctuation, the initial value of which at the instant t_0 is δx_0. Note that the absolute value of a fluctuation, for a given initial value, may either decrease or increase in time. However, *on average* it usually (but not always) decreases for any given initial value. The correlation function is the value of $\delta x_1 \langle \delta x(t_2) | \delta x_1, t_1 \rangle$ averaged over the 'initial' values δx_1.

Let us select those systems of the ensemble in which the fluctuation at an instant t_1 is close to a definite value δx_1. After t_1, eventually the fluctuations in different systems (of the initially selected ones) will be quite different (they are randomized) and their distribution will be the same as that of the fluctuations in the entire ensemble. However, the mean value of the fluctuation for the entire ensemble is zero. Thus, the conditional average value of the fluctuation tends to zero:

$$\lim_{(t_2 - t_1) \to \infty} \langle \delta x(t_2) | \delta x_1, t_1 \rangle = 0.$$

In the case of a Gaussian random process the conditional mean fluctuation $\langle \delta x(t) | \delta x_1, 0 \rangle$ and the correlation function are connected by a simple equation. Substitution of $P(\delta x_2, t_2 | \delta x_1, t_1) = w_2(\delta x_1, t_1; \delta x_2, t_2)/w_1(x_1, t_1)$ for a Gaussian process (Eq. (1.1.16)) into Eq. (1.2.5) yields after integration over δx_2:

$$\frac{\langle \delta x(t) | \delta x_0, 0 \rangle}{\delta x_0} = \frac{\psi_x(t)}{\langle (\delta x)^2 \rangle}. \tag{1.2.7}$$

This relationship means that, for a Gaussian process, the time-dependence of the conditional mean fluctuation is the same as that of the correlation function at *any* initial fluctuation δx_0. If $\psi_x(t)$ is a monotonically decreasing function of $|t|$, the fluctuation $|\delta x(t)|$ at $t > 0$ *on average* decreases at any, arbitrarily small, $|\delta x(0)|$. Of course, this behavior takes place only on average, the absolute value of a random fluctuation may grow and exceed its value at a preceding instant of time.

Several random quantities $x_\alpha(t)$ $(\alpha = 1, \ldots, M)$ are often coupled (correlated) with one another. The correlation function is then a matrix:

$$\psi_{\alpha\beta}(t_1, t_2) = \langle \delta x_\alpha(t_1) \delta x_\beta(t_2) \rangle. \tag{1.2.8}$$

Any function with $\alpha = \beta$ is called an auto-correlation function, those with $\alpha \neq \beta$ are the cross-correlation functions of different physical quantities.

The above definition of the correlation function is meaningful only if the quantum-mechanical effects are insignificant and the random quantity $x(t)$ may be considered as a classical quantity (c-number). In quantum mechanics, a quantum-mechanical operator $\hat{x}$ is associated with any physical quantity x. Several representations (pictures) of quantum mechanics are known. They are, of course, equivalent. In the Schrödinger representation the wave functions of the system's states depend on time but the operators do not.

In the Heisenberg representation, on the contrary, the wave functions (state vectors) are independent of time, and the entire dependence on time t is transferred to the operators $\hat{x}(t)$, which obey the Heisenberg equation of motion:

$$\frac{\partial \hat{x}(t)}{\partial t} = \frac{i}{\hbar}(\hat{H}\hat{x} - \hat{x}\hat{H}). \qquad (1.2.9)$$

Here, $\hat{H}$ is the operator of the system's energy (Hamiltonian) which is also expressed in terms of Heisenberg operators depending on the same time t, $\hbar = h/2\pi$, h is Planck's constant.

The correlation function $\psi_x(t_1, t_2)$ is more conveniently written in terms of Heisenberg operators of the fluctuating quantities. Obviously, it is expressed in terms of the product of two Heisenberg operators taken at instants t_1 and t_2, i.e., $\hat{x}(t_1)\hat{x}(t_2)$. However, the correlation function is an observable (measurable) physical quantity, and its values must be real. According to general rules of quantum mechanics, the corresponding operator of the correlation function must be Hermitian. The symmetrical combination,

$$\frac{1}{2}\{\hat{x}(t_1), \hat{x}(t_2)\} \equiv \frac{1}{2}(\hat{x}(t_1)\hat{x}(t_2) + \hat{x}(t_2)\hat{x}(t_1)), \qquad (1.2.10)$$

meets this condition (Ekstein & Rostoker, 1955).

As in classical systems, where the correlation function is an average of the product $x(t_1)x(t_2)$, for quantum-mechanical systems it is equal to the expectation (mean value) of the operator given by Eq. (1.2.10). In general, the state of a quantum-mechanical system is determined by a density matrix $\hat{\rho}$, not by a wave function as a system in a 'pure' quantum-mechanical state. The expectation of a physical quantity, the operator of which is $\hat{x}$, equals $\langle \hat{x}(t) \rangle = \mathrm{Tr}\,(\rho \hat{x}(t))$. Here, Tr denotes the trace (sum of the diagonal matrix elements) of the operator in brackets. In these notations, the correlation function may be written in the form:

$$\psi_x(t_1, t_2) = \frac{1}{2}\langle\{\hat{x}(t_1), \hat{x}(t_2)\}\rangle. \qquad (1.2.11)$$

It follows from the very definition of $\psi_x(t_1, t_2)$ that it is a symmetrical function of the instants t_1 and t_2, i.e.,

$$\psi_x(t_1, t_2) = \psi_x(t_2, t_1). \qquad (1.2.12)$$

If the system is a stationary one, the correlation function is a function of the difference $t_1 - t_2$ only (see above). According to Eq. (1.2.12), it is an even function of this difference.

In a more general case when there are several coupled (correlated) quantities x_α ($\alpha = 1, \ldots, M$), the correlation functions form a matrix (compare with

Eqs. (1.2.8) and (1.2.11)):

$$\psi_{\alpha\beta}(t_1, t_2) = \frac{1}{2}\langle(\hat{x}_\alpha(t_1)\hat{x}_\beta(t_2) + \hat{x}_\beta(t_2)\hat{x}_\alpha(t_1))\rangle. \tag{1.2.13}$$

Its elements obviously satisfy the relation

$$\psi_{\alpha\beta}(t_1, t_2) = \psi_{\beta\alpha}(t_2, t_1), \tag{1.2.14}$$

i.e., in a stationary system

$$\psi_{\alpha\beta}(t_1 - t_2) = \psi_{\beta\alpha}(t_2 - t_1). \tag{1.2.15}$$

There is another, more profound and less obvious, relationship between the correlation functions $\psi_{\alpha\beta}(t_1, t_2)$. It follows, as does the Onsager's principle of symmetry of kinetic coefficients (Onsager, 1931), from the symmetry of the equations of motion of any system of particles under the time reversal $t \rightarrow -t$. If the system under consideration is placed in a magnetic field **B**, this symmetry takes place if simultaneously the magnetic field is also reversed, i.e., **B** replaced by $-$**B**. This relationship, when applied to stationary systems, takes the form:

$$\psi_{\alpha\beta}(t_1 - t_2; \mathbf{B}) = \pm\psi_{\alpha\beta}(t_2 - t_1; -\mathbf{B}). \tag{1.2.16}$$

The plus sign corresponds to the case when the physical quantities $x_\alpha(t)$ and $x_\beta(t)$ are either both invariant under reversal of time t, or both change their signs simultaneously. An example of a quantity that changes its sign under time inversion is the velocity of particles. The minus sign corresponds to the case when only one of the two quantities, x_α or x_β, changes its sign.

Obviously, in the absence of magnetic field (**B** $= 0$) Eq. (1.2.16) becomes simpler. The correlation functions then are either strictly even or strictly odd functions of the difference $t_1 - t_2$.

Applying Eq. (1.2.16) to small times $t \equiv t_1 - t_2$, one obtains for **B** $= 0$ two relationships for the positive and negative signs in the r.h.s. of Eq. (1.2.16), respectively. The first one is

$$\dot{\psi}_{\alpha\beta}(0) = 0, \tag{1.2.17}$$

where $\dot{\psi}$ is the time derivative of ψ, while the second one is

$$\psi_{\alpha\beta}(0) = 0. \tag{1.2.18}$$

Let $\psi_{\alpha\beta}(\omega)$ be the Fourier transform of the correlation function of a stationary system:

$$\psi_{\alpha\beta}(\omega) = \int_{-\infty}^{+\infty} d(t_1 - t_2)e^{i\omega(t_1 - t_2)}\psi_{\alpha\beta}(t_1 - t_2). \tag{1.2.19}$$

Because the correlation functions are real quantities, their Fourier transforms satisfy a relationship:

$$\psi_{\alpha\beta}^{*}(\omega) = \psi_{\alpha\beta}(-\omega). \tag{1.2.20}$$

Two more relationships for $\psi_{\alpha\beta}(\omega)$ follow immediately from the relationships in Eqs.(1.2.15) and (1.2.16):

$$\psi_{\alpha\beta}(\omega) = \psi_{\beta\alpha}(-\omega) = \psi_{\beta\alpha}^{*}(\omega), \tag{1.2.21}$$

$$\psi_{\alpha\beta}(\omega; \mathbf{B}) = \pm\psi_{\alpha\beta}(-\omega; -\mathbf{B}) = \pm\psi_{\alpha\beta}^{*}(\omega; -\mathbf{B}). \tag{1.2.22}$$

The second Eq. (1.2.21) follows from Eq. (1.2.20). This implies that the matrix $\psi_{\alpha\beta}(\omega)$ is a Hermitian one. Equation (1.2.22) means that in the absence of a magnetic field the functions $\psi_{\alpha\beta}(\omega)$ are either purely real or purely imaginary, depending on the symmetry of the quantities $\hat{x}_{\alpha}(t)$ and $\hat{x}_{\beta}(t)$ under time reversal, see above.

The correlation of fluctuations at two different points $\mathbf{r}_1$ and $\mathbf{r}_2$ in a medium is negligible if the distance $R = |\mathbf{r}_1 - \mathbf{r}_2|$ between these points significantly exceeds some characteristic length L_c, which is called the correlation length. At $R \gg L_c$ the correlation function is proportional to $\exp(-R/L_c)$. A system, the dimensions L of which are many times greater than L_c, can be called macroscopic. This definition of a macroscopic system differs from the common one: a system with $L \simeq L_c$ or even with $L \ll L_c$ may consist of a huge number of atoms and be macroscopic from the point of view of atomic physics.

Macroscopic systems with dimensions $L \gg L_c$ possess an important property: the correlation function of relative (fractional) fluctuations, $\psi_x(t_1, t_2)/\langle x \rangle^2$, is inversely proportional to the system's volume V. If only two dimensions or one dimension are much greater than L_c, this correlation function is inversely proportional to the area A or the length L of the system, respectively. To prove this statement, one has to thoughtfully divide the system into blocks each of $\simeq L_c$ dimension. Then the fluctuations in different blocks are uncorrelated, and their contributions to the total fluctuation partially cancel one another. If $x(t) = \sum_i \xi_i(t)$, where $\xi_i(t)$ is the fluctuating quantity in the i-th block, then the mean value $\langle x \rangle = \sum_i \langle \xi_i \rangle = N \langle \xi \rangle$ (all N blocks are assumed to be identical). The fluctuation is also a sum over N blocks, $\delta x = \sum_i \delta \xi_i$. The correlation function then equals:

$$\psi_x(t_1, t_2) = \sum_{ij} \langle \delta\xi_i(t_1)\delta\xi_j(t_2) \rangle = \sum_i \psi_{\xi_i}(t_1, t_2) = N\psi_\xi(t_1, t_2). \tag{1.2.23}$$

The cross-terms with $i \neq j$ dropped out owing to the absence of correlation between different blocks. The correlation function of relative fluctuations,

$$\frac{\psi_x(t_1, t_2)}{\langle x \rangle^2} = \frac{1}{N} \frac{\psi_\xi(t_1, t_2)}{\langle \xi \rangle^2}, \tag{1.2.24}$$

is inversely proportional to the number of blocks, i.e., to V.

In the case of resistance fluctuations, the calculations are different because the total resistance is, in general, not a sum of the resistances of all blocks, but the result for a macroscopic resistor is the same: $\psi_R(t_1, t_2)/R^2 \propto 1/V$.

Usually only the lowest-order correlation functions are studied, both experimentally and theoretically. The higher-order correlations may also be of interest. As was mentioned in Sec. 1.1, in the case of Gaussian random processes these correlations are expressed in terms of pair correlations defined by Eq. (1.2.1). Specifically, the fourth-order correlation function equals:

$$\langle \delta x(t_1) \delta x(t_2) \delta x(t_3) \delta x(t_4) \rangle_{\text{Gauss}} = \langle \delta x(t_1) \delta x(t_2) \rangle \langle \delta x(t_3) \delta x(t_4) \rangle$$
$$+ \langle \delta x(t_1) \delta x(t_3) \rangle \langle \delta x(t_2) \delta x(t_4) \rangle + \langle \delta x(t_1) \delta x(t_4) \rangle \langle \delta x(t_2) \delta x(t_3) \rangle. \tag{1.2.25}$$

The deviation from this relation indicates that the random process under investigation is not a Gaussian one (Sec. 1.3).

1.3 Spectral density of noise

In order to understand the meaning of the spectral density of noise and to derive the relationship between this quantity and the correlation function, we examine a basic circuit arrangement (Fig. 1.1). Consider a resistor connected to a spectrum analyzer which measures the voltage fluctuations across the resistor. A spectrum analyzer contains a band-pass adjustable-frequency filter with a narrow bandwidth and an output detector that responds to the mean square of the signal. In a real experiment the fluctuation signal is amplified, but we ignore these details. Let the central frequency and the bandwidth of the filter be $\bar{f}$ and Δf, respectively. Let the input fluctuation signal and the signal at the output of the filter be $\delta x(t)$ and $\delta x(t|\bar{f}, \Delta f)$, respectively. The latter contains only the frequencies from $f_1 = \bar{f} - \Delta f/2$ to $f_2 = \bar{f} + \Delta f/2$.

The random noise signal measured in a long time $t_m \to \infty$, i.e., the input signal of the analyzer, can be represented as a Fourier integral:

$$\delta x(t) = \int_{-\infty}^{+\infty} \frac{d\omega}{2\pi} \delta x(\omega) e^{-i\omega t}. \tag{1.3.1}$$

Because $\delta x(t)$ is a real quantity, the Fourier amplitude $x(-\omega) = x^*(\omega)$. Taking into account this relationship, one can write the Fourier integral as

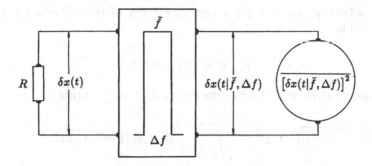

Fig. 1.1 Schematic diagram of a spectral analyzer for spectral density of noise measurement. R is the source of time-dependent fluctuations $\delta x(t)$. The rectangular box in the middle denotes a bandwidth filter with pass-band Δf and central frequency $\bar{f}$. $\delta x(t|\bar{f}, \Delta f)$ is the output (filtered) random signal. $[\delta x(t|\bar{f}, \Delta f)]^2$ is the power of the filtered signal.

an integral over positive frequencies only:

$$\delta x(t) = \int_0^\infty \frac{d\omega}{2\pi} [\delta x(\omega) e^{-i\omega t} + \delta x^*(\omega) e^{i\omega t}]. \tag{1.3.2}$$

Similarly, the filter output signal

$$\delta x(t|\bar{f}, \Delta f) = \int_{\bar{\omega} - \Delta\omega/2}^{\bar{\omega} + \Delta\omega/2} \frac{d\omega'}{2\pi} [\delta x(\omega') e^{-i\omega' t} + \delta x^*(\omega') e^{i\omega' t}]. \tag{1.3.3}$$

This signal squared, which is called noise <u>power</u>, depends on time and randomly fluctuates around its mean value $\overline{[\delta x(t|\bar{f}, \Delta f)]^2}$ which is time-independent. Assuming that the averaging over the time of measurement, t_m, and over an ensemble of identical noise sources and spectral analyzers give identical results (Sec. 1.4), one can write:

$$\langle [\delta x(t|\bar{f}, \Delta f)]^2 \rangle = \int_{\bar{\omega} - \Delta\omega/2}^{\bar{\omega} + \Delta\omega/2} \frac{d\omega'}{2\pi} \frac{d\omega''}{2\pi} \langle [\delta x(\omega') e^{-i\omega' t} + \delta x^*(\omega') e^{i\omega' t}]$$
$$\times [\delta x(\omega'') e^{-i\omega'' t} + \delta x^*(\omega'') e^{i\omega'' t}] \rangle. \tag{1.3.4}$$

To find the result of averaging let us examine in the same way Eq. (1.2.1) for the correlation function of the same random quantity $\delta x(t)$:

$$\psi_x(t_1 - t_2) = \langle x(t_1) x(t_2) \rangle = \int_{-\infty}^{+\infty} \frac{d\omega' d\omega''}{(2\pi)^2} \langle \delta x(\omega') \delta x(\omega'') \rangle e^{-i\omega' t_1 - i\omega'' t_2}. \tag{1.3.5}$$

Because the correlation function of a stationary random process depends only on the difference $t_1 - t_2$, the mean value $\langle \delta x(\omega') \delta x(\omega'') \rangle$ should be nonzero only for $\omega' + \omega'' = 0$. In other words, it is proportional to the delta-function $\delta(\omega' + \omega'')$. The coefficient at this delta-function is obviously

$2\pi\psi_x(\omega')$, where $\psi_x(\omega)$ is the Fourier transform of the correlation function $\psi_x(t_1 - t_2)$. Thus,

$$\langle \delta x(\omega') \delta x(\omega'') \rangle = 2\pi\psi_x(\omega')\delta(\omega' + \omega''). \tag{1.3.6}$$

Upon substituting Eq. (1.3.6) into Eq. (1.3.4) one obtains:

$$\overline{[\delta x(t|\bar{f}, \Delta f)]^2} = 2 \int_{\bar{f}-\Delta f/2}^{\bar{f}+\Delta f/2} df \psi_x(\omega) \approx S_x(\bar{f})\Delta f. \tag{1.3.7}$$

The last equation means that the mean squared signal transmitted through a filter with a sufficiently narrow passband is proportional to the bandwidth Δf. The quantity $S_x(f)$ introduced in the r.h.s. of Eq. (1.3.7), i.e., the mean random signal squared (noise power) per unit frequency band is called the spectral density of noise (or of fluctuations), or simply the spectrum of fluctuations. It follows from Eq. (1.3.7) that:

$$S_x(f) = 2 \int_{-\infty}^{+\infty} d(t_1 - t_2)e^{i\omega(t_1-t_2)}\psi_x(t_1 - t_2) \equiv 2\psi_x(\omega), \tag{1.3.8}$$

i.e., the spectral density is twice the Fourier transform of the correlation function. This relationship is called the Wiener–Khintchine theorem (Wiener, 1930; Khintchine, 1934).

The equation for the spectral density may be represented in a different but equivalent form. Let $\delta x(t)$ be a record of the fluctuations measured during a sufficiently long time interval t_m, say, from $t = -t_m/2$ to $t = t_m/2$. Then:

$$S_x(f) = 2 \lim_{t_m \to \infty} \frac{1}{t_m} \left| \int_{-t_m/2}^{t_m/2} dt\, e^{i\omega t}\delta x(t) \right|^2. \tag{1.3.9}$$

In the case of several coupled, i.e., correlated, fluctuating quantities x_α, where $\alpha = 1,\ldots,M$, the spectral density is obviously a matrix, $S_{\alpha\beta}(f)$, which may be defined in two ways. The first is a generalization of Eq. (1.3.9). Let $\delta x_\alpha(t)$ and $\delta x_\beta(t)$ be the records of two different random quantities during the time t_m. Then

$$S_{\alpha\beta}(f) = 2 \lim_{t_m \to \infty} \frac{1}{t_m} \int_{-t_m/2}^{t_m/2} dt_1 e^{i\omega t_1}\delta x_\alpha(t_1) \times \int_{-t_m/2}^{t_m/2} dt_2 e^{-i\omega t_2}\delta x_\beta(t_2)$$

$$= 2\psi_{\alpha\beta}(\omega). \tag{1.3.10}$$

By this definition, the spectral density matrix is twice the Fourier transform of the correlation matrix, and its properties are, obviously, the same as those of $\psi_{\alpha\beta}(\omega)$ (Eqs. (1.2.20–22)). For the reader's convenience, these relationships are reproduced here:

$$S_{\alpha\beta}^*(f) = S_{\alpha\beta}(-f), \tag{1.3.11}$$

$$S_{\alpha\beta}(f) = S_{\beta\alpha}^*(f), \tag{1.3.12}$$

$$S_{\alpha\beta}(f; \mathbf{B}) = \pm S_{\alpha\beta}^*(f; -\mathbf{B}). \tag{1.3.13}$$

These spectral densities are defined on the entire frequency axis, i.e., both for positive and negative frequencies. The nondiagonal elements of $S_{\alpha\beta}(f)$ are, in general, complex quantities.

The spectral density may be defined also in a way analogous to Eq. (1.3.7) as the spectral density of the mean product of filtered signals $\delta x_\alpha(\bar{f}, \Delta f)$ and $\delta x_\beta(\bar{f}, \Delta f)$:

$$\bar{S}_{\alpha\beta}(f) = \lim_{\Delta f \to 0} (\Delta f)^{-1} \overline{\delta x_\alpha(t|\bar{f}, \Delta f) \delta x_\beta(t|\bar{f}, \Delta f)}. \tag{1.3.14}$$

One can easily prove that $\bar{S}_{\alpha\beta}(f)$ is a symmetrical combination:

$$\bar{S}_{\alpha\beta}(f) = \psi_{\alpha\beta}(\omega) + \psi_{\beta\alpha}(\omega) = \frac{1}{2}[S_{\alpha\beta}(f) + S_{\beta\alpha}(f)]$$

$$= \frac{1}{2}[S_{\alpha\beta}(f) + S_{\alpha\beta}(-f)] = \frac{1}{2}[S_{\alpha\beta}(f) + S_{\alpha\beta}^*(f)]. \tag{1.3.15}$$

Although this spectral density, $\bar{S}_{\alpha\beta}(f)$, is defined for positive frequencies, it may be considered as an even function of f (Eq. (1.3.15)). It is always, even in a magnetic field $\mathbf{B} \neq 0$, a real symmetric matrix.

It follows from Eq. (1.3.7) that in the limit of very wide bandwidth which encompasses all significant frequencies, the mean squared signal is just the variance of the fluctuations:

$$\langle(\delta x)^2\rangle = \psi_x(t_1 - t_2 = 0) = \int_0^{+\infty} df\, S_x(f) \tag{1.3.16}$$

It means that the integral of the spectral density over all positive frequencies is exactly the variance of the noise.

The correlation function $\psi_x(t_1 - t_2)$ and the corresponding spectral density $S_x(f)$ are of the lowest order. The measurement of higher-order correlation functions and spectral densities may be useful as well, especially in the case of non-Gaussian random processes, when these functions can not be expressed in terms of $\psi_x(t_1 - t_2)$ and $S_x(f)$, respectively. As was mentioned above, the noise power, $[\delta x(t|\bar{f}, \Delta f)]^2$, of the noise signal, transmitted through a filter with bandwidth Δf and central frequency $\bar{f}$, randomly fluctuates in time around its mean value

$$\overline{[\delta x(t|\bar{f}, \Delta f)]^2} = \int_{\bar{f}-\Delta f/2}^{\bar{f}+\Delta f/2} df\, S_x(f). \tag{1.3.17}$$

The correlation function of these fluctuations may be called 'the second correlation function' $\psi_x^{(2)}(t_1, t_2 | \bar{f}, \Delta f)$ (the term 'second' may be used only for simplicity, this function is expressed in terms of a product of *four* fluctuations $\delta x(t)$). The corresponding spectral density is called 'the second spectral density':

$$
\begin{aligned}
&S_x^{(2)}(f | \bar{f}, \Delta f) \\
&= 2 \lim_{t_m \to \infty} \frac{1}{t_m} \left| \int_{-t_m/2}^{t_m/2} dt\, e^{i\omega t} \left[[\delta x(t | \bar{f}, \Delta f)]^2 - \overline{[\delta x(t | \bar{f}, \Delta f)]^2} \right] \right|^2.
\end{aligned}
\tag{1.3.18}
$$

This spectral density can be represented in terms of the fourth-order correlation function $\psi_x^{(4)}$ as follows:

$$
\begin{aligned}
S_x^{(2)}(f | \bar{f}, \Delta f) = 16\pi \int_{\bar{\omega} - \Delta\omega/2}^{\bar{\omega} + \Delta\omega/2} \frac{d\omega_1 d\omega_2}{(2\pi)^3} \Big[&\psi_x^{(4)}(\omega, \omega_1, \omega_2) \\
&- 2\pi \delta(\omega) \psi_x(\omega_1) \psi_x(\omega_2) \Big].
\end{aligned}
\tag{1.3.19}
$$

Here

$$
\begin{aligned}
\psi_x^{(4)}(\omega, \omega_1, \omega_2) = \int_{-\infty}^{+\infty} d\theta d\theta_1 d\theta_2\, e^{i[\omega\theta + (\omega_1 - \omega/2)\theta_1 + (\omega_2 + \omega/2)\theta_2]} \\
\times \langle \delta x(t_1) \delta x(t_1') \delta x(t_2) \delta x(t_2') \rangle,
\end{aligned}
\tag{1.3.20}
$$

and $\theta = (t_1 + t_1' - t_2 - t_2')/2$, $\theta_1 = t_1 - t_1'$, $\theta_2 = t_2 - t_2'$.

If the random process is Gaussian, the fourth-order correlation function can be expressed in terms of the lowest-order (ordinary) one by decoupling the mean product of four fluctuations (Eq. (1.3.20)) in all possible mean products of pairs of fluctuations (Eq. (1.2.25)). For Gaussian processes the correlation function of $[\delta x(t | \bar{f}, \Delta f)]^2$ and the spectral density $S^{(2)}$ equal, respectively,

$$
\begin{aligned}
\left[\psi_x^{(2)}(t_1, t_2 | \bar{f}, \Delta f) \right]_{\text{Gauss}} &= 2 \left[\overline{\delta x(t_1 | \bar{f}, \Delta f) \delta x(t_2 | \bar{f}, \Delta f)} \right]^2, \\
\left[S_x^{(2)}(f | \bar{f}, \Delta f) \right]_{\text{Gauss}} &= 2 \int_{\bar{f} - \Delta f/2 + f/2}^{\bar{f} + \Delta f/2 - f/2} dv\, S_x(v + f/2) S_x(v - f/2).
\end{aligned}
\tag{1.3.21}
$$

By the very meaning of the second spectrum, its frequencies are much smaller than the lowest limit of the frequency range of the first spectrum, i.e., $f/2 \ll f_1 = \bar{f} - \Delta f/2$. Hence, as follows from Eq. (1.3.21), the second spectral density of a Gaussian random process is independent of frequency f, i.e., its second spectrum is always 'white'. Thus, if $S^{(2)}(f)$ depends on f, this dependence is directly connected with deviations of the noise statistics from purely Gaussian (Weissman, 1988, 1993).

The difference

$$\tilde{S}_x^{(2)}(f|\bar{f}, \Delta f) - 1 \equiv \frac{S_x^{(2)}(f|\bar{f}, \Delta f)}{[S_x^{(2)}(f|\bar{f}, \Delta f)]_{\text{Gauss}}} - 1, \qquad (1.3.22)$$

is a measure of the deviation of the random process from a Gaussian one. The measurements of the second spectral density have been successfully employed to find out the statistical properties of $1/f$ noise (Weissman, 1988, 1993 and Secs. 8.2.2 and 8.6).

1.4 Ergodicity and nonergodicity of random processes

Each random quantity and each random process are characterized by some mean (average), nonrandom quantities. They include the moments (mean value, variance, etc.), the distribution functions, the correlation functions and so on. As we have already mentioned above in Secs. 1.1 and 1.2, two methods of averaging random quantities and obtaining these characteristics are possible. According to the first method, the data (records) of the random quantity measurement during a sufficiently long time t_m are averaged. Just this method is usually used in experiments. The equations for the mean value $\bar{x}$ and for the correlation function found by this method are given by Eq. (1.1.18) and (1.2.2), respectively.

The second method is averaging over an ensemble of identical systems, that is, systems having identical conditions for the random process considered and identical methods of its measurement. The equations for the mean value $\langle x \rangle$ and the correlation function $\psi_x(t_1, t_2)$ using this method of averaging are Eqs. (1.1.19) and (1.2.1), respectively.

Does each of these procedures of averaging yield a definite mean value of the random quantity being averaged? Are the results obtained by the two methods identical? The systems and random processes for which the answers are positive are called ergodic. Otherwise the system or the random process is nonergodic. It is important to understand what physical phenomena underlie the property of ergodicity of many physical systems, and why some other systems, or even the same systems under different conditions, are nonergodic.

The thermal motion of any statistical system may be viewed as incessant transitions between the microstates of this system. If the interaction of the system with surrounding bodies is sufficiently weak, each such microstate is an eigenstate of the Hamiltonian of the system itself. Owing just to the weak interaction of the system with other bodies, having often much greater thermal capacity ('thermal bath'), the energies E_n of the states which the system happens to visit in the course of its thermal motion are, in general, different.

Let us consider the simplest case of an ensemble of identical equilibrium systems each of which is interacting with a 'thermal bath' with absolute temperature T. The probability that a system belonging to this ensemble happens to be in its m-th state is given by the well known Gibbs distribution:

$$w_m = \frac{\exp(-E_m/k_B T)}{\sum_n \exp(-E_n/k_B T)} \tag{1.4.1}$$

Here k_B is the Boltzmann constant. The mean value of a random quantity x is equal to:

$$\langle x \rangle = \sum_m w_m x_m, \tag{1.4.2}$$

where x_m is the value of x in the m-th microstate of the system, the sum is over all such microstates. Obviously, the probability of finding the system in states, the energies E_m of which significantly differ from the mean energy $\langle E \rangle$ of the system by an amount which exceeds the fluctuations of energy, is very small, and these microstates practically do not contribute to the mean value $\langle x \rangle$. If the fluctuations of system's energy are neglected, its motion can be considered as confined to a surface of constant energy in the space of microstates (this 'space' is called phase space). The averaging in Eq. (1.4.2) is then over microstates belonging to this surface only (over so called microcanonical ensemble).

The existence of a definite single mean value of averaging and the equivalence of the results of averaging over time and over an ensemble ($\bar{x} = \langle x \rangle$), are based on a definite assumption. It is assumed that during the time t_m of measurement the system visits, if not all the allowed microstates (it is, perhaps, not necessary), a representative enough sample of microstates which make up the major contribution to the mean value. According to this assumption, the measurement during such a finite time t_m gives with sufficient accuracy the same result as for much longer times of measurement, even $t_m \to \infty$. This assumption is really valid if the time t_m is much greater than the characteristic times τ of passing through all groups of representative states. The inequality

$$t_m \gg \tau \tag{1.4.3}$$

is just the condition of ergodicity of the system.

However, there are many systems for which the surface of allowed states in the phase space consists of several or even a great number of disconnected parts. These parts are separated by some barriers (in particular, high energy barriers) which cannot be surmounted during the given time t_m of experiment and, moreover, during any reasonable time of experiment. The parts of the energy surface and the states close to this surface (within thermal fluctuations of the energy) are called 'valleys', or 'components' (Palmer, 1982). If the barriers separating the valleys are sufficiently high, the time of intervalley transitions τ_{interv} is much greater than any possible time t_m of the experiment. In this case the system is confined within only one valley in which it happened to be 'placed' by the procedure of the system's preparation. The measured mean value $\bar{x}$ of the random quantity is a characteristic of this valley only. The result of averaging may be different if the measurements are performed on a system which happened to be in a different valley. Let the quantity x be averaged over an ensemble which includes systems in different valleys, for instance, over the Gibbs ensemble, Eq. (1.4.1), which takes into account all valleys. If the measured mean quantities $\bar{x}$ corresponding to different valleys are different, the ensemble mean value $\langle x \rangle$ obviously differs from any specific measured $\bar{x}$, and the ergodicity is broken.

A simple example is a ferromagnet (Palmer, 1982). It is well known to be in a paramagnetic state at temperatures T higher than the critical Curie temperature T_c. The relaxation ('intermixing') times in this state are sufficiently small, therefore the ergodicity condition is easily satisfied. However, below T_c a spontaneous magnetic moment $\mathbf{M}$ is established. In the simplest case, the ferromagnet can be in two states with identical free energies and absolute values $|\mathbf{M}|$ but with opposite directions of the magnetic moment $\mathbf{M}$. The transition from one state to another can be achieved if a domain wall crosses the magnet. The free energy of formation of the domain wall is proportional to its surface. The energy which is required to overturn one elementary spin magnetic moment is of the order of $k_B T_c \sim 10^{-2}$ eV. If the domain surface contains N spins, the energy of domain wall formation is $\sim N k_B T_c$. The probability of such fluctuation is proportional to $\exp(-N k_B T_c / k_B T)$, where the actual temperature $T < T_c$. Even for a magnet of microscopic dimensions ($N \sim 10^3$) the time τ_{interv} for domain wall formation and for crossing the barrier between the two states of the magnet becomes extremely long and may be greater by many orders than even the age of the Universe. The magnetic moment measured in a sample of a ferromagnet in any accessible time t_m will be $\mathbf{M} \neq 0$. But the mean value $\langle \mathbf{M} \rangle$ obtained by averaging over the full Gibbs ensemble, Eq. (1.4.1), in which systems with $\mathbf{M}$ of opposite signs, having identical energies, are equally probable, is zero (this ensemble corresponds to $t_m \rightarrow \infty$). Strictly speaking, this system is nonergodic. However, such kind of nonergodicity is easily removed by averaging over an ensemble of

Fig. 1.2 Schematic diagram of a continuous spectrum of relaxation times τ. τ_{min} and τ_{max} are the minimal and maximum relaxation times in the system. t_m is the time of noise measurement (averaging).

systems with only one direction of **M**, i.e., over a 'restricted ensemble' (Palmer, 1982).

Another kind of nonergodicity is encountered in disordered macroscopic systems. A known example is glass, e.g., amorphous SiO_2. It is in a metastable state (according to Eq. (1.4.1), at low temperatures it should exist in a crystalline form). The relaxation time of its transition to a stable equilibrium crystalline state is finite but huge. However, for many kinds of measurement, a macroscopic sample of glass may be considered as an ergodic system.

Another example is spin glasses (Sec. 8.6 and references to Ch. 8). Their study has shown that the phase space of these systems consists of a large or even a huge number of valleys. In disordered systems these valleys are separated by energy barriers of various heights: from small microscopic barriers up to barriers the heights of which are macroscopic energies, i.e., a hierarchy of barriers exists. As the time of intervalley transition τ_{interv} depends on the height of the barrier exponentially, there is an extremely wide ('exponentially wide') continuous spectrum of the times τ_{interv} from some minimal τ_{min} up to some τ_{max} which, as in the above mentioned ferromagnets, may exceed the age of the Universe by many orders of magnitude (Fig. 1.2). The condition $t_m \gg \tau_{min}$ is usually easily satisfied. However, it is absolutely impossible to satisfy the condition $t_m \gg \tau_{max}$. At any accessible time of the experiment, t_m is within the continuous spectrum of relaxation times. The system is an equilibrium and ergodic one for those states, the transitions between which require time $\tau < t_m$, but is nonergodic for states with $\tau \gg t_m$. Since there are such groups of states between which equilibrium is not reached, these systems are, in principle, not only nonergodic but also nonequilibrium. Some of the measured quantities depend therefore on the duration t_m of the measurement procedure and on the 'waiting' time t_w between establishing the conditions of the experiment and the start of the measurements.

The problem of $1/f$ (flicker) noise observed in many systems is connected with the problem of nonergodicity of these systems (Ch. 8).

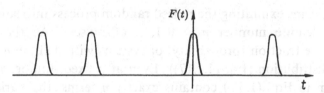

Fig. 1.3 Sequence of random pulses. $F(t)$ is the shape of a pulse.

1.5 Random pulses and shot noise

Random sequences of pulses, or pulsed random processes, are encountered very often. In the simplest case, the pulses have identical form, amplitude, and duration, and the only random quantity is the instant at which a pulse starts (Fig. 1.3). One well known example is the current in the anode circuit of the thermionic tube. Each electron emitted from the cathode and moving in the vacuum space between cathode and anode creates in the anode circuit a pulse $i(t)$, where t is time. The integral of the function $i(t)$ over time is equal to the electron charge e. Other examples are the impacts of gas molecules on the wall of the container, or the impacts of the molecules of the liquid on the macroscopic (Brownian) particle (Secs. 1.1, 1.9, and the books by Rytov, 1976 and Rytov *et al.*, 1987, 1988).

Let the index k number the pulses, and t_k be the instant corresponding to the start of the k-th pulse (or to its maximum). Each pulse can be described by some function $F(t-t_k; \mathbf{a}_k)$ which, in general, depends, if the pulses are not identical, on a finite set of random parameters $\mathbf{a}_k = (a_{k1}, \ldots, a_{km})$ considered as a m-component vector. These parameters include the amplitude of the pulse, its duration, etc. Let the measured random quantity be the sum:

$$x(t) = \sum_k F(t - t_k; \mathbf{a_k}). \tag{1.5.1}$$

The random quantities t_k and $\mathbf{a}_k$ are usually assumed to have the following statistical properties:

1. The random quantities t_k and $\mathbf{a}_k$ are statistically independent for different k and their distribution functions do not depend on k.
2. The probability of the time t_k being in the range between t and $t + dt$ is proportional to dt, independent of t, and equal to $n_1 dt$, where n_1 is the constant mean frequency of pulses.

Suppose that the measurement of the pulsed random process lasts a time t_m. The number n of pulses in this time interval is a random quantity fluctuating around the mean value $\bar{n} = n_1 t_m$. Let us split the entire ensemble of identical systems exhibiting the pulsed random process into subensembles, each with a definite number $n = 0, 1, \ldots$ of pulses. As the pulses are independent, the fraction (probability) of systems with definite n is given by the Poisson distribution (Eq. (1.1.17)). Even at a fixed number n of pulses, when the sum in Eq. (1.5.1) contains exactly n terms, the variable x is a random quantity because the sequences of pulses are random. According to Eq. (1.1.11), its characteristic function is a product of n characteristic functions, each of which equals:

$$\phi_{x_k}(u) = \overline{\exp(iux_k)} = \int da w_a(\mathbf{a}) \int_{-t_m/2}^{t_m/2} \frac{dt_k}{t_m} \exp[iuF(t - t_k; \mathbf{a})]. \qquad (1.5.2)$$

Here $w_a(\mathbf{a})$ is the probability density function of the parameters $\mathbf{a}$ of the pulses. The second integral in Eq. (1.5.2) is over instants t_k at which the pulses start. The value of the integral is independent of both the number k and the time t.

Both the characteristic function $\phi_x(u|n) = [\phi_{x_k}(u)]^n$ and the corresponding probability density $w(x|n)$ are *conditional* ones. They each relate to a subensemble (see above) with a definite number n of pulses in the time interval between $-t_m/2$ and $t_m/2$. The total, i.e., unconditional characteristic function is obtained by averaging over all subensembles by using the Poisson probability (Eq. (1.1.17)) for the number of pulses that occur in the time t_m. The result reads:

$$\phi_x(u) = \exp\{n_1 \int da w_a(\mathbf{a}) \int_{-\infty}^{+\infty} d\theta [\exp(iuF(\theta, \mathbf{a})) - 1]\}. \qquad (1.5.3)$$

Here $\theta = t - t'$. The replacement of the limits $t \pm t_m/2$ by $\pm\infty$ uses the steep enough drop of the pulse $F(t - t_k; \mathbf{a})$ at its edges.

The mean value and the variance can be expressed in terms of the characteristic function (Eq. (1.5.3)):

$$\langle x \rangle = \int_{-\infty}^{+\infty} dx \, x w(x) = \left(\frac{\partial \phi_x(u)}{\partial iu}\right)_{u=0} = n_1 \int da w_a(\mathbf{a}) \int_{-\infty}^{+\infty} d\theta F(\theta, \mathbf{a}),$$

$$\langle (\delta x)^2 \rangle = \langle x^2 \rangle - \langle x \rangle^2 = \left(\frac{\partial^2 \ln \phi_x(u)}{\partial (iu)^2}\right)_{u=0} \qquad (1.5.4)$$

$$= n_1 \int da w_a(\mathbf{a}) \int_{-\infty}^{+\infty} d\theta F^2(\theta, \mathbf{a}).$$

When all pulses are identical there is no integration over $\mathbf{a}$. For this case Eq. (1.5.4) has been derived by Campbell (1909).

The correlation function of the stationary pulsed random process can be found in the same way as the characteristic function, Eq. (1.5.3):

$$\psi_x(t_1 - t_2) = \langle \delta x(t_1) \delta x(t_2) \rangle = n_1 \int da w_a(\mathbf{a}) \int_{-\infty}^{+\infty} d\theta F(\theta; \mathbf{a}) F(\theta + t_1 - t_2; \mathbf{a}).$$
(1.5.5)

Obviously, this equation at $t_1 \to t_2$ reduces to Eq. (1.5.4) for the noise variance.

According to the Wiener–Khintchine theorem (Eq. (1.3.8)), the corresponding spectral density is

$$S_x(f) = 2n_1 \int da w_a(\mathbf{a}) |F(\omega; \mathbf{a})|^2,$$
(1.5.6)

where $F(\omega; \mathbf{a})$ is the Fourier transform of $F(t; \mathbf{a})$. For identical pulses, when no integration over $\mathbf{a}$ has to be performed, this equation was derived by Carson (1931).

The Fourier transform $F(\omega)$ at $\omega = 0$ is equal to the integral over the pulse. It may be denoted by q because in the case of current pulses it is exactly the charge of a particle. Hence the spectral density at low frequencies, many times lower than the inverse duration of each pulse, equals:

$$S_x(0) = 2q^2 n_1. \qquad q = \int_{-\infty}^{+\infty} dt F(t).$$
(1.5.7)

As was mentioned above, the current in the anode circuit of a thermionic tube is a pulsed random process. Schottky (1918), who was the first to examine it, called it shot noise. The pulses are uncorrelated because so are the events of electron emission from the cathode. The integral over a single pulse is equal to the absolute value of the electron charge e. The frequency of the pulses n_1 is the number flow, J, of particles per unit time, $J = I/e$, where I is the mean current. In the case of shot noise, the spectral density at low frequencies equals (Schottky, 1918):

$$S_I(0) = 2eI = 2e^2 J.$$
(1.5.8)

Measuring $S_I(0)$ and the mean current I one is able to determine the electron charge e. A number of measurements of e by this method are known.

Let us write down the equation for the spectral density in the case when the pulses are identical and rectangular. If the duration of a pulse is τ, its height is q/τ. In this case

$$F(\omega) = \frac{q}{\tau} \int_{-\tau/2}^{\tau/2} dt e^{i\omega t} = q \frac{\sin(\omega\tau/2)}{\omega\tau/2}.$$
(1.5.9)

According to Eq. (1.5.7),

$$S_x(f) = 2n_1 q^2 \left[\frac{\sin(\omega\tau/2)}{\omega\tau/2} \right]^2 \tag{1.5.10}$$

The spectral density significantly deviates from its low-frequency value at frequencies $f \simeq 1/\tau$. Obviously, the same is true even if the form of a single pulse deviates from the rectangular one: the characteristic duration of a pulse then plays the role of τ.

1.6 Markov processes; general theory

The notion of Markov, or Markovian, random processes, as well as the Markov approach in the theory of random processes are widely used (including this book) and applied to many systems. This notion and this approach are based on an approximation. As any well substantiated physical approximation it is, in its turn, based on definite properties of random processes in real physical systems (for comprehensive accounts of Markov processes for physicists see Lax, 1960; Rytov, 1976; Rytov *et al.*, 1987, 1988).

Let us examine, for instance, the temperature fluctuations in a macroscopic specimen due to heat exchange with the surrounding bodies. Since the temperature determines the distribution in energies of the specimen's states (through the Gibbs distribution), its fluctuation is accompanied by a change of a huge number of microscopic parameters characterizing the specimen: parameters of the electron system, phonons, and so on. Therefore the characteristic times of temperature measurement and monitoring (the duration of one measurement, the intervals between consecutive measurements) must be and, in fact, are many times greater than the times that are necessary for the microscopic parameters to adjust the change of the temperature. As on the time scale of temperature measurements the microscopic parameters can be considered as completely adjusted to the instantaneous value of the temperature, the conditional probability of the temperature fluctuation at instant t_2 taking a value δT_2, if at the preceding measurement at instant t_1 it happened to be δT_1, depends only on this last quantity and does not depend on the previous history of temperature fluctuations before t_1, which is 'forgotten'.

Some defects and impurities in solids are known to be mobile. They are moving by hops. Each such hop usually requires surmounting of an activation barrier. Therefore, the rate of hops of any individual defect is many times smaller than the frequency v of attempts to surmount the barrier ($v \sim 10^{11} - 10^{14}$ s^{-1}). After each hop, the crystal and the defect in its new position release the activation energy to the lattice vibrations (phonons) and reach an equilibrium state. The equilibration time, which can be considered as a 'microscopic' one, is usually smaller by many orders of magnitude than the time between two consequent hops. That is why the defect 'forgets' its previous history long before the next hop occurs, the probability of which depends therefore on its present position only. Thus, the hopping motion of defects in a solid is usually a Markov random process.

Let us assume that the random function of time $x(t)$ is known (e.g., as a result of measurements) at some n instants $t_1 < t_2 < \ldots < t_n$, and the time intervals between these instants are many times greater than the times that can be considered as microscopic for the random process $x(t)$. A random process $x(t)$ can be characterized by the probability $w_n(x_1, t_1; \ldots; x_n, t_n) \times dx_1 \cdots dx_n$ that the random quantity $x(t)$ is at time t_1 in the range $(x_1, x_1 + dx_1)$, at time t_2 in the range $(x_2, x_2 + dx_2)$, and so on (Eq. (1.1.3)). The probability density functions w_n and w_{n-1} may be related by an equation which is, in fact, a definition of the conditional probability function P_n (Eq. (1.2.4)):

$$w_n(x_1, t_1; \ldots; x_n, t_n)$$
$$= w_{n-1}(x_1, t_1; \ldots; x_{n-1}, t_{n-1}) P_n(x_n, t_n | x_1, t_1; \ldots; x_{n-1}, t_{n-1}).$$
(1.6.1)

The quantity $P_n dx_n$ is the conditional probability that $x(t_n)$ is in the range $(x_n, x_n + dx_n)$ *if* at preceding instants t_i ($i = 1, \ldots, n-1$) the random quantity took the given values x_i. Thus, in general, the conditional probability P_n of the random quantity to be in the state x_n at instant t_n depends on the entire preceding history of the process (on the entire path it has run), that is, on a great number of its previous states.

However, real random processes often possess a fundamental property: the conditional probability P_n at any number n depends not on all values $x_1, \ldots, x_{n-1}$ at all previous times $t_1, \ldots, t_{n-1}$ but only on the value x_{n-1} at the last instant t_{n-1} which precedes t_n:

$$P_n(x_n, t_n | x_1, t_1; \ldots; x_{n-1}, t_{n-1})$$
$$= P_2(x_n, t_n | x_{n-1}, t_{n-1}) \equiv P(x_n, t_n | x_{n-1}, t_{n-1}).$$
(1.6.2)

This equation means that for those methods of monitoring the random quantity which are used in the experiment and are appropriate to this random quantity, the system which determines the randomness of $x(t)$ has enough time to forget the previous history of this random quantity. In other words, the system has no long-term memory. The processes, for which the conditional probability P_n satisfies the condition Eq. (1.6.2), have been first investigated by A.A. Markov, and are called Markov (or Markovian) random processes, or Markov chains, or processes without after-effect. Such processes have been illustrated above.

The conditional probability function $P(x, t|x_0, t_0)$ is called the transition probability. In the case of a homogeneous Markov process it depends only on the difference $t - t_0$. The homogeneous process may be nonstationary, but for stationary processes the homogeneity condition is always satisfied. Let us restrict ourselves to stationary processes only.

The definition of a Markov random process may be easily generalized: there may be several, say M, random quantities in the system. This multitude of random quantities may be denoted as an M-component vector $\mathbf{x}(t)$. In this case the transition probability can be written as $P(\mathbf{x}, t|\mathbf{x}_0, t_0)$.

The transition probability $P(\mathbf{x}, t|\mathbf{x}_0, t_0)$ at $t > t_0$ must satisfy a definite consistency condition. Let t' be any intermediate point of time between t_0 and t, i.e., $t_0 < t' < t$. The probability of the transition $\mathbf{x}_0 \to \mathbf{x}$ is obviously a sum of probabilities of such transitions through all possible values of $\mathbf{x}'$ at the instant t'. Therefore the following equation holds:

$$P(\mathbf{x}, t|\mathbf{x}_0, t_0) = \int d\mathbf{x}' P(\mathbf{x}, t|\mathbf{x}', t') P(\mathbf{x}', t'|\mathbf{x}_0, t_0), \qquad (1.6.3)$$

where $d\mathbf{x}' = dx'_1 \cdots dx'_M$. It is called the Smoluchowski equation.

This equation imposes a substantial restriction on the possible form of the transition probability: integration over $\mathbf{x}'$ of the product of two such probabilities has to result also in a transition probability, and this result is independent of the intermediate time t'.

As the total probability of transition to *all* states of the system is equal to unity, the transition probability must satisfy the condition:

$$\int d\mathbf{x} P(\mathbf{x}, t|\mathbf{x}_0, t_0) = 1. \qquad (1.6.4)$$

Let $w(\mathbf{x})$ be the stationary probability density function. If the transition probability $P(\mathbf{x}, t|\mathbf{x}_0, t_0)$ is averaged over all initial values $\mathbf{x}_0$ with the density function $w(\mathbf{x}_0)$, the result must be the value of the probability density function $w(\mathbf{x})$:

$$\int d\mathbf{x}_0 P(\mathbf{x}, t|\mathbf{x}_0, t_0) w(\mathbf{x}_0) = w(\mathbf{x}). \qquad (1.6.5)$$

The physical meaning of $P(\mathbf{x}, t|\mathbf{x}_0, t_0)$ implies that if the difference $t - t_0$ is small as compared with the characteristic time of the random process, the random quantity $\mathbf{x}(t)$ has no time to deviate from the initial value $\mathbf{x}_0$. Therefore

$$\lim_{(t-t_0)\to 0} P(\mathbf{x}, t|\mathbf{x}_0, t_0) = \delta(\mathbf{x} - \mathbf{x}_0). \qquad (1.6.6)$$

Here $\delta(\mathbf{x} - \mathbf{x}_0)$ is the pulse (delta-) function which is nonzero only at $\mathbf{x} = \mathbf{x}_0$.

If, on the contrary, the time $t - t_0$ is much longer than the relaxation time of the random process, the system has enough time to 'forget' the initial condition $x(t_0) = x_0$, and, irrespective of the value of x_0, the transition probability tends to the probability density at x:

$$\lim_{(t-t_0)\to\infty} P(x, t|x_0, t_0) = w(x). \qquad (1.6.7)$$

This property of P makes it more convenient to use instead of P another function:

$$p(x, t|x_0, t_0) = P(x, t|x_0, t_0) - w(x), \qquad (1.6.8)$$

which falls off to *zero* as $(t - t_0) \to \infty$. It follows from Eqs. (1.6.4) and (1.6.5) that this function satisfies the relations:

$$\int dx\, p(x, t|x_0, t_0) = 0, \qquad \int dx_0\, p(x, t|x_0, t_0) w(x_0) = 0. \qquad (1.6.9)$$

The Smoluchowski equation for p has the same form as the one for P (Eq. (1.6.3)).

Let us find the differential equations which govern the kinetics of the transition probabilities as functions of time (Kolmogorov equations). In the Smoluchowski equation for p we take the maximum time equal to $t + \Delta t$, the intermediate time $t' = t$:

$$p(x, t + \Delta t|x_0, t_0) = \int dx'\, p(x, t + \Delta t|x', t) p(x', t|x_0, t_0). \qquad (1.6.10)$$

Taking into account Eqs. (1.6.6) and (1.6.8) one may, at $\Delta t \to 0$, expand $p(x, t + \Delta t|x', t)$ and represent it in the form:

$$p(x, t + \Delta t|x', t) = \delta(x - x') - w(x) - \lambda(x, x')\Delta t. \qquad (1.6.11)$$

The meaning of $\lambda(x, x')$ will become obvious below.

Let us substitute the last equation into Eq. (1.6.10). According to the first Eq. (1.6.9), the term with $w(x)$ yields zero. Therefore

$$\frac{p(x, t + \Delta t|x_0, t_0) - p(x, t|x_0, t_0)}{\Delta t} = -\int dx'\, \lambda(x, x') p(x', t|x_0, t_0). \qquad (1.6.12)$$

In the limit $\Delta t \to 0$ Eq. (1.6.12) yields the Kolmogorov equation:

$$\frac{d}{dt} p(x, t|x_0, t_0) = -\int dx'\, \lambda(x, x') p(x', t|x_0, t_0). \qquad (1.6.13)$$

The quantities $-\lambda(x, x')$ are the derivatives of $p(x, t|x', t')$ with respect to the first time argument. At $x \neq x'$ they may be interpreted, with some reservations, as probabilities of transition per unit time from the state $\mathbf{x}'$ to the state $\mathbf{x}$. By virtue of Eqs. (1.6.9) and (1.6.11),

$$\int d\mathbf{x}\lambda(\mathbf{x}, \mathbf{x}') = 0, \qquad \int d\mathbf{x}'\,\lambda(\mathbf{x}, \mathbf{x}')w(\mathbf{x}') = 0. \qquad (1.6.14)$$

Our next goal is to derive equations for the correlation functions of the fluctuations, $\psi_{\alpha\beta}(t_1 - t_2)$, given, in general, by Eq. (1.2.8). These equations should express these functions in terms of one-time correlation functions $\psi_{\alpha\beta}(0)$ (variances). However, at first one has to find, using the same methods, how a small perturbation, given at an initial instant, relaxes in time. Let, at the initial instant t_0, the quantities $\mathbf{x}(t_0) = \mathbf{x}_0$. The expected conditional mean value of a component x_α at the instant $t > t_0$ equals:

$$\langle x_\alpha(t)|\mathbf{x}_0, t_0\rangle = \int d\mathbf{x}\,x_\alpha P(\mathbf{x}, t|\mathbf{x}_0, t_0). \qquad (1.6.15)$$

Since the mean value of this component $\langle x_\alpha \rangle = \int d\mathbf{x}\,x_\alpha w(\mathbf{x})$, the expected deviation of the random quantity x_α from its mean value equals:

$$\langle \Delta x_\alpha(t)|\mathbf{x}_0, t_0\rangle = \int d\mathbf{x}\,x_\alpha p(\mathbf{x}, t|\mathbf{x}_0, t_0) = \int d\mathbf{x}\,\Delta x_\alpha p(\mathbf{x}, t|\mathbf{x}_0, t_0). \qquad (1.6.16)$$

According to the Kolmogorov equation, the time derivative of this quantity equals:

$$\frac{d}{dt}\langle \Delta x_\alpha(t)|\mathbf{x}_0, t_0\rangle = -\int d\mathbf{x}'d\mathbf{x}\Delta x_\alpha\lambda(\mathbf{x}, \mathbf{x}')p(\mathbf{x}', t|\mathbf{x}_0, t_0). \qquad (1.6.17)$$

The deviations $\Delta\mathbf{x}'$ from the mean values $\langle \mathbf{x}'\rangle$ may also be considered as small. Therefore the integral $\int d\mathbf{x}\,\Delta x_\alpha\lambda(\mathbf{x}, \mathbf{x}')$, which depends on $\mathbf{x}'$, may be represented as a linear function of $\Delta\mathbf{x}'$:

$$\int d\mathbf{x}\Delta x_\alpha\,\lambda(\mathbf{x}, \mathbf{x}') = \Lambda_{\alpha\beta}\Delta x'_\beta. \qquad (1.6.18)$$

Here $\Lambda_{\alpha\beta}$ is a matrix, the eigenvalues of which have the meaning of inverse relaxation times of the system, and summation over repeated indices is implied. From the last two equations one immediately obtains:

$$\frac{d}{dt}\langle \Delta x_\alpha(t)|\mathbf{x}_0, t_0\rangle = -\Lambda_{\alpha\beta}\langle \Delta x_\beta(t)|\mathbf{x}_0, t_0\rangle. \qquad (1.6.19)$$

The correlation function is easily expressed in terms of the conditional value of the fluctuation (Eq. (1.6.16)). Let us first represent $\psi_{\alpha\beta}(t_1 - t_2)$ as a sum of 'one-sided' parts:

$$\psi_{\alpha\beta}(t_1 - t_2) = \psi_{\alpha\beta}^+(t_1 - t_2) + \psi_{\alpha\beta}^-(t_1 - t_2), \qquad (1.6.20)$$

where

$$\psi_{\alpha\beta}^+(t_1 - t_2) = \Theta(t_1 - t_2)\langle \delta x_\alpha(t_1)\delta x_\beta(t_2)\rangle$$
$$\psi_{\alpha\beta}^-(t_1 - t_2) = \Theta(t_2 - t_1)\langle \delta x_\beta(t_2)\delta x_\alpha(t_1)\rangle. \tag{1.6.21}$$

Here $\Theta(t)$ is the known Heaviside step function:

$$\Theta(t) = \begin{cases} 1, & \text{if } t \geq 0; \\ 0, & \text{if } t < 0. \end{cases} \tag{1.6.22}$$

According to Eqs. (1.2.4), (1.6.8), and (1.6.16):

$$\psi_{\alpha\beta}^+(t) = \Theta(t)\int dx dx' x_\alpha p(\mathbf{x}, t|\mathbf{x}', 0)x'_\beta w(\mathbf{x}')$$

$$= \Theta(t)\int dx dx' \delta x_\alpha p(\mathbf{x}, t|\mathbf{x}', 0)\delta x'_\beta w(\mathbf{x}')$$

$$= \Theta(t)\int dx' \langle \delta x_\alpha(t)|\mathbf{x}_0, 0\rangle \delta x'_\beta w(\mathbf{x}'), \tag{1.6.23}$$

$$\psi_{\alpha\beta}^-(t) = \psi_{\beta\alpha}^+(|t|).$$

The time derivative of $\psi_{\alpha\beta}^+(t)$, where $t = t_1 - t_2$, consists of two parts. The derivative of the step function is $\Theta'(t) = \delta(t)$. The corresponding coefficient is the integral in Eq. (1.6.23), taken at $t_1 = t_2$, i.e., $\psi_{\alpha\beta}(0)$. The time-derivative of $p(\mathbf{x}, t_1|\mathbf{x}', t_2)$ is found using the Kolmogorov equation (1.6.13). One of the integrals is given by Eq. (1.6.18). The result is a system of equations for $\psi_{\alpha\beta}^+$:

$$\frac{\partial}{\partial t}\psi_{\alpha\beta}^+(t) + \Lambda_{\alpha\gamma}\psi_{\gamma\beta}^+(t) = \psi_{\alpha\beta}(0)\delta(t). \tag{1.6.24}$$

Perhaps, the most convenient way to solve this system is to diagonalize the matrix $\Lambda_{\alpha\beta}$ (we assume that it is possible). Let λ_m $(m = 1, \dots, M)$ and $\chi_\alpha^{(m)}$ be, respectively, the eigenvalues and eigenvectors of the $M \times M$ matrix $\Lambda_{\alpha\beta}$. Similarly, $\phi_\alpha^{(m)}$ is the eigenfunction of the conjugated matrix $\Lambda_{\alpha\beta}^+ = \Lambda_{\beta\alpha}^*$ which corresponds to its eigenvalue λ_m^* of the last matrix:

$$\Lambda_{\alpha\beta}\chi_\beta^{(m)} = \lambda_m\chi_\alpha^{(m)}, \qquad \Lambda_{\alpha\beta}^+\phi_\beta^{(m)} = \lambda_m^*\phi_\alpha^{(m)}. \tag{1.6.25}$$

Let us assume also that all eigenvalues λ_m are different. The normalized eigenfunctions $\chi_\alpha^{(m)}$ and $\phi_\alpha^{(m)}$ satisfy the following orthogonality conditions (Gantmacher, 1977; Pease, 1965):

$$\sum_\alpha \chi_\alpha^{(m)}\phi_\alpha^{(n)*} = \delta_{nm}. \tag{1.6.26}$$

We can now introduce M linear combinations of fluctuations δx_α, the correlation functions of which decay in time exponentially, each with a definite relaxation time λ_m^{-1}:

$$\Phi_{mn}^+(t) = \sum_{\alpha\beta} \phi_\alpha^{(m)*} \psi_{\alpha\beta}^+(t) \chi_\beta^{(n)}. \tag{1.6.27}$$

It follows from Eqs. (1.6.24)–(1.6.27) that $\Phi_{mn}(t)$ satisfy the equations:

$$\frac{d\Phi_{mn}^+(t)}{dt} + \lambda_m \Phi_{mn}^+(t) = \Phi_{mn}(0)\delta(t). \tag{1.6.28}$$

The solutions to these equations are:

$$\Phi_{mn}^+(t) = \Theta(t)\Phi_{mn}(0)e^{-\lambda_m t}. \tag{1.6.29}$$

Using Eqs. (1.6.26) and (1.6.27) one can find the original correlation functions:

$$\psi_{\alpha\beta}^+(t) = \sum_{mn} \chi_\alpha^{(m)} \Phi_{mn}^+(t) \phi_\beta^{(n)*} = \Theta(t) \sum_{m\gamma} e^{-\lambda_m t} \chi_\alpha^{(m)} \phi_\gamma^{(m)*} \psi_{\gamma\beta}(0). \tag{1.6.30}$$

The matrix of spectral densities can be found from Eqs. (1.3.8) and (1.6.21):

$$S_{\alpha\beta}(f) = 2\sum_{m\gamma} \left\{ \frac{\chi_\alpha^{(m)} \psi_{\gamma\beta}(0) \phi_\gamma^{(m)*}}{\lambda_m - i\omega} + \frac{\chi_\beta^{(m)} \psi_{\gamma\alpha}(0) \phi_\gamma^{(m)*}}{\lambda_m + i\omega} \right\}. \tag{1.6.31}$$

If there is only one quantity $\lambda = \tau^{-1}$ (τ is the relaxation time), the spectral density has a simple Lorentzian form:

$$S_x(f) = 4\langle(\delta x)^2\rangle\tau(1 + \omega^2\tau^2)^{-1}. \tag{1.6.32}$$

According to Eqs. (1.6.21) and (1.6.30) the correlation functions decay at $(t_1 - t_2) > 0$ and $(t_1 - t_2) < 0$ as linear superpositions of exponential functions $\exp(-\lambda_m|t_1 - t_2|)$. It means that in the Markov approximation the slopes of the correlation functions at $(t_1 - t_2) \to +0$ and at $(t_1 - t_2) \to -0$ are obviously different and are not zero, in contradiction to the general condition Eq. (1.2.17). According to this condition the correlation function is an even function of time with zero derivative at $t = 0$. This inconsistency of the Markov approximation has been very clearly explained by Lax (1960). This approximation is valid only at times much greater than the microscopic ones. The true time-dependence of the correlation function satisfies Eq. (1.2.17). The difference between the true behaviour of the correlation function and its approximate, Markovian, dependence is significant only in a very small time interval $|t| \leq\sim \tau_c \ll \tau$, where τ is the relaxation time of the fluctuations considered, and τ_c is a microscopic forgetting time of the nonmacroscopic variables which has different meaning in different systems. For instance, in a gas it is the duration of the particles' collisions that, at low density of the particles, is much smaller than the time between successive collisions of a

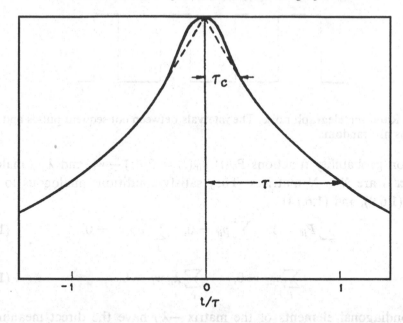

Fig. 1.4 Correlation function $\psi(t)$ of a random process with a single relaxation time τ. Solid curve: real decay of the correlation function. Dashed curve: purely exponential correlation function $\exp(-|t|/\tau)$ obtained in the Markov approximation. It deviates from the real one at small times $|t| < \tau_c$ which in Markov approximation are 'microscopic' times. From Lax (1960).

particle. The true correlation function and its Markov approximation are schematically shown in Fig. 1.4 (Lax, 1960).

1.7 Discrete Markov processes; random telegraph noise

A very important special case, especially in solid state physics, is random processes in which the random quantity takes only discrete values $x_i(i = 1,\ldots,N)$ and randomly switches between these values. It may be a defect, the transitions of which between two states modulate the resistance of the conductor with this defect. The noise generated by these transitions is illustrated in Fig. 1.5. It is called random telegraph noise (RTN), or random telegraph signals. The time between two successive transitions is usually many times greater than the time of relaxation (equilibration) of the crystal after each such transition. Then this process is a typical Markov one: the probability of transition depends only on the system's present state and does not depend on its history.

In the case of a discrete random quantity the unconditional probability w_i of finding the system in its i-th 'state' depends on the discrete index i. The

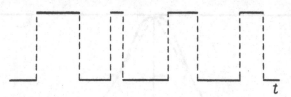

Fig. 1.5 Random telegraph noise. The intervals between consequent pulses and their durations are random.

transition probability functions $P_{ii'}(t), p_{ii'}(t) = P_{ii'}(t) - w_i$, and $\lambda_{ii'}$ (analogue of $\lambda(\mathbf{x}, \mathbf{x}'))$ are $N \times N$ matrices. They satisfy conditions analogous to Eqs. (1.6.4), (1.6.9), and (1.6.14):

$$\sum_i P_{ii'} = 1, \quad \sum_i p_{ii'} = 0, \quad \sum_{i'} p_{ii'} w_{i'} = 0, \tag{1.7.1}$$

$$\sum_i \lambda_{ii'} = 0, \quad \sum_{i'} \lambda_{ii'} w_{i'} = 0. \tag{1.7.2}$$

The nondiagonal elements of the matrix $-\lambda_{ij}$ have the direct meaning of transition probabilities per unit time from j to i and, for clarity, can be denoted as $W(j \to i)$. The diagonal elements can be expressed in terms of the sum of nondiagonal elements using the first Eq. (1.7.2):

$$\lambda_{ii} = -\sum_{j(\neq i)} \lambda_{ji}. \tag{1.7.3}$$

The Kolmogorov equations for a discrete Markov process take the form:

$$\frac{d}{dt} p_{ij}(t) = -\sum_k \lambda_{ik} p_{kj}(t). \tag{1.7.4}$$

The initial conditions are: $p_{ij}(0) = \delta_{ij} - w_i$.

The solutions to the Kolmogorov equations are exponential functions of the type $\exp(-Wt)$, where W are the eigenvalues of the matrix λ_{ij}, i.e., the roots of the equation of degree N:

$$\det \|\lambda_{ij} - W \delta_{ij}\| = 0. \tag{1.7.5}$$

In a stable system, the roots are either positive real or form complex pairs with a positive real part.

As in the general case, the correlation function of a discrete Markov random process can be represented in the form of Eq. (1.6.20). The equation for $\psi^+(t)$ follows from Eq. (1.6.23):

$$\psi^+(t) = \Theta(t) \sum_{ij} x_i p_{ij}(t) x_j w_j. \tag{1.7.6}$$

In order to find the explicit form of $\psi(t)$ at given transition rates, one has to solve the Kolmogorov equations (1.7.4) for $p_{ij}(t)$ and then find the probabilities w_i by solving the last Eq. (1.7.2) which, actually, is a system of equations in w_i. In the simplest case when $N = 2$ (the quantity x randomly switches between two states x_1 and x_2) the only nonzero eigenvalue is $W \equiv \tau^{-1} = W(1 \to 2) + W(2 \to 1)$, i.e., is equal to the sum of the rates of transitions back and forth. The probabilities of the two states are:

$$w_1 = \frac{W(2 \to 1)}{W(1 \to 2) + W(2 \to 1)}, \quad w_2 = \frac{W(1 \to 2)}{W(1 \to 2) + W(2 \to 1)}, \quad (1.7.7)$$

$w_1 + w_2 = 1$.

The matrix $p_{ij}(t)$ in the case of $N = 2$ equals:

$$\hat{p}(t) = \begin{pmatrix} w_2 & -w_1 \\ -w_2 & w_1 \end{pmatrix} e^{-Wt}. \quad (1.7.8)$$

The one-sided correlation functions

$$\psi^+(t) = \Theta(t)w_1 w_2 (x_1 - x_2)^2 e^{-Wt}, \quad \psi^-(t) = \psi^+(|t|). \quad (1.7.9)$$

Thus, according to Eq. (1.3.8), the spectral density equals:

$$S(f) = 4w_1 w_2 (x_1 - x_2)^2 \tau / (1 + \omega^2 \tau^2). \quad (1.7.10)$$

The spectrum of the simplest telegraph noise is Lorentzian (Machlup, 1954).

1.8 Quasi-continuous (diffusion-like) Markov processes

In some Markov processes, the random quantities $x(t)$ vary in time quasi-continuously, that is, by relatively very small steps which, moreover, are randomly positive and negative. Any considerable change of the random quantity is in this case a result of a very great number of such steps. Such processes are called continuous or, quite naturally, diffusion-like Markov processes.

Let us find the equation for the conditional probability distribution $P(x, t|x_0, t_0)$ for this kind of random process. The starting point is the Smoluchowski equation (1.6.3) in which the maximum time is taken to be $t + \Delta t$, and the intermediate time equals t (as in Eq. (1.6.10)). Let us multiply both sides of the equation by an arbitrary function $h(x)$, which is assumed to be zero at the boundaries of the interval of x variation, and integrate over x. Upon interchanging x and x' in the integrand one obtains (Wang & Uhlenbeck, 1945; Rytov, 1976; Rytov et al., 1987, 1988):

$$\int dx h(x) P(x, t + \Delta t|x_0, t_0) = \int dx dx' h(x') P(x', t + \Delta t|x, t) P(x, t|x_0, t_0). \quad (1.8.1)$$

At small values of Δt the quantity $P(\mathbf{x}', t + \Delta t | \mathbf{x}, t)$ is significant only for small differences $|\mathbf{x}' - \mathbf{x}|$ (quasi-continuity). Since $\mathbf{x}$ is quasi-continuous and $\Delta t \to 0$, only terms up to the second order in $\mathbf{x}' - \mathbf{x}$ should be retained. Taking into account the normalization condition for P, Eq. (1.6.4), one obtains:

$$\int d\mathbf{x} h(\mathbf{x})[P(\mathbf{x}, t + \Delta t | \mathbf{x}_0, t_0) - P(\mathbf{x}, t | \mathbf{x}_0, t_0)]$$

$$= \int d\mathbf{x} h'_\alpha(\mathbf{x}) P(\mathbf{x}, t | \mathbf{x}_0, t_0) \int d\mathbf{x}'(x'_\alpha - x_\alpha) P(\mathbf{x}', t + \Delta t | \mathbf{x}, t) \qquad (1.8.2)$$

$$+ \frac{1}{2} \int d\mathbf{x} h''_{\alpha\beta}(\mathbf{x}) P(\mathbf{x}, t | \mathbf{x}_0, t_0) \int d\mathbf{x}'(x'_\alpha - x_\alpha)(x'_\beta - x_\beta) P(\mathbf{x}', t + \Delta t | \mathbf{x}, t).$$

Here $h' \equiv \partial h / \partial x_\alpha$, $h''_{\alpha\beta} \equiv \partial^2 h / \partial x_\alpha \partial x_\beta$.

In this equation, $P(\mathbf{x}, t + \Delta t | \mathbf{x}, t)$ can be expanded in powers of Δt. The zero-th order term, i.e., $\delta(\mathbf{x}' - \mathbf{x})$ (Eq. (1.6.6)) yields a zero contribution to the integrals over $\mathbf{x}'$. In the r.h.s. only terms linear in Δt remain. By dividing both sides by Δt one obtains in the l.h.s. a time derivative. Integration over $\mathbf{x}$ by parts, taking into account that $h(\mathbf{x}) = 0$ at the lower and upper limits, yields an equation in integrals over $\mathbf{x}$ with the function $h(\mathbf{x})$ as a factor in the integrand. Since the equation is valid at arbitrary $h(\mathbf{x})$, one obtains the equation which is called by physicists the Fokker–Planck (Fokker, 1914; Planck, 1917), or Einstein–Fokker–Planck, equation, and in mathematical text-books it is called the second (or forward) Kolmogorov equation:

$$\frac{\partial P(\mathbf{x}, t | \mathbf{x}_0, t_0)}{\partial t} + \frac{\partial}{\partial x_\alpha} [A_\alpha(\mathbf{x}, t) P(\mathbf{x}, t | \mathbf{x}_0, t_0)]$$

$$- \frac{\partial^2}{\partial x_\alpha \partial x_\beta} [D_{\alpha\beta}(\mathbf{x}, t) P(\mathbf{x}, t | \mathbf{x}_0, t_0)] = 0. \qquad (1.8.3)$$

Here summation over repeated indices is implied and the following notations have been introduced:

$$A_\alpha(\mathbf{x}, t) = \int d\mathbf{x}'(x'_\alpha - x_\alpha) \dot{P}(\mathbf{x}', t | \mathbf{x}, t),$$

$$D_{\alpha\beta}(\mathbf{x}, t) = \frac{1}{2} \int d\mathbf{x}'(x'_\alpha - x_\alpha)(x'_\beta - x_\beta) \dot{P}(\mathbf{x}', t | \mathbf{x}, t), \qquad (1.8.4)$$

In the last equations

$$\dot{P}(\mathbf{x}', t | \mathbf{x}, t) = \left[\frac{\partial P(\mathbf{x}', t' | \mathbf{x}, t)}{\partial t'} \right]_{t'=t}.$$

The solution to Eq. (1.8.3) must be nonnegative and must satisfy the normalization condition (1.6.4) and the initial condition (1.6.6).

The Fokker–Planck equation has the form of a continuity equation in which the role of the flow in x-space is played by the quantity:

$$J_\alpha(\mathbf{x}, t) = A_\alpha(\mathbf{x}, t)P(\mathbf{x}, t|\mathbf{x}_0, t_0) - \frac{\partial}{\partial x_\beta}[D_{\alpha\beta}(\mathbf{x}, t)P(\mathbf{x}, t|\mathbf{x}_0, t_0)]. \qquad (1.8.5)$$

The first part of J_α has the meaning of a drift flow, the second one, proportional to the gradient, is a diffusion flow. In fact, the quantity $\mathbf{A}(\mathbf{x}, t)$ is nonzero only if there is an asymmetry of transitions, those with $x' - x > 0$ and with $x' - x < 0$. The quantity $D_{\alpha\beta}$ has an obvious meaning of the tensor of diffusion coefficients. Depending on the meaning of the variables $\mathbf{x}$, it may be not only the diffusion in the configuration (common) space but also diffusion in the space of velocities, diffusion of energy, diffusion of the phase of oscillations and so on. The Fokker–Planck equation is widely used in the theory of random processes.

If at the initial time t_0, not a definite state $\mathbf{x}_0$ of the system is given but a probability density function $w(\mathbf{x}_0 t_0)$, it is more appropriate to use an equation for the probability density function $w(\mathbf{x}t)$. One has to take into account Eq. (1.2.4) for the two-dimensional density function,

$$w_2(\mathbf{x}t; \mathbf{x}_0 t_0) = P(\mathbf{x}t|\mathbf{x}_0 t_0)w(\mathbf{x}_0 t_0),$$

and Eq. (1.1.4), which expresses the function $w(\mathbf{x}t)$ in terms of $w_2(\mathbf{x}t; \mathbf{x}_0 t_0)$. Then

$$w(\mathbf{x}t) = \int d\mathbf{x}_0 P(\mathbf{x}t|\mathbf{x}_0 t_0)w(\mathbf{x}_0 t_0).$$

Multiplying the Fokker–Planck Eq. (1.8.3) by $w(\mathbf{x}_0 t_0)$ and integrating over $\mathbf{x}_0$ one obtains the equation for $w(\mathbf{x}t)$:

$$\frac{\partial w(\mathbf{x}t)}{\partial t} + \frac{\partial}{\partial x_\alpha}[A_\alpha(\mathbf{x}t)w(\mathbf{x}t)] - \frac{\partial^2}{\partial x_\alpha \partial x_\beta}[D_{\alpha\beta}(\mathbf{x}t)w(\mathbf{x}t)] = 0. \qquad (1.8.6)$$

Summation over repeated indices is implied.

To have some idea of how the Fokker-Planck equation 'works', it is instructive to discuss its application to a simple problem. For randomly moving Brownian particles (Sec. 1.9) the vectors $\mathbf{J}_\mathbf{r}(\mathbf{r}\mathbf{v}t)$ and $\mathbf{J}_\mathbf{v}(\mathbf{r}\mathbf{v}t)$ are flow densities in common space ($\mathbf{r}$ is the radius-vector) and in the space of the particles' velocities $\mathbf{v}$, respectively, at definite $\mathbf{r}$, $\mathbf{v}$, and time t. According to the definition, Eq. (1.8.4), $\mathbf{A}_\mathbf{r}$ is the speed of the radius-vector's change,

$$\mathbf{A}_\mathbf{r} = \lim_{\Delta t \to 0} \int d\mathbf{r}' d\mathbf{v}' \frac{(\mathbf{r}' - \mathbf{r})}{\Delta t}[P(\mathbf{r}'\mathbf{v}', t + \Delta t|\mathbf{r}\mathbf{v}t) - P(\mathbf{r}'\mathbf{v}', t|\mathbf{r}\mathbf{v}t)] = \mathbf{v}, \qquad (1.8.7)$$

that is, the velocity of a particle. $D_{r_\alpha r_\beta}$ is the speed of growth of the product of the particle's displacements along the directions α and β, i.e., the tensor of diffusion coefficients.

According to the same definition, Eq. (1.8.4),

$$\mathbf{A_v}(\mathbf{v}) = \lim_{\Delta t \to 0} \int d\mathbf{r}' d\mathbf{v}' \frac{\mathbf{v}' - \mathbf{v}}{\Delta t} [P(\mathbf{r}'\mathbf{v}', t + \Delta t | \mathbf{rvt}) - P(\mathbf{r}'\mathbf{v}', t | \mathbf{rvt})] \qquad (1.8.8)$$

is the acceleration of the particle. Similarly, $D_{v_\alpha v_\beta}$ determines the spreading of particles' velocities, i.e., diffusion in the space of velocities.

The acceleration $\mathbf{A_v}(\mathbf{v})$ consists of two parts. If a force $-\nabla U$, where U is the potential energy, is acting on each particle it yields an acceleration $(-\nabla U)/M$, where M is the mass of a particle. The second part is due to the hindering of the particle's motion by collisions with the molecules of the medium. If the particle's velocity v_α is at instant t equal to zero, it can change (due to collisions) by the instant $t + \Delta t$ to $v'_\alpha > 0$ and $v'_\alpha < 0$ with equal probability. Integration over $\mathbf{v}'$ in Eq. (1.8.8) yields zero. Assume that the component $v_\alpha > 0$. Then the relative velocity of the particle and the molecules moving in the opposite direction is higher, and that of the particle and molecules moving in the same direction is lower than in the case $v_\alpha = 0$. The impacts of the former are on average stronger, and the probability of a negative change of velocity $v'_\alpha - v_\alpha < 0$ is higher than that of a positive change. It means that at small velocities $|\mathbf{v}|$ this second part of $\mathbf{A_v}$ may be represented by $-\mathbf{v}/\tau$. The time τ is the relaxation time of the directional velocity of a particle. Therefore

$$\mathbf{A_v} = \frac{1}{M}(-\nabla U) - \frac{\mathbf{v}}{\tau}. \qquad (1.8.9)$$

The quantities

$$n(\mathbf{r}, t) = \int d\mathbf{v} w(\mathbf{rvt}), \quad \mathbf{j}(\mathbf{r}, t) = \int d\mathbf{v} \mathbf{J_r} w(\mathbf{rvt}) \qquad (1.8.10)$$

are the total density and flow density of particles, respectively, (with any velocity) at the point $\mathbf{r}$ and at instant t. After integration of Eq. (1.8.6) over $\mathbf{v}$, the term with divergence of $\mathbf{J_v}$ drops out. One obtains the continuity equation:

$$\frac{\partial n}{\partial t} + \operatorname{div} \mathbf{j} = 0. \qquad (1.8.11)$$

The flow density is a sum of drift and diffusion flows. In an isotropic medium the diffusion coefficient $D_\mathbf{r}$ is isotropic:

$$\mathbf{j} = \langle \mathbf{v} \rangle n(\mathbf{rt}) - D \nabla n(\mathbf{rt}), \quad \langle \mathbf{v} \rangle n(\mathbf{rt}) = \int d\mathbf{v} \, \mathbf{v} w(\mathbf{rvt}). \qquad (1.8.12)$$

These relations are used, in particular, in the theory of Brownian motion.

1.9 Brownian motion

In 1827 the English botanist Robert Brown discovered and thoroughly studied a phenomenon called after him 'Brownian motion'. Fine particles, in particular, spores of mushrooms or pollen suspended in a liquid, are moving quickly, randomly, and incessantly (Brown, 1828). As the dimensions of the particles are of the order of 1 μm, they are visible only by microscope. R. Brown was not the first to observe this motion (see the early history in the book by Brush, 1976).

'Brown's discovery, therefore, was not his observation of the motion of microscopic particles in fluids; that observation had been made many times before; instead, it was his emancipation from the previously current notion that such movements had a specifically organic character. What Brown showed was that almost any kind of matter, organic or inorganic, can be broken into fine particles that exhibit the same kind of dancing motion; thus he removed the subject from the realm of biology into the realm of physics' (Brush, 1976).

His experiments also proved that this motion is not a result of effects which may be related to the imperfectness of the experiment, like convection under inhomogeneous heating of the liquid, but a fundamental physical phenomenon.

Brownian motion was properly explained only in the first decade of the twentieth century when Einstein (1905, 1906, 1907) and Smoluchowski (1906a, 1906b, 1915) developed the theory of this phenomenon. Most likely, the reader knows that the Brownian particles move because they participate in the thermal motion of the molecules of the liquid. Brownian motion may be also viewed as being caused by uncompensated impacts of the liquid's molecules.

According to statistical mechanics, the mean random velocity squared of a particle with mass M in a medium with absolute temperature T is equal to $\langle v^2 \rangle = 3k_B T/M$, where k_B is the Boltzmann constant. For a particle with radius $a \sim 1$ μm and with a mass density of the order of the water density at $T = 300$ K, the mean velocity is ~ 0.1 cm/s, that is, of the order of thousands of radii of the particle per second. The velocities seen under microscope are smaller by many times. The theory of Brownian motion developed by A. Einstein and M. von Smoluchowski had not only qualitatively explained the phenomenon and eliminated the paradoxes, in particular, the one just mentioned, but also established its quantitative laws. The experimental verification of these laws became one of the most important arguments in favor of the molecular structure of matter and kinetic nature of thermal phenomena.

Over the decades after the publication of the theory by A. Einstein and
M. von Smoluchowski, many fluctuation phenomena similar to Brownian
motion have been discovered. Therefore at present this term is understood
in a broad sense. Besides the motion of fine particles suspended in a liquid,
the rotational Brownian motion of molecules, random (Brownian) motion
of charge carriers in conductors that produces electrical noise, Brownian
motion of domain walls in ferromagnets and ferroelectrics, random changes
of the phase of an oscillator, and many other similar phenomena are known.

The presentation of the fundamental properties of Brownian motion given
below is based on the theory of diffusion-like Markov processes (Sec. 1.8).
The Langevin approach to Brownian motion is presented in the next section.

The Brownian particle is moving randomly under the impacts of the
surrounding molecules. As the mass m of the molecules is many times
smaller than the mass M of the Brownian particle ($m \ll M$), the change $|\Delta \mathbf{P}|$
of the momentum of the latter is relatively very small. In fact, $|\Delta \mathbf{P}| \leq 2p$,
where $\mathbf{p}$ is the momentum of the incident molecule. The equality $|\Delta \mathbf{P}| = 2p$
corresponds to the motion of the molecule after the elastic collision exactly
in the backward direction. As $p \sim \sqrt{2mk_B T}$ and $P \sim \sqrt{2Mk_B T}$, the relative
changes of the momentum and velocity v of the Brownian particle are

$$|\Delta \mathbf{P}|/P = |\Delta \mathbf{v}|/v \leq 2\sqrt{m/M} \ll 1. \tag{1.9.1}$$

The duration of one impact may be estimated, assuming that the forces
acting between the molecule and the surface of the Brownian particle in
the process of collision are atomic forces. Therefore, the duration of the
interaction is expected to be of the order of a period of short-wavelength
vibrations of atoms in condensed matter, i.e., $\sim 10^{-14} - 10^{-12}$ s. According
to the theory of random pulsed processes (Sec. 1.5) the correlation time τ_c
of such a process is of the order of the duration of one pulse, that is,

$$\tau_c \sim 10^{-14} - 10^{-12} \text{ s}. \tag{1.9.2}$$

By this time, the Brownian particle moving with thermal velocity $v_T =
\sqrt{3k_B T/M}$ traverses a very small path $\sim v_T \tau_c$.

These estimates show that the changes of both the position $\mathbf{r}$ and velocity
$\mathbf{v}$ of the Brownian particle occur by very small steps, or in other words,
considerable changes of $\mathbf{r}$ and $\mathbf{v}$ occur as a result of a huge number of
collisions with the molecules of the liquid. It allows us to consider Brownian
motion as a quasi-continuous (diffusion-like) Markov random process which
is governed by the Fokker–Planck equation (1.8.3) (or Eq. (1.8.6)).

Let us derive the fundamental relations of the theory of Brownian motion.
Suppose that a force $-\nabla U$ is acting on each Brownian particle, where U is

the potential energy of a particle. It may be either the force of gravity $M\mathbf{g}$, where $\mathbf{g}$ is the free fall acceleration, or the force $q\mathbf{F}$ of an electric field $\mathbf{F}$ if each particle has a charge q. Consider the case of a spatially uniform system of particles with probability density $w(\mathbf{v}, t)$ independent of $\mathbf{r}$. Multiplying Eq. (1.8.6) by $\mathbf{v}$ and integrating over $\mathbf{v}$, one obtains the equation for the mean (drift) velocity $\langle\mathbf{v}\rangle$ of the particles. With account of Eq. (1.8.9) it takes the form:

$$\frac{\partial}{\partial t}\langle\mathbf{v}\rangle = \frac{1}{M}(-\nabla U) - \frac{1}{\tau}\langle\mathbf{v}\rangle. \tag{1.9.3}$$

Under stationary conditions

$$\langle\mathbf{v}\rangle = \frac{\tau}{M}(-\nabla U). \tag{1.9.4}$$

In the case of an electric field $-\nabla U = q\mathbf{F}$, the coefficient $\mu = q\tau/M$ is the mobility of the particles. The quantity $k_f = M/\tau$ may be considered as the friction coefficient of the Brownian particle. In a simple model of a spherical particle of radius a, according to the Stokes equation, $k_f = 6\pi\eta a$, where η is the dynamic viscosity of the liquid.

Let us examine two problems. The first is the distribution of the particles in space. According to Eqs. (1.8.12) and (1.9.4) the density of the particles' flow is at each point given by

$$\mathbf{J} = \frac{\tau}{M}n(\mathbf{rt})(-\nabla U) - D\nabla n(\mathbf{r}, t). \tag{1.9.5}$$

In equilibrium, that is, in the absence of a net flow of particles (in the field of gravity the fall of particles is prevented by a reflecting wall or Earth's surface)

$$\frac{1}{n}\nabla n = \frac{\tau}{MD}(-\nabla U) = \frac{\mu}{qD}(-\nabla U). \tag{1.9.6}$$

As is known from statistical mechanics, the equilibrium distribution of the concentration in a field is the Boltzmann distribution:

$$n(\mathbf{r}) = n(0)\exp\left[-\frac{U(\mathbf{r}) - U(0)}{k_B T}\right] \tag{1.9.7}$$

Comparing Eqs. (1.9.6) and (1.9.7), A. Einstein found the fundamental relation between the diffusion coefficient and the mobility:

$$D = \frac{k_B T}{q}\mu, \quad \text{or} \quad D = \frac{k_B T\tau}{M}. \tag{1.9.8}$$

In the absence of any external forces, the density of the Brownian particles obeys the diffusion equation :

$$\frac{\partial n}{\partial t} - D\Delta n = 0. \tag{1.9.9}$$

Suppose the particles at the initial instant $t_0 = 0$ are confined to the point $r_0 = 0$, that is, their density is $n(r,0) = \delta(r)$. The solution to the diffusion equation which satisfies this condition reads:

$$n(r,t) = [4\pi Dt]^{-d/2} \exp\left[-\frac{r^2}{4Dt}\right]. \tag{1.9.10}$$

Here d is the Euclidean dimension of the space in which the spreading of the initially confined particles is observed. The mean displacement squared of the particles from their initial position in any direction α is equal to

$$\langle x_\alpha^2 \rangle = \int dr x_\alpha^2 n(r,t) = 2Dt. \tag{1.9.11}$$

It is the famous Einstein's equation for the variance of the displacement of diffusing particles.

The second problem is the relaxation of the Brownian particle's velocity and the correlation function of the particle's velocities. According to Eq. (1.9.3), in the absence of external forces the mean velocity at instant t, under the condition that at time t_0 it equals v_0, is given by the following equation:

$$\langle v(t) \rangle_{v_0} = v_0 \exp(-t/\tau). \tag{1.9.12}$$

It means that τ is, in fact, the relaxation time of the directional velocity of a particle.

Let us find the correlation function $\psi_{v\alpha\beta}(t_1 - t_2)$ of the components $v_\alpha(t_1)$ and $v_\beta(t_2)$ of a particle's velocity. According to the general theory of Markov processes (Sec. 1.6), one has to find first the one-sided correlation function:

$$\begin{aligned}
\psi_{v\alpha\beta}^+(t_1 - t_2) &= \Theta(t_1 - t_2)\psi_{v\alpha\beta}(t_1 - t_2) \\
&= \Theta(t_1 - t_2) \int dr_1 dv_1 dv_2 v_{1\alpha} P(r_1 v_1 t_1 | r_2 v_2 t_2) v_{2\beta} w(v_2).
\end{aligned} \tag{1.9.13}$$

Here $w(v)$ is the probability density function of the velocities.

Multiply the Fokker–Planck equation (1.8.3) by $v_{1\alpha} v_{2\beta} w(v_2)$ and integrate all terms over r and velocities. Using Eq. (1.8.9) for A_v one obtains a simple equation:

$$\frac{\partial}{\partial t}\psi_{v\alpha\beta}^+(t) + \frac{1}{\tau}\psi_{v\alpha\beta}^+(t) = \delta_{\alpha\beta} \langle v_\alpha^2 \rangle \delta(t)$$

Using also the equation for $\psi^-_{\mathrm{v}\alpha\beta}(t)$ one obtains:

$$\psi_{\mathrm{v}\alpha\beta}(t_1 - t_2) = \delta_{\alpha\beta}\langle v_\alpha^2\rangle \exp[-|t_1 - t_2|/\tau]. \qquad (1.9.14)$$

According to the Wiener–Khintchine relation (1.3.8), the corresponding spectral density is

$$S_{\mathrm{v}\alpha\beta}(f) = \delta_{\alpha\beta}4\langle v_\alpha^2\rangle\tau(1 + \omega^2\tau^2)^{-1}. \qquad (1.9.15)$$

Using Eq. (1.9.8) for the diffusion coefficient of Brownian particles, one can relate the low-frequency ($\omega \ll \tau^{-1}$) spectral density and the diffusion coefficient:

$$S_{\mathrm{v}\alpha\beta}(0) = 4D_{\alpha\beta} = 4D\delta_{\alpha\beta}. \qquad (1.9.16)$$

This relation, which may be called the fluctuation–diffusion relation, means that the low-frequency spectral density of the particles' random velocities differs from their diffusion coefficient only by a constant factor 4. It is important that the temperature T of the system does not appear in this relation. A similar fluctuation–diffusion relation is valid even in some nonequilibrium systems in which the distribution of particles in energy strongly differs from an equilibrium one. An example is a gas of hot electrons in semiconductors when the interelectron interaction is negligible (Sec. 3.5).

Let Brownian particles with radii $a = 1$ μm move in water at $T = 300$ K and their mass density ρ be close to the density of water. The dynamic viscosity of water equals $\eta \approx 10^{-3}$ Pa·s. The velocity relaxation time is then

$$\tau = 2\rho a^2/9\eta \sim 2 \cdot 10^{-7}\mathrm{s}. \qquad (1.9.17)$$

The trajectory of a Brownian particle under visual observation is, in fact, averaged over time intervals of the order of the time of the eye's persistence of vision t_v. In other words, of all frequencies f of particle's velocity fluctuations only a low-frequency band with $f \leq t_v^{-1}$ is visually recorded. The mean square velocity measured by visual observations is of the order of $S_v(0)\Delta f$. As $t_v \gg \tau$ the measured path-length is smaller than the real one and the measured velocity is smaller than the thermal velocity $\sqrt{3k_BT/M}$ by a factor τ/t_v. The difference between the real path and the visible, smoothed, one is schematically shown in Fig. 1.6.

1.10 The Langevin approach to the kinetics of fluctuations

In this section a very general approach to the calculation of correlation properties of random processes is presented. Its idea was suggested by

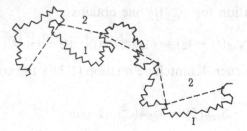

Fig. 1.6 Schematic diagram of a trajectory of a Brownian particle. Dashed curve: visible trajectory smoothed over time intervals equal to the time of the eye's persistence of vision, t_v. Solid curve: trajectory closer to the real one.

Langevin (1908) in his theory of Brownian motion. The principal difference between the Langevin approach and the Markov one (Secs. 1.6–1.8) is the following. The correlation functions $\psi_{\alpha\beta}(t_1, t_2)$ and the transition probabilities $P(\mathbf{x}, t | \mathbf{x}_0, t_0)$ are definite, not random, functions, and the equations for the one-sided correlation function $\psi_{\alpha\beta}^+$ and for P in the Markov approach are equations for definite, not random, quantities which are just the sought-for characteristics of the random process. In contrast with this approach, in the Langevin method equations are derived and solved for the actual random variables $\mathbf{x}(t)$ or their fluctuations $\delta\mathbf{x}(t) = \mathbf{x}(t) - \langle\mathbf{x}\rangle$, and the correlation function is then calculated by averaging the product of fluctuations $\delta\mathbf{x}(t)$ taken at different instants of time (Eqs. (1.2.1) and (1.2.8)):

$$\psi_{\alpha\beta}(t_1, t_2) = \langle \delta x_\alpha(t_1) \delta x_\beta(t_2) \rangle.$$

Let us examine the correlation properties of the Brownian particle's velocity $\mathbf{v}(t)$ using the Langevin approach. As was shown in Sec. 1.9, the relaxation time of the velocity, $\tau = M/k_f \gg \tau_c$, where τ_c is the correlation time of random forces acting on the Brownian particle and is of the order of the duration of an impact of a molecule onto the particle. Using this inequality, $\mathbf{v}(t)$ can be smoothed over time intervals Δt, which satisfy two inequalities: $\tau_c \ll \Delta t \ll \tau$, i.e., they must be much longer than τ_c but much shorter than τ. Let us derive the equation of motion for this smoothed $\mathbf{v}(t)$. In the case of Brownian motion this equation is Newton's equation.

The force acting on the Brownian particle is set up by the impacts of surrounding molecules. This force falls, however, into two parts, and this is the main Langevin's idea. The first part is explicitly expressed in terms of the random velocity $\mathbf{v}$ itself, which, as was mentioned above, is smoothed (averaged) over time intervals Δt. The latter must be long enough so that during this time the particle experiences a huge number of impacts. Hence, this part of the force can be treated macroscopically. The time intervals must be also so short that the particle's velocity remains almost unchanged during Δt. At small $|v|$ this 'friction' force equals $F_f = -k_f v = -(M/\tau)v$ (Eq. (1.9.3)). It results from the fact that the impacts of molecules moving in the opposite direction are on average stronger than the impacts of the ones moving in the direction of the particle's motion.

However, at each instant t there is also another part of the force created by the random impacts of the liquid's molecules. This part of the force is nonzero even at those instants at which $v = 0$. Its variation in time is very frequent, and its correlation time is smaller by many orders of magnitude than the relaxation time τ of the velocity $v(t)$. This force divided by mass M is denoted here by $y(t)$. Its mean value $\langle y \rangle = 0$. According to the theory of random pulses (Sec. 1.5), the correlation function of $y(t)$ is a narrow peak with a width of the order of τ_c. It means that, when only processes smoothed over time intervals $\Delta t \gg \tau_c$ are considered, the correlation function of $y(t)$ may be taken as proportional to the delta-function of the time difference:

$$\psi_y(t_1 - t_2) = \frac{1}{2}S_y(0)\delta(t_1 - t_2).\tag{1.10.1}$$

Here the coefficient at $\delta(t_1 - t_2)$ is expressed, according to the Wiener–Khintchine relationship (1.3.8), in terms of the spectral density at zero frequency:

$$S_y(f) = S_y(0) = \text{const.}, \qquad f \ll \tau_c^{-1}.\tag{1.10.2}$$

Thus, the equation of motion of the Brownian particle takes the form suggested by Langevin (1908):

$$\frac{dv}{dt} + \frac{v}{\tau} = y(t).\tag{1.10.3}$$

The r.h.s. can not be expressed in terms of $v(t)$. It plays the role of the source of the Brownian particle's random motion. Such a source is called a Langevin source, and Eq. (1.10.2) is called the Langevin equation.

Turning to Fourier transforms $v(\omega)$ and $y(\omega)$, one obtains from Eq. (1.10.3):

$$v(\omega) = (-i\omega + \tau^{-1})^{-1}y(\omega)\tag{1.10.4}$$

As was shown above (Sec. 1.3), the spectral density of a random quantity $x(t)$ is proportional to $\langle x(\omega)x^*(\omega)\rangle$. Therefore,

$$S_{v_\alpha}(f) = (-i\omega + \tau^{-1})^{-1}(i\omega + \tau^{-1})^{-1}S_{y_\alpha}(f) = (\omega^2 + \tau^{-2})^{-1}S_{y_\alpha}(0).\tag{1.10.5}$$

Thus, in the Langevin approach the spectral density of the fluctuations is expressed in terms of the spectral density of the fluctuations' source, which is independent of frequency, i.e., is a constant, at all frequencies $f \ll \tau_c^{-1}$. This constant has to be found.

The last problem is especially simple in equilibrium systems. The variance of the fluctuations (Sec. 1.2 and Eq. (1.3.16)), i.e.,

$$\psi_x(0) = \langle (\delta x)^2 \rangle = \int_0^\infty df \, S_x(f), \tag{1.10.6}$$

is a thermodynamic, not a kinetic, quantity (it does not depend on the kinetics of the system, in particular, on its relaxation times), and is known from statistical mechanics or thermodynamics. In the case of Brownian particles

$$\langle v_\alpha^2 \rangle = k_B T / M. \tag{1.10.7}$$

On the other hand,

$$\langle v_\alpha^2 \rangle = \int_0^\infty df \, S_{v_\alpha}(f) = S_{y_\alpha} \int_0^\infty df(\omega^2 + \tau^{-2})^{-1} = \frac{\tau}{4} S_{y_\alpha}. \tag{1.10.8}$$

Comparing Eqs. (1.10.8) and (1.10.7) one obtains:

$$S_{y_\alpha} = \frac{4 \, k_B T}{\tau \; M}. \tag{1.10.9}$$

Upon substitution of this expression into Eq. (1.10.5) one obtains an equation for the spectral density of random velocity which coincides with Eq. (1.9.15) derived by a different (Markov, Fokker–Plank) method.

Equation (1.7.10) for the spectral density $S_x(f)$ of a random variable $x(t)$, which is randomly switching between two values x_1 and x_2, has been derived using the Markov method. It is instructive to derive the same equation by using the Langevin approach.

The fluctuations $\delta x(t)$ can be viewed as originating from the fluctuations of the occupancy numbers, $n_1(t)$ and $n_2(t) = 1 - n_1(t)$, of the two states of the system around their mean values $\langle n_1 \rangle = w_1$ and $\langle n_2 \rangle = w_2 = 1 - w_1$. Then

$$\delta x(t) = x_1 \delta n_1(t) + x_2 \delta n_2(t) = (x_1 - x_2) \delta n_1(t). \tag{1.10.10}$$

The Langevin kinetic equation for $\delta n_1(t)$ reads:

$$\frac{\partial \delta n_1}{\partial t} + W(1 \to 2) \delta n_1(t) - W(2 \to 1) \delta n_2(t) = \delta J_{2 \to 1}(t) - \delta J_{1 \to 2}(t). \tag{1.10.11}$$

Here $W(1 \to 2)$ and $W(2 \to 1)$ are, as in Sec. 1.7, the probabilities, per unit time, of transitions between the two states, $\delta J_{2 \to 1}(t)$ and $\delta J_{1 \to 2}(t)$ are the Langevin sources (flows) corresponding to these transitions. Using the equation $\delta n_2 = -\delta n_1$, one obtains the final form of the Langevin equation:

$$\frac{\partial \delta n_1}{\partial t} + \frac{1}{\tau}\delta n_1(t) = \delta J_{2\to1}(t) - \delta J_{1\to2}(t), \qquad \tau^{-1} = W(1 \to 2) + W(2 \to 1).$$

(1.10.12)

The random process is assumed to be a Markov one: any transition of the system depends only on the its present state, all previous transitions are 'forgotten'. It implies that different transitions are uncorrelated. Therefore the correlation functions of the Langevin sources are proportional to the mean rates of the transitions (compare with Eq. (1.5.8) for the shot noise):

$$\langle \delta J_{1\to2}(t_1)\delta J_{1\to2}(t_2)\rangle = \delta(t_1 - t_2)J_{1\to2}, \qquad S_{J_{1\to2}}(f) = 2J_{1\to2},$$
$$\langle \delta J_{2\to1}(t_1)\delta J_{2\to1}(t_2)\rangle = \delta(t_1 - t_2)J_{2\to1}, \qquad S_{J_{2\to1}}(f) = 2J_{2\to1}, \qquad (1.10.13)$$
$$\langle \delta J_{1\to2}(t_1)\delta J_{2\to1}(t_2)\rangle = 0.$$

Under stationary conditions, the mean rates are equal: $J_{1\to2} = J_{2\to1} = W(1 \to 2)w_1 = W(2 \to 1)w_2$.

The Fourier transform of $\delta n_1(t)$ is found from Eq. (1.10.12):

$$\delta n_1(\omega) = (-i\omega + \tau^{-1})^{-1}[\delta J_{2\to1}(\omega) - \delta J_{1\to2}(\omega)].$$

(1.10.14)

The spectral density of a random variable $x(t)$ is proportional to $\langle x(\omega) \times x^*(\omega)\rangle$. Therefore

$$S_{n_1}(f) = \frac{S_{J_{1\to2}} + S_{J_{2\to1}}}{\omega^2 + \tau^{-2}} = \frac{2(J_{2\to1} + J_{1\to2})}{\omega^2 + \tau^{-2}} = \frac{4J_{1\to2}\tau^2}{1 + \omega^2\tau^2}.$$

(1.10.15)

Since $S_x(f) = (x_1 - x_2)^2 S_{n_1}(f)$ and $W(1 \to 2)\tau = w_2$, the equation for $S_x(f)$ exactly coincides with Eq. (1.7.10).

The Langevin equations for fluctuations in other systems are derived in the same way. They have the form:

$$\hat{L}(t, x)\delta x(t) = y(t).$$

(1.10.16)

Here $\hat{L}(t, x)$ is a linear operator acting on the fluctuation $\delta x(t)$. The r.h.s. $y(t)$, as in Eq. (1.10.3), is the Langevin source of fluctuations. Depending on the physical meaning of both the fluctuating quantity $x(t)$ and Eq. (1.10.16), the Langevin source may be a random force, random flux of particles, random current, etc.

It is assumed that the fluctuation $\delta x(t)$ is smoothed (averaged) over some time intervals Δt that satisfy two conditions (see below). The kinetics of such a smoothed fluctuation is the same as the kinetics of a small (linear) deviation $\langle \Delta x(t)\rangle$ of the same quantity from its stationary (e.g., equilibrium) value under a given, nonrandom perturbation. That is why the l.h.s. of the

Langevin equation for $\delta x(t)$ coincides with the linearized equation for the linear response $\langle \Delta x(t) \rangle$ to a small perturbation.

On one hand, the intervals of time, Δt, should be sufficiently short. Specifically, the expected change of the random variable, $\delta x(t + \Delta t) - \delta x(t)$, must be so small that time-derivatives can be used in the operator $\hat{L}(t, x)$. On the other hand, Δt should be much greater than the correlation time of the Langevin source, so that the correlation between $y(t + \Delta t)$ and $y(t)$ is negligible: the Langevin source randomly changes its sign many times during Δt. Thus, the Langevin approach, like the Markov approach (Sec. 1.6), is based on the existence of the same large gap between the relaxation time of the random quantity under consideration and the much smaller relaxation times, τ_c, of other random variables in the same system which can be considered, comparatively, as 'microscopic' ones. The meaning of the term 'microscopic' used here may differ from the meaning of this term in atomic physics: the quantity $x(t)$, from the standpoint of atomic physics, may also be considered as a microscopic quantity.

The Langevin source of fluctuations can be considered as delta-correlated, that is, its correlation function can be approximated by $\delta(t_1 - t_2)$ times a constant coefficient. If the system is in an equilibrium state, this coefficient is found, as in the example of Brownian motion, by equating the variance of the fluctuations (integral of the spectral density over all positive frequencies) to the corresponding expression known from statistical mechanics. In many nonequilibrium systems the source of fluctuations is similar to a sequence of uncorrelated random events like uncorrelated pulses (see Eq. (1.10.13)). Then Eq. (1.5.6) for the spectral density of this random process can be used.

Several advantages of the Langevin approach can be stated:

1. If the system is out of equilibrium, the one-time correlations (variances and covariances) $\psi_{\alpha\beta}(0)$, in terms of which the correlation functions $\psi_{\alpha\beta}^+(t)$ are expressed if the Markov approach is used (Sec. 1.6), can not be calculated from statistical-mechanical equations. In this case one has to derive and solve the equations for one-time correlation functions $\psi_{\alpha\beta}(0)$ apart from equations for $\psi_{\alpha\beta}^+(t)$. At the same time the Langevin source, as was mentioned above, is often similar to a sequence of uncorrelated random events. Its low-frequency spectral density, which has to be known if the Langevin approach is used, is proportional to the mean rate of these events (Eq. (1.5.6)). This spectral density can be found for all those systems, including nonequilibrium ones, for which the problem of mean quantities can be solved. Thus, for nonequilibrium systems the Langevin approach may be much simpler.

2. The Langevin approach can be applied to nonstationary systems if the char-

acteristic time of the system's variation is much larger than the correlation time τ_c of the Langevin sources.

3. If there are N coupled random variables x_α $(\alpha = 1, \ldots, N)$, one has to solve N^2 equations for the functions $\psi_{\alpha\beta}^+(t)$ (Eq. (1.6.23)), and, in the case of nonequilibrium systems, also the equations for $\psi_{\alpha\beta}(0)$. Using the Langevin method, one has to solve only N inhomogeneous equations for N fluctuations $\delta x_\alpha(t)$ (Lax, 1966).

4. Consider two coupled systems in which the fluctuations are caused by internal Langevin sources. If the coupling between these systems is varied, the correlation functions of the fluctuations may change drastically. However, the spectral densities of the Langevin sources may remain unchanged (Lax, 1966).

5. If the random variable $x(\mathbf{r}t)$ depends not only on the time t but also on the radius-vector $\mathbf{r}$, its correlation function depends on the boundary conditions for the fluctuations. However, because the correlation function of the bulk Langevin sources, $\langle y(\mathbf{r}_1 t_1) y(\mathbf{r}_2 t_2) \rangle$, is proportional to $\delta(\mathbf{r}_1 - \mathbf{r}_2)$, it is independent of the boundary conditions. One and the same expression for the correlation function of the bulk Langevin sources can be used in problems with different boundary conditions, and, consequently, quite different correlation functions $\langle x(\mathbf{r}_1 t_1) x(\mathbf{r}_2 t_2) \rangle$.

In this book the Langevin method is widely used.

2
Fluctuation–dissipation relations in equilibrium systems

In any physical system the dependence of the fluctuations' correlation function on time or, equivalently, the frequency dependence of the spectral density, on one hand, and the response of the same system to external perturbation, on the other hand, are governed by the same kinetic processes, and one can expect that there is some relationship between the two kinetic characteristics of the system. For example, the velocity correlation function of a Brownian particle $\psi_v(t_1 - t_2)$ decays exponentially with $|t_1 - t_2|$ (Sec. 1.9). The corresponding relaxation time τ depends on the viscosity of the liquid, on the mass and linear dimensions of the particle. If the Brownian particle is brought into motion by an external perturbation (e.g., by an electric field if the particle is charged) the particle's stationary velocity and the time of its acceleration and deceleration after switching off the force are determined by the same parameters and, consequently, by the same relaxation time τ.

Such qualitative considerations are usually true for any physical system. However, for equilibrium systems an exact relationship holds between the spectral density of fluctuations at any given frequency f and that part of the linear response of the same system to an external perturbation of the same frequency f, which corresponds to the dissipation of the power of the perturbation. This fundamental relation is called the fluctuation–dissipation relation (FDR), or theorem (FDT). It was first derived by Nyquist in 1928 for fluctuations of current and voltage in quasi-stationary electric circuits, and subsequently, in a far more general form, by Callen & Welton (1951); Callen & Green (1952); Green & Callen (1952), and Callen, Barasch & Jackson (1952). We present below in Sec. 2.1 a general derivation of the FDR. In the next sections we apply it to some specific systems.

2.1 Derivation of fluctuation–dissipation relations

Consider a system which weakly interacts with a thermal bath with a temperature T and is in equilibrium with this thermal bath. The probability of the equilibrium system being in its m-th state is given by the Gibbs distribution:

$$w_m = \exp\left(\frac{F - E_m}{k_B T}\right). \tag{2.1.1}$$

Here E_m is the energy of the m-th eigenstate of the system, F is the free energy of the system which is found using the condition that the sum of all probabilities w_m is equal to unity. The density matrix ρ of any equilibrium system is known to be diagonal in the basis of the eigenstates Ψ_m of its Hamiltonian, and the diagonal elements of ρ are just the Gibbs probabilities of these eigenstates w_m.

Let us first derive the formal quantum-mechanical equation for the spectral density of fluctuations of a physical quantity $x(t)$ in an equilibrium system. In accordance with Eqs. (1.2.11) and (1.3.8),

$$S_x(f) = \int_{-\infty}^{+\infty} d(t_1 - t_2) \text{Tr}\{\rho(\hat{x}(t_1)\hat{x}(t_2) + \hat{x}(t_2)\hat{x}(t_1))\} e^{i\omega(t_1 - t_2)}$$
$$= \int_{-\infty}^{+\infty} d(t_1 - t_2) \sum_{mn} w_m (x_{mn}(t_1)x_{nm}(t_2) + x_{mn}(t_2)x_{nm}(t_1)) e^{i\omega(t_1 - t_2)}. \tag{2.1.2}$$

Here $x_{mn}(t)$ is the matrix element of the Heisenberg operator $\hat{x}(t)$ between the states Ψ_m and Ψ_n. The equation for it,

$$\frac{\partial x_{mn}(t)}{\partial t} = i\omega_{mn} x_{mn}(t), \tag{2.1.3}$$

follows from Eq. (1.2.9) for Heisenberg operators (for brevity, the notation $\hbar\omega_{mn} = E_m - E_n$ was introduced). The solution to Eq. (2.1.3) equals:

$$x_{mn}(t) = e^{i\omega_{mn}t} x_{mn}(0), \tag{2.1.4}$$

The matrix element $x_{mn}(0)$ which is independent of time may be considered as a matrix element of a common (Schrödinger) operator and denoted simply by x_{mn}. Equation (2.1.2) then reads:

$$S_x(f) = 2\pi \sum_{mn} w_m |x_{mn}|^2 [\delta(\omega_{nm} - \omega) + \delta(\omega_{nm} + \omega)]. \tag{2.1.5}$$

In deriving this equation, the integral representation for the delta-function was used:

$$\delta(\omega) = \int_{-\infty}^{+\infty} \frac{dt}{2\pi} e^{i\omega t}. \tag{2.1.6}$$

The spectral density of fluctuations of the quantity $x(t)$ at frequency f, as is evident from Eq. (2.1.5), is determined by the matrix elements of the operator $\hat{x}$ of this quantity between states the energies of which differ by $E_m - E_n = \pm hf$.

The r.h.s. of Eq. (2.1.5) consists of two sums with different delta-functions. Interchanging m and n in the second sum, one easily obtains an equivalent equation for the spectral density:

$$S_x(f) = 2\pi \sum_{mn} (w_m + w_n)|x_{mn}|^2 \delta(\omega_{nm} - \omega). \qquad (2.1.7)$$

Let us substitute Eq. (2.1.1) into Eq. (2.1.7). In the case of equilibrium systems, the ratio of probabilities of two states m and n depends only on the difference of the corresponding energies $E_n - E_m \equiv \hbar\omega_{nm}$:

$$w_n = \exp\left(-\frac{\hbar\omega_{nm}}{k_B T}\right) w_m. \qquad (2.1.8)$$

Due to the delta-function of frequency in Eq. (2.1.5), $\hbar\omega_{nm} = \hbar\omega = hf$. Therefore

$$S_x(f) = 2\pi \left[1 + \exp\left(-\frac{hf}{k_B T}\right)\right] \sum_{mn} w_m |x_{mn}|^2 \delta(\omega_{nm} - \omega). \qquad (2.1.9)$$

Equation (2.1.9) has to be compared with the equation for the response of the system to external perturbation. This response must be expressed in terms of the same quantities: w_m, x_{mn}, and ω_{nm}. Suppose that the quantity x interacts with the field $F(t)$. Let $\hat{x}$ be the quantum-mechanical operator of x and the interaction energy be $\hat{H}_F = -\hat{x}F$. In particular, if x is a component p_α of the system's dipole moment, then F is the corresponding component of the electric field F_α acting on the system. Let $F(t)$ be a periodic 'force' with angular frequency $\omega = 2\pi f$ and amplitude F_0. It must be real:

$$F(t) = \frac{1}{2}\left(F_0 e^{-i\omega t} + F_0^* e^{i\omega t}\right) \qquad (2.1.10)$$

Under this 'force' the system undergoes transitions between its states. It absorbs the energy of the perturbation $F(t)$ and dissipates it. According to the rules of quantum mechanics the rate of transitions from a state m to a state n under a periodic perturbation equals:

$$W_{mn} = (\pi/2\hbar^2)|F_0|^2 |x_{mn}|^2 [\delta(\omega_{nm} - \omega) + \delta(\omega_{mn} - \omega)]. \qquad (2.1.11)$$

Each such transition is accompanied by absorption or emission of an energy quantum $\hbar\omega$ (the first and second terms in rectangular brackets, respectively).

The power transferred to the system is obtained by averaging over the initial states, taking into account their probabilities w_m:

$$\bar{Q} = \sum_{mn} w_m W_{mn} \hbar \omega_{nm}. \tag{2.1.12}$$

Using the Gibbs distribution for equilibrium systems (Eq. (2.1.1)) and Eq. (2.1.11) one obtains:

$$\bar{Q} = (\pi\omega/2\hbar)|F_0|^2 \left[1 - \exp\left(-\frac{hf}{k_B T}\right)\right] \sum_{mn} w_m |x_{mn}|^2 \delta(\omega_{nm} - \omega). \tag{2.1.13}$$

This power may be expressed also in terms of the response of the quantity x to a given external force $F(t)$. The expected linear in F change of x in a stationary system may be written as:

$$\langle x(t) \rangle = \int_{-\infty}^{t} dt' A(t - t') F(t'). \tag{2.1.14}$$

Here the real function $A(t)$ may be called the response function. By virtue of the causality principle the response $\langle x(t) \rangle$ is determined by the force $F(t')$ at instants t' preceding t, that is why the upper limit of integration in Eq. (2.1.8) is t.

If $F(t)$ is a harmonic function of time given by Eq. (2.1.10),

$$\langle x(t) \rangle = \frac{1}{2} \left\{ F_0 A(\omega) e^{-i\omega t} + F_0^* A(-\omega) e^{i\omega t} \right\}, \tag{2.1.15}$$

where

$$A(\omega) = \int_0^{\infty} dt e^{i\omega t} A(t) = A^*(-\omega) \tag{2.1.16}$$

is the frequency dependent susceptibility of the entire system.

The power absorbed by the system and dissipated by it, after averaging over a period $2\pi/\omega$, equals:

$$\bar{Q} = -\overline{\langle x(t) \rangle \dot{F}} = \overline{\langle \dot{x}(t) \rangle F(t)}$$

$$= \frac{1}{4} \left[-i\omega F_0 A(\omega) e^{-i\omega t} + i\omega F_0^* A^*(\omega) e^{i\omega t} \right] \left[F_0 e^{-i\omega t} + F_0^* e^{i\omega t} \right]. \tag{2.1.17}$$

Here the upper dot denotes time derivative, the bar and overline denote averaging over time. Only those products of exponential functions which are independent of time contribute to the dissipated power. Therefore,

$$\bar{Q} = \frac{1}{4} |F_0|^2 i\omega [A^*(\omega) - A(\omega)] = \frac{1}{2} |F_0|^2 \omega A''(\omega). \tag{2.1.18}$$

Here $A''(\omega) \equiv \text{Im} A(\omega)$ is the imaginary part of the complex susceptibility $A(\omega)$.

Comparing Eq. (2.1.9) for $S_x(f)$ with Eqs. (2.1.13) and (2.1.18) for $\bar{Q}$, one obtains the FDR, that is, the relationship between the spectral density of noise $S_x(f)$ and that part of the response at the same frequency $(A''(\omega))$ which determines the dissipated power:

$$S_x(f) = 2\hbar \coth\left(\frac{hf}{2k_B T}\right) A''(\omega). \tag{2.1.19}$$

In many cases it is more convenient to use not the susceptibility $A(\omega)$ but the generalized conductance $G(\omega)$ which is defined as the response of the time derivative of x, i.e., $\dot{x}$, to the 'force' $F(t)$. Obviously, $G(\omega) = -i\omega A(\omega)$. Therefore $G' \equiv \mathrm{Re}\ G = \omega A''$. The spectral density of the quantity $\dot{x}$ is $S_{\dot{x}}(f) = \omega^2 S_x(f)$. Then:

$$S_{\dot{x}}(f) = 2hf \coth\left(\frac{hf}{2k_B T}\right) G'(\omega). \tag{2.1.20}$$

The FDR can be generalized to systems with several coupled (correlated) fluctuating quantities x_α $(\alpha = 1,\ldots, M)$ (Callen, Barasch & Jackson, 1952; Callen & Green, 1952). In equilibrium systems, for which Eq. (2.1.1) holds, the matrix of spectral densities $S_{\alpha\beta}(f)$, given by Eq. (1.3.10), can be represented in the form analogous to Eq. (2.1.9):

$$S_{x_\alpha x_\beta}(f) = 2\pi \left[1 + \exp\left(-\frac{hf}{k_B T}\right)\right] \sum_{mn} w_m(x_\alpha)_{mn}(x_\beta)_{nm}\delta(\omega_{nm} - \omega). \tag{2.1.21}$$

The energy of interaction of the system with the generalized forces (or fields) F_α, where $\alpha = 1,\ldots, M$ equals:

$$\hat{H}_F = -\hat{x}_\alpha F_\alpha, \tag{2.1.22}$$

where summation over α is implied. If the forces $F_\alpha(t)$ are periodic functions of time with amplitudes $F_{0\alpha}$, the rate of transitions of the system from a state m to state n is given by an equation similar to Eq. (2.1.11), and the dissipated power by an equation similar to Eq. (2.1.12):

$$\bar{Q} = (\pi\omega/2\hbar)F_{0\alpha}F_{0\beta}^*\left[1 - \exp\left(-\frac{hf}{k_B T}\right)\right] \sum_{mn} w_m(x_\beta)_{mn}(x_\alpha)_{nm}\delta(\omega_{nm} - \omega). \tag{2.1.23}$$

The equation for the variations of the quantities x_α, linear in the forces F_β, is a generalization of Eq. (2.1.14):

$$\langle x_\alpha(t)\rangle = \int_{-\infty}^{t} dt' A_{\alpha\beta}(t - t')F_\beta(t'). \tag{2.1.24}$$

The quantities $A_{\alpha\beta}(t - t')$ are elements of the response matrix, and

$$A_{\alpha\beta}(\omega) = \int_0^\infty dt e^{i\omega t} A_{\alpha\beta}(t) = A_{\alpha\beta}^*(-\omega) \qquad (2.1.25)$$

are elements of the matrix of susceptibilities. The time derivative of the system's energy equals (a dot on the top of a symbol denotes time derivative):

$$\dot{U} = -\langle x_\alpha(t) \rangle \dot{F}_\alpha(t). \qquad (2.1.26)$$

In full analogy with Eqs. (2.1.17) and (2.1.18), the dissipated power is

$$\bar{Q} = -\overline{\langle x_\alpha(t) \rangle \dot{F}_\alpha(t)} = \overline{\langle \dot{x}_\alpha(t) \rangle F_\alpha(t)} = \frac{\omega}{4i} F_{0\alpha} F_{0\beta}^* [A_{\beta\alpha}(\omega) - A_{\alpha\beta}^*(\omega)]. \qquad (2.1.27)$$

Comparing Eq. (2.1.21) with Eqs. (2.1.23) and (2.1.27), one obtains the generalized fluctuation–dissipation relations (compare with Eqs. (2.1.19) and (2.1.20)):

$$\begin{aligned} S_{x_\alpha x_\beta}(f) &= \hbar \frac{1}{i} [A_{\alpha\beta}(\omega) - A_{\beta\alpha}^*(\omega)] \coth\left(\frac{hf}{2k_B T}\right), \\ S_{\dot{x}_\alpha \dot{x}_\beta}(f) &= hf [G_{\alpha\beta}(\omega) + G_{\beta\alpha}^*(\omega)] \coth\left(\frac{hf}{2k_B T}\right). \end{aligned} \qquad (2.1.28)$$

Here $G_{\alpha\beta}(\omega) = -i\omega A_{\alpha\beta}(\omega)$ is the generalized conductance.

FDRs yield the spectral densities of such quantities as polarization and current in the absence of any forces, e.g., fields, voltages. These fluctuations may be considered as intrinsic, and they play the role of Langevin sources (Sec. 1.10). For instance, a fluctuation of current may create in the conductor a fluctuation of electric field, i.e., a 'force' $F_\alpha^{tot}(t)$. In general, the total fluctuation of the quantity $x_\alpha(t)$ equals $x_\alpha^{tot}(\omega) = x_\alpha(\omega) + A_{\alpha\beta}(\omega) F_\beta^{tot}(\omega)$, where the spectral density of $x_\alpha(t)$ is given by Eq. (2.1.28). The second term in the r.h.s. is the fluctuation of x_α induced by the fluctuation of force. One can represent the total fluctuation of 'force' as $F_\alpha^{tot}(\omega) = y_\alpha(\omega) + A_{\alpha\beta}^{-1}(\omega) x_\beta^{tot}(\omega)$, where

$$x_\alpha(\omega) = -A_{\alpha\beta}(\omega) y_\beta(\omega), \qquad y_\alpha(\omega) = -A_{\alpha\beta}^{-1}(\omega) x_\beta(\omega). \qquad (2.1.29)$$

Here $\hat{A}^{-1}$ is the reciprocal of the tensor $\hat{A}$. The 'force' fluctuation y_α may be considered as a Langevin source on the same footing as x_α, and in some problems its use is more convenient than the use of x_α.

The spectral density of the quantities y_α is easily obtained from Eqs.

(2.1.28) and (2.1.29):

$$S_{y_\alpha y_\beta}(f) = A_{\alpha\alpha'}^{-1} A_{\beta\beta'}^{-1*} S_{x_{\alpha'} x_{\beta'}}(f) = i\hbar [A_{\alpha\beta}^{-1}(\omega) - A_{\beta\alpha}^{-1*}(\omega)] \coth\left(\frac{hf}{2k_B T}\right)$$

$$= hf [G_{\alpha\beta}^{-1}(\omega) + G_{\beta\alpha}^{-1*}(\omega)] \coth\left(\frac{hf}{2k_B T}\right).$$

$$(2.1.30)$$

FDRs are frequently applied to fluctuations of polarization, current density, and the electric field in homogeneous dielectrics and conductors. The fluctuation of the specific polarization, averaged over volume V, is $\delta\mathbf{P}(t) = \delta\mathbf{p}(t)/V$, where $\delta\mathbf{p}$ is the fluctuation of the total dipole moment and corresponds to x in Eq. (2.1.19). Therefore, $S_\mathbf{P} = S_\mathbf{p}/V^2$. The fluctuation of the current density is $\delta\mathbf{j}(t) = \partial\delta\mathbf{P}/\partial t$. In the case of a homogeneous system, $A_{\alpha\beta} = V\chi_{\alpha\beta}$, where $\chi_{\alpha\beta}$ is the tensor of specific susceptibility. Taking into account these equations and the general relations (2.1.28), one obtains the FDR for the fluctuations of current density in the absence of electric field fluctuations and the fluctuations of electric field in the absence of current fluctuations:

$$S_{j_\alpha j_\beta}(f) = \frac{1}{V} hf (\sigma_{\alpha\beta}(\omega) + \sigma_{\beta\alpha}^*(\omega)) \coth\left(\frac{hf}{2k_B T}\right), \qquad (2.1.31)$$

$$S_{F_\alpha F_\beta}(f) = \frac{1}{V} hf (\rho_{\alpha\beta}(\omega) + \rho_{\beta\alpha}^*(\omega)) \coth\left(\frac{hf}{2k_B T}\right). \qquad (2.1.32)$$

Here $\sigma_{\alpha\beta}$ and $\rho_{\alpha\beta}$ are tensors of conductivity and resistivity, respectively.

In the derivation of Eqs. (2.1.31) and (2.1.32) the linear dimensions of the volume V over which the fluctuations are averaged have been implicitly assumed to be much larger than the correlation length (or correlation radius) of the fluctuations, e.g., larger than the free path length. However, one can write down the FDR for a more general case (Sec. 2.3).

The fluctuations of current density and electric field, the spectral densities of which are given by Eqs. (2.1.31) and (2.1.32), are intrinsic fluctuations, and play the role of Langevin sources (see above). The total fluctuation of the current density is the sum of the intrinsic fluctuation and the fluctuation produced by the fluctuation of the electric field, $\delta\mathbf{F}(t)$. The Fourier components of the total current density fluctuation equal:

$$\delta j_\alpha(\omega) = \sigma_{\alpha\beta}(\omega)\delta F_\beta(\omega) + \delta j_\alpha^{\text{int}}(\omega). \qquad (2.1.33)$$

The fluctuation of the field is determined by the conditions imposed on the system. In the case of fluctuations in a conductor, the electric field fluctuation is zero if the conductor is shorted out.

The combination of generalized polarizabilities $[A_{\alpha\beta} - A_{\beta\alpha}^*]/i$ is obviously a Hermitian matrix, i.e., it remains invariant under simultaneous complex conjugation and interchange of indexes. It means that the spectral density matrix given by the FDR is Hermitian in accordance with the general property of the spectral density matrix (Eq. 1.3.12).

The Onsager's principle of kinetic coefficients' symmetry is a consequence of the symmetry of the equations of the particles' motion under time reversal and simultaneous change of the sign of the magnetic field **B** in which the system is placed. According to this principle,

$$A_{\alpha\beta}(\omega; \mathbf{B}) = \pm A_{\beta\alpha}(\omega; -\mathbf{B}). \tag{2.1.34}$$

The plus sign applies when the quantities $\hat{x}_\alpha$ and $\hat{x}_\beta$ are either both invariant under time reversal or both change sign under time reversal. The minus sign applies when only one of the two quantities changes its sign under time reversal. An example of a quantity that changes its sign under time reversal is a particle's velocity.

It follows from Eq. (2.1.34) that the combination of kinetic coefficients in the r.h.s. of Eq. (2.1.28) satisfies the following condition:

$$A_{\alpha\beta}(\omega; \mathbf{B}) - A_{\beta\alpha}^*(\omega; \mathbf{B}) = \pm[A_{\beta\alpha}(\omega; -\mathbf{B}) - A_{\alpha\beta}^*(\omega; -\mathbf{B})]. \tag{2.1.35}$$

Owing to Eq. (2.1.35) the spectral density matrix given by the FDR satisfies the relation:

$$S_{x_\alpha x_\beta}(f; \mathbf{B}) = \pm S_{x_\alpha x_\beta}^*(f; -\mathbf{B}). \tag{2.1.36}$$

The last relation means that in the absence of a magnetic field, the elements of the spectral density matrix given by Eq. (2.1.28) are either purely real, if both quantities $\hat{x}_\alpha$ and $\hat{x}_\beta$ are of the same time reversal symmetry, or otherwise they are purely imaginary (compare with Eq. (1.3.13)).

The universal function of frequency f and absolute temperature T in the FDR can be represented as:

$$\frac{hf}{2} \coth\left(\frac{hf}{2k_B T}\right) = hf\left[\bar{N}(f, T) + \frac{1}{2}\right]. \tag{2.1.37}$$

Here

$$\bar{N}(f, T) = \left[\exp\left(\frac{hf}{k_B T}\right) - 1\right]^{-1}. \tag{2.1.38}$$

is the mean number of quanta (discrete excitations) of an oscillator with frequency f at temperature T. The r.h.s. of Eq. (2.1.37) is the mean energy of this oscillator. It consists of ground state energy $hf/2$ (it is called zero-point energy, or the energy of zero-point vibrations) and the mean energy of the oscillator's excitations. At absolute zero temperature only the ground state energy remains.

The expression in the r.h.s. of Eq. (2.1.37) can be represented by $\bar{N}+1/2 = [\bar{N} + (\bar{N} + 1)]/2$. It is known from quantum mechanics that the rate of processes in which the oscillator loses a quantum of energy (it is gained by the system with which the oscillator interacts) is proportional to $\bar{N}(f, T)$. These processes are impossible at $T = 0$ when $\bar{N}(f, T) = 0$. Similarly, the rate of processes in which the oscillator gains a quantum from another system, e.g., the measuring device, is proportional to $\bar{N}(f, T) + 1$. At $T = 0$ only processes in which the system under consideration is excited are possible, and they correspond to the last unity.

At the opposite extreme case $k_B T \gg hf$, the quasi-classical approximation holds. Even at $T = 1$ K, the ratio $hf/2k_B T > 1$ at $f > 4 \cdot 10^{10}$ Hz only. In this, quasi-classical, limit $\coth(hf/2k_B T) \approx 2k_B T/h$ and

$$S_{\dot{x}}(f) = 4k_B T G'(f). \qquad (2.1.39)$$

The spectral density in an equilibrium system can be expressed also in terms of the relaxation function of the same system. Let F_0 be a small force acting on the system from the remote past up to $t = 0$ when this force is switched off. Under the action of this force, the mean value of the quantity $\hat{x}$ was, at $t < 0$, a nonzero constant, linear in F_0, which we denote as $x_0 \equiv \langle \hat{x}(0) \rangle$. Owing to the same kinetic processes that produce fluctuations, the quantity $\langle \hat{x}(t) \rangle$ decays to zero after switching off the force ($t > 0$). The ratio $\Phi_x(t) = \langle \hat{x}(t) \rangle / F_0$ at $t > 0$ can be called the relaxation function. At small F_0 it is independent of F_0.

The expectation value of the quantity $\hat{x}$ can be represented by the average value of the linear in $F(t)$ change of the Heisenberg operator $\hat{x}(t)$. According to Eqs. (1.2.9) and (5.5.5),

$$\langle \hat{x}(t) \rangle = \frac{i}{\hbar} \int_{-\infty}^{t} dt' \langle [\hat{H}_F(t'), \hat{x}(t)] \rangle. \qquad (2.1.40)$$

We have to substitute into this equation the expression for $\hat{H}_F = -\hat{x}(t)F(t)$, and to take into account that $F(t) = F_0$ up to $t = 0$. As the function $\langle [\hat{x}(t'), \hat{x}(t)] \rangle$ depends on $t - t'$ only, one obtains:

$$\Phi_x(t) = \frac{i}{\hbar} \int_{t}^{\infty} dt' \langle [\hat{x}(t'), \hat{x}(0)] \rangle. \qquad (2.1.41)$$

It is easy to show that

$$\int_0^\infty dt \sin(\omega t)\left(-\frac{d\Phi_x(t)}{dt}\right) = \frac{1}{2\hbar}\int_{-\infty}^{+\infty} dt e^{i\omega t}\langle[\hat{x}(t), \hat{x}(0)]\rangle$$

$$= \frac{\pi}{\hbar}\left[1 - \exp\left(-\frac{hf}{k_B T}\right)\right]\sum_{mn} w_m |x_{mn}|^2 \delta(\omega_{nm} - \omega). \qquad (2.1.42)$$

Comparing Eqs. (2.1.42) and (2.1.9) one obtains the sought relation:

$$S_x(f) = 2\hbar \coth\left(\frac{hf}{2k_B T}\right)\int_0^\infty dt \sin(\omega t)\left(-\frac{d\Phi_x(t)}{dt}\right). \qquad (2.1.43)$$

FDRs express the spectral density of fluctuations in the equilibrium state of *any* system in terms of the dissipative part of the linear response of the same system to a small enough force (or in terms of the relaxation function). FDRs hold even if the response of this system to a strong force is arbitrarily nonlinear. In this sense FDRs are quite general.

FDRs (Eqs. (2.1.19), (2.1.28), and (2.1.43)) may be applied only to equilibrium systems. Of course, the spectral densities of the Langevin sources can be found, in principle, also for nonequilibrium systems. However, in these systems the spectral densities are not directly related to the linear response of the corresponding system. It means that in nonequilibrium systems the linear response and spectral density of noise yield *independent* information on the kinetics of the system and complement each other. The theory of fluctuations in nonequilibrium gases is presented in Ch. 3.

A 'force' applied to the system removes it from its equilibrium state and, at the same time, reveals such fluctuation processes which, due to their symmetry, are not observable in an equilibrium noise. Often this nonequilibrium noise exceeds the equilibrium one even when the 'force' is so small that the deviation from equilibrium is negligible in all other respects.

2.2 Equilibrium noise in quasi-stationary circuits. Nyquist theorem

The thermal motion of charge carriers in any conductor is a source of random fluctuations of current (de Haas-Lorentz, 1913; Johnson, 1928; Nyquist, 1928). Due to the randomness of the thermal motion, the current created by those charge carriers that are moving in a given direction is, in general, not counterbalanced by the current of those moving in the opposite direction. The net current is therefore not zero even in the absence of any external current sources, i.e., electromotive forces. The mean value of this random, fluctuational, current is, of course, zero. The random current stems from the randomness of the thermal motion of charge carriers, which is analogous to Brownian motion (de Haas-Lorentz, 1913). This noise is universal, i.e., exists in any conductor, irrespective of the type of conduction (electronic, ionic).

The random current, $\delta I(t)$, that passes through the conductor in a circuit and can be, in principle, measured, differs, in general, from the random current that is directly induced by the thermal motion of charge carriers and may be called intrinsic current, $\delta I^{\text{int}}(t)$. Let us denote the impedance of the conductor under consideration by $Z(f)$, and that of the circuit that closes the conductor (load) by $Z_L(f)$. Let us also assume that the only sources of fluctuations are the charge carriers in the conductor. The current $\delta I(t)$ induces a voltage across the load impedance. Its Fourier component is

$$\delta U_L(\omega) = Z_L(f)\delta I(\omega), \tag{2.2.1}$$

where $\omega = 2\pi f$ is the angular frequency.

Let us consider quasi-stationary circuits, the linear dimensions of which $L \ll \lambda$, where $\lambda = c/f$ is the wavelength of the electromagnetic wave with frequency f and c is the speed of light. According to the Kirchhoff law, the fluctuation of voltage across the electrodes of the conductor is $\delta U(\omega) = -\delta U_L(\omega)$. The current in the conductor and, hence, in the quasi-stationary circuit is the sum of the source current $\delta I^{\text{int}}(\omega)$ and the current $Z^{-1}(f)\delta U(\omega)$ induced by the voltage $\delta U(\omega)$, i.e.,

$$\delta I(\omega) = -[Z_L(f)/Z(f)]\delta I(\omega) + \delta I^{\text{int}}. \tag{2.2.2a}$$

It means that

$$\delta I(\omega) = [1 + Z_L(f)/Z(f)]^{-1}\delta I^{\text{int}}(\omega),$$
$$\delta U(\omega) = -Z_L(\omega)[1 + Z_L(f)/Z(f)]^{-1}\delta I^{\text{int}}(\omega). \tag{2.2.2}$$

In a short-shunted conductor ($Z_L \to 0$) the current $\delta I = \delta I^{\text{int}}$ and, obviously, $\delta U = 0$. In a disconnected conductor, on the contrary, $\delta I = 0$, $\delta U(\omega) = -\delta U^{\text{int}}(\omega)$, where the voltage

$$\delta U^{\text{int}}(\omega) \equiv Z(f)\delta I^{\text{int}}(\omega) \tag{2.2.3}$$

can be viewed as a random electromotive force applied across the conductor. It can be considered and used in calculations, on the same footing as δI^{int}, as the source of equilibrium noise.

The equations for the spectral densities of the current $\delta I^{\text{int}}(t)$ and emf $\delta U^{\text{int}}(t)$ have been derived in a very fundamental and elegant way by Nyquist (1928). He analyzed the experiments by Johnson (1928), who was the first to observe the equilibrium noise in conductors and to prove experimentally that the spectral density of δU^{int} is proportional to the resistance of the conductor and to its absolute temperature. Now one can derive the Nyquist theorem, as Nyquist's equation for the spectral density of equilibrium current and voltage noise is called, from the fluctuation–dissipation relations (Sec. 2.1). However, one must keep in mind that these relations may be viewed as a generalization of the Nyquist theorem.

Since the dissipated power in the conductor is $\bar{Q} = \overline{IU}$ (Eq. (2.1.17)), the FDR in the form of Eq. (2.1.20) must be used. In our case the conductance $G(f)$ is the inverse impedance $Z^{-1}(f)$:

$$S_I^{\text{int}}(f) = 2hf \coth\left(\frac{hf}{2k_B T}\right) \text{Re}\, Z^{-1}(f),$$

$$S_U^{\text{int}}(f) = |Z(f)|^2 S_I^{\text{int}}(f) = 2hf \coth\left(\frac{hf}{2k_B T}\right) \text{Re}\, Z(f).$$

$$(2.2.4)$$

It is very instructive to recover Eq. (2.2.4) using the original Nyquist's ideas. Consider two conductors with impedances $Z_1(f)$ and $Z_2(f)$, respectively, connected by perfect conductors. An ideal filter that passes only fluctuations in a definite frequency band can be placed between the conductors. Both conductors, which are assumed to have identical temperatures, $T_1 = T_2 \equiv T$, are sources of current fluctuations $\delta I_1^{\text{int}}(t)$ and $\delta I_2^{\text{int}}(t)$, respectively. The first one induces a current $\delta I_1(t)$ in the circuit and a voltage $\delta U_{1\to 2}(t)$ across the second conductor. The electric power dissipated by the first conductor in the second one is the average of the product of these two quantities:

$$W_{1\to 2} = \langle \delta U_{1\to 2}(t)\delta I_1(t) \rangle. \qquad (2.2.5)$$

Substituting the Fourier expansions of the fluctuations and using Eqs. (2.2.1), (2.2.2), and the relation $Z(-\omega) = Z^*(\omega)$, one obtains:

$$W_{1\to 2} = \int_0^\infty df \frac{\text{Re}\, Z_2^{-1}(f)}{|Z_1^{-1}(f) + Z_2^{-1}(f)|^2} S_{I1}^{\text{int}}(f). \qquad (2.2.6)$$

Similarly, the second conductor dissipates in the first one a power

$$W_{2\to 1} = \int_0^\infty df \frac{\text{Re}\, Z_1^{-1}(f)}{|Z_1^{-1}(f) + Z_2^{-1}(f)|^2} S_{I2}^{\text{int}}(f). \qquad (2.2.7)$$

The two powers must be identical. If they were not identical the power would be transferred from one body to another one with the same temperature, violating the second law of thermodynamics. Not only must the integrals over frequency in Eqs. (2.2.6) and (2.2.7) be identical, but also the integrands, because the identity holds also when both conductors are

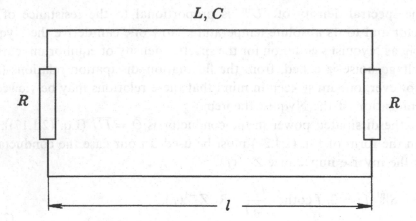

Fig. 2.1 The scheme of resistors connected by a transmission line used in the Nyquist's derivation of the fluctuation–dissipation relation. R are the resistances at the ends of the transmission line, L and C are the inductance and capacity of the line.

connected by a filter with any passband. Thus,

$$\frac{S_{I1}^{int}(f)}{\operatorname{Re} Z_1^{-1}(f)} = \frac{S_{I2}^{int}(f)}{\operatorname{Re} Z_2^{-1}(f)}. \tag{2.2.8}$$

This equation holds for any pair of conductors. It means that the ratio,

$$\frac{S_I^{int}(f)}{\operatorname{Re} Z^{-1}(f)} \equiv F(f, T), \tag{2.2.9}$$

is a universal function of frequency f and temperature. The expression for it can be found by considering any simple model.

Nyquist (1928) considered a system of two identical resistors with resistance R, both at a common temperature T, connected by an ideal non-dissipative transmission line with the same wave resistance, i.e., the line and the two resistors at its ends are matched (Fig. 2.1). The electric fluctuations generated in the resistors are exciting electromagnetic waves into the line and thus the resistors exchange power. As the line and the resistors are matched, the excited waves are not reflected back. If the line is abruptly shorted out at its ends the waves are trapped and become standing waves. According to quantum statistical mechanics, the average energy per mode with a frequency f is equal to the mean energy of one quantum oscillator,

$$\bar{E}_f = hf\left(\bar{N}(f, T) + \frac{1}{2}\right), \tag{2.2.10}$$

where (see Eq. (2.1.38))

$$\bar{N}(f,T) = \left[\exp\left(\frac{hf}{k_B T}\right) - 1\right]^{-1} \tag{2.2.11}$$

is the mean number of excitations (quanta) in the mode.

The number of modes (standing waves) in the line, shorted out at its ends, in a frequency range Δf is $(2l/c)\Delta f$, where l is the length of the line and c is the phase speed of the waves. The average energy of the waves in the line in this frequency range is $(2l/c)\bar{E}_f \Delta f$. It is equal to the energy which was sent by the two resistors over a time $\tau = l/c$ that is necessary for the waves to reach the opposite end of the line. Hence, the power sent by each resistor is just $\bar{E}_f \Delta f$. On the other hand, according to Eqs. (2.2.6) and (2.2.9), this power is $(1/4)RS_I^{int}(f)\Delta f = (1/4)F(f,T)\Delta f$. Comparing the two expressions, one concludes that:

$$F(f,T) = 4\bar{E}_f, \tag{2.2.12}$$

in accordance with Eq. (2.2.4).

In some problems one has to find the flow of the thermal electric fluctuations' power from one conductor to another one. Due to the presence of 1/2 in the r.h.s. of Eq. (2.2.10) this flow always consists partly of the flow of zero-point (ground state) energy of quantum oscillators. This part of the flow is always exactly compensated by an identical but opposite flow from other conductors, even if these conductors are at absolute zero temperature. Hence, when considering the exchange of fluctuations' power between different conductors, the zero-point energy term in the spectral density of equilibrium fluctuations may be always dropped out of all energy fluxes (flows). This can be done not only when the conductors are matched, as in the model considered by Nyquist (see above), but in a general case when the waves generated by the equilibrium noise are partly reflected (for a discussion of these problems see the book by Levin & Rytov, 1967). After the zero-point energy term is dropped out, the energy radiation from a conductor at $T = 0$ becomes zero, and seeming paradoxes disappear. Indeed, a nonzero energy radiation from a body, which at $T = 0$ is in its ground state, seems to be paradoxical.

The noise temperature T_N multiplied by the Boltzmann constant k_B is, by definition, the power, per unit frequency band around a frequency f, which is dissipated by a resistor with impedance $Z(f)$ in a matched load, i.e., in a load with impedance $Z^*(f)$. Using Eqs. (2.2.4) and (2.2.6) one obtains for an equilibrium resistor in the quasi-classical regime, $hf \ll k_B T$:

$$T_N = T. \tag{2.2.13}$$

It means that the absolute temperature of a conductor can be measured by measuring its noise in equilibrium.

If the noise frequency $f \ll k_B T / h$, Eq. (2.2.4) can be simplified:

$$S_I^{int}(f) = 4k_B T \operatorname{Re} Z^{-1}(f), \qquad S_U^{int}(f) = 4k_B T \operatorname{Re} Z(f). \tag{2.2.14}$$

In fact, even at $T = 1$ K the ratio $hf/k_B T$ is unity only at a rather high frequency $f \approx 20$ GHz. The quantum noise, i.e., noise at $hf \gg k_B T$, is difficult to observe directly. The main obstacle is, apart from low temperatures, the cutoff frequency of the circuitry: $\operatorname{Re} Z(f)$ and $\operatorname{Re} Z^{-1}(f)$ fall off at high frequencies. In the case of an $R-C$ and $R-L$ circuit the cutoff frequency is $f_{RC} = 1/2\pi RC$ and $f_{RL} = R/2\pi L$, respectively. However, the quantum noise was observed in Josephson junctions (Koch, Van Harlingen & Clarke, 1982) due to the strong nonlinearity of this device: the high-frequency quantum noise in the resistor was mixed-down to low frequencies (Sec. 9.1.3).

Owing to the Nyquist relationship, the measurement or calculation of the equilibrium noise can not give any additional information besides that which is obtained by measurement or calculation of the real part of the impedance or real part of the inverse impedance.

As was emphasized above, the equilibrium current noise is generated by the random motion back and forth of the charge carriers, i.e., by the fluctuations of the directions of the charge carriers' velocities. In the same equilibrium systems many other physical quantities fluctuate: the temperature, the number of charge carriers, the number of scattering centers and their orientation in the crystal, the distribution of charge carriers in energy, and so on. However, these fluctuations, due to their symmetry, are not measured (revealed) in the current or voltage noise unless a mean current, generated by an external current source, is passed through the conductor. The magnitude of the mean current which is sufficient to reveal the 'hidden' fluctuations and generate a measurable 'excess', nonequilibrium noise is often so small that the conductor in all other respects can be considered as an equilibrium one. This sensitivity of the noise is the basis of fluctuation spectroscopy, i.e., of studying the kinetic properties by measurements of noise spectra.

2.3 Fluctuations of electromagnetic fields in continuous media

The random motion of the charged particles in any medium creates, in accordance with the Maxwell equations, a random electromagnetic field, i.e., electromagnetic fluctuations. In its turn, this field affects the motion of the charged particles. Using the fluctuation–dissipation relations (FDR, Sec. 2.1) one is able to relate, in a very general way, the spectral densities of the electric and magnetic fields with that part of the dielectric permittivity of the medium which corresponds to the dissipation of the electromagnetic field's energy. These spectral densities have been found by Rytov (1953). A straightforward method of their derivation from the general FDR was developed by Landau & Lifshitz (1957) (see also Levin & Rytov, 1967; Lifshitz & Pitaevskii, 1980). By solving the Maxwell equations at given boundary conditions, one can express the electromagnetic field fluctuations in terms of the Langevin current sources. If the correlation function of the latter is known, the sought correlation functions and spectral densities of the electromagnetic fields are easily found.

The random electromagnetic field satisfies the Maxwell equations. The current density fluctuation $\delta \mathbf{j}(\mathbf{r}t)$, which appears in these equations, may be represented as a sum of two random currents. One of them is the response of the charges in the medium to the random electromagnetic field $\delta \mathbf{E}, \delta \mathbf{H}$, the second is the Langevin, intrinsic, current $\delta \mathbf{j}^{\text{int}}(\mathbf{r}t)$ which plays the role of the source of fluctuations (Sec. 1.10). In the electrodynamics of continuous media, the current density is usually split in two parts: the polarization current $\partial \mathbf{P}/\partial t$ and the magnetization current $c\,\mathbf{curl}\,\mathbf{M}$, where $\mathbf{P}$ and $\mathbf{M}$ are the specific dielectric polarization and magnetization, respectively, and c is the speed of light. The Langevin current may be split in the same way:

$$\delta \mathbf{j}^{\text{int}} = \frac{\partial \delta \mathbf{P}^{\text{int}}}{\partial t} + c\,\mathbf{curl}\,\delta \mathbf{M}^{\text{int}}. \tag{2.3.1}$$

Each Fourier amplitude of electric or magnetic induction fluctuations consists of a part which is the response to the fluctuation of the corresponding field and a part which is directly connected to the corresponding part of the Langevin current:

$$\delta D_\alpha(\mathbf{r}\omega) = \int d\mathbf{r}'\,\epsilon_{\alpha\beta}(\mathbf{r}\mathbf{r}'\omega)\delta E_\beta(\mathbf{r}'\omega) + 4\pi\delta P_\alpha^{\text{int}}(\mathbf{r}\omega),$$
$$\delta B_\alpha(\mathbf{r}\omega) = \int d\mathbf{r}'\,\mu_{\alpha\beta}(\mathbf{r}\mathbf{r}'\omega)\delta H_\beta(\mathbf{r}'\omega) + 4\pi\delta M_\alpha^{\text{int}}(\mathbf{r}\omega). \tag{2.3.2}$$

Here $\epsilon_{\alpha\beta}(\mathbf{r}\mathbf{r}'\omega)$ and $\mu_{\alpha\beta}(\mathbf{r}\mathbf{r}'\omega)$ are the dielectric and magnetic permittivity, respectively. In general, each of them depends on two radius-vectors: the first determines the point at which the induction is measured, the second is the point where the field is taken.

The Maxwell equations can be written as follows:

$$\frac{1}{c}\frac{\partial(\epsilon\delta\mathbf{E})}{\partial t} - \mathbf{curl}\,\delta H + \frac{4\pi}{c}\frac{\partial\delta\mathbf{P}^{\text{int}}}{\partial t} = 0,$$

$$\frac{1}{c}\frac{\partial(\mu\delta\mathbf{H})}{\partial t} + \mathbf{curl}\,\delta E + \frac{4\pi}{c}\frac{\partial\delta\mathbf{M}^{\text{int}}}{\partial t} = 0. \tag{2.3.3}$$

Here $(\epsilon\delta\mathbf{E})$ and $(\mu\delta\mathbf{H})$ are the first terms in the r.h.s. of the first and second Eq. (2.3.2), respectively.

Multiplying all terms of the first equation by $\delta\mathbf{E}$, the second by $\delta\mathbf{H}$ and summing both equations, one obtains the continuity equation for the electromagnetic field energy. The term that corresponds to the rate of dissipation of this energy equals:

$$\dot{U} = -\int d\mathbf{r}\left[\delta\mathbf{E}(\mathbf{r}t)\delta\dot{\mathbf{P}}^{\text{int}}(\mathbf{r}t) + \delta\mathbf{H}(\mathbf{r}t)\delta\dot{\mathbf{M}}^{\text{int}}(\mathbf{r}t)\right]. \tag{2.3.4}$$

Here a dot over a symbol denotes time derivative. This equation has the same form as Eq. (2.1.26) if the integral over the volume is substituted by a sum over small but finite and macroscopic volumes ΔV, and the fields and the Langevin sources are substituted by their mean values averaged over each of these small volumes (Landau & Lifshitz, 1957). Comparing this sum with Eq. (2.1.26), one concludes that the field fluctuations correspond to the quantities x_α and the Langevin sources to the 'forces' y_α:

$$x_\alpha = \{\delta\mathbf{E}(\mathbf{r}_i)\Delta V, \Delta\mathbf{H}(\mathbf{r}_i)\Delta V\}; \quad y_\alpha = \{\delta\mathbf{P}^{\text{int}}(\mathbf{r}_i), \delta\mathbf{M}^{\text{int}}(\mathbf{r}_i)\}. \tag{2.3.5}$$

Here each $\mathbf{r}_i$ corresponds to one of the small volumes ΔV.

To obtain the specific FDR one has to know the relation between the quantities x_α and y_α (Eq. (2.1.29)):

$$y_\alpha(\omega) = -A_{\alpha\beta}^{-1}(\omega)x_\beta(\omega), \tag{2.3.6}$$

that is, to know the coefficients $A_{\alpha\beta}^{-1}(\omega)$. In the case of electromagnetic fluctuations the role of the linear relations, Eq. (2.3.6), is played by the Maxwell equations, which may be represented in the form:

$$\delta\mathbf{P}^{\text{int}}(\mathbf{r}\omega) = -\frac{1}{4\pi}\sum_{\beta\mathbf{r}'}\epsilon_{\alpha\beta}(\mathbf{r}\mathbf{r}'\omega)\delta E_\beta(\mathbf{r}'\omega)\Delta V + \frac{ic}{4\pi\omega}\mathbf{curl}\,\delta\mathbf{H}(\mathbf{r}\omega),$$

$$\delta M_\alpha^{\text{int}}(\mathbf{r}\omega) = -\frac{1}{4\pi}\sum_{\beta\mathbf{r}'}\mu_{\alpha\beta}(\mathbf{r}\mathbf{r}'\omega)\delta H_\beta(\mathbf{r}'\omega)\Delta V - \frac{ic}{4\pi\omega}\mathbf{curl}\,\delta\mathbf{E}(\mathbf{r}\omega). \tag{2.3.7}$$

In accordance with the transition from the integration over coordinates to summation over finite volumes ΔV, the differential operation **curl** has to be substituted by finite differences.

Let us assign the first $3N$ components x_α to the fluctuations of the electric field (three components of the vector $\delta\mathbf{E}$ at N points $\mathbf{r}_i$), and the next $3N$ components to the fluctuations of the magnetic field $\delta\mathbf{H}(\mathbf{r}_i)$. Respectively, the first $3N$ components y_α are $\delta\mathbf{P}^{\mathrm{int}}(\mathbf{r}_i)$, the second $3N$ components are the fluctuations $\delta\mathbf{M}^{\mathrm{int}}(\mathbf{r}_i)$. By virtue of the fact that the coefficients at the **curl** $\delta\mathbf{E}$ and **curl** $\delta\mathbf{H}$ in the Maxwell equations are purely imaginary and have opposite signs, the coefficients $A_{\alpha\beta}^{-1}$ with $\alpha \le 3N$, $\beta > 3N$ satisfy the relation $A_{\alpha\beta}^{-1} = A_{\beta\alpha}^{-1*}$. It follows from the FDR (2.1.28) that the corresponding spectral density is zero. It means that there is no mutual correlation between the fluctuations $\delta\mathbf{P}^{\mathrm{int}}$ and $\delta\mathbf{M}^{\mathrm{int}}$.

The equations for the spectral densities of the Langevin sources taken at different points $\mathbf{r}_1$ and $\mathbf{r}_2$ follow immediately from the FDR (Eq. (2.1.28)) and Eq. (2.3.7):

$$S_{P_\alpha P_\beta}^{\mathrm{int}}(\mathbf{r}_1, \mathbf{r}_2, f) = \frac{\hbar}{4\pi i}\left[\epsilon_{\alpha\beta}(\mathbf{r}_1, \mathbf{r}_2, \omega) - \epsilon_{\beta\alpha}^*(\mathbf{r}_2, \mathbf{r}_1, \omega)\right]\coth\left(\frac{hf}{2k_B T}\right),$$

$$S_{M_\alpha M_\beta}^{\mathrm{int}}(\mathbf{r}_1, \mathbf{r}_2, f) = \frac{\hbar}{4\pi i}\left[\mu_{\alpha\beta}(\mathbf{r}_1, \mathbf{r}_2, \omega) - \mu_{\beta\alpha}^*(\mathbf{r}_2, \mathbf{r}_1, \omega)\right]\coth\left(\frac{hf}{2k_B T}\right). \tag{2.3.8}$$

If the nonlocal effects are negligible,

$$\epsilon_{\alpha\beta}(\mathbf{r}_1, \mathbf{r}_2, \omega) = \epsilon_{\alpha\beta}(\mathbf{r}_1, \omega)\delta(\mathbf{r}_1 - \mathbf{r}_2), \quad \mu_{\alpha\beta}(\mathbf{r}_1, \mathbf{r}_2, \omega) = \mu_{\alpha\beta}(\mathbf{r}_1, \omega)\delta(\mathbf{r}_1 - \mathbf{r}_2), \tag{2.3.9}$$

where $\epsilon_{\alpha\beta}(\mathbf{r}, \omega)$ and $\mu_{\alpha\beta}(\mathbf{r}, \omega)$ are the common macroscopic dielectric and magnetic permittivities. If, moreover, the system is not placed in a magnetic field, the simple form of the Onsager relations holds, i.e.,

$$\epsilon_{\alpha\beta}(\omega) = \epsilon_{\beta\alpha}(\omega), \qquad \mu_{\alpha\beta}(\omega) = \mu_{\beta\alpha}(\omega). \tag{2.3.10}$$

In this case, the spectral densities of the Langevin sources of the random electromagnetic field are proportional to the imaginary parts $\epsilon_{\alpha\beta}''(\omega)$ and $\mu_{\alpha\beta}''(\omega)$ of the permittivities.

It is well known that the electrodynamic properties of any medium may be described either by $\epsilon_{\alpha\beta}$ or by the conductivity $\sigma_{\alpha\beta}$. They are related by the equation:

$$\epsilon_{\alpha\beta}(\omega) = \delta_{\alpha\beta} + \frac{4\pi i}{\omega}\sigma_{\alpha\beta}(\omega). \tag{2.3.11}$$

The time derivative of the polarization $\dot{\mathbf{P}}$ is the current density (less the magnetization current). Therefore the spectral density of the Langevin currents equals:

$$S^{int}_{\dot{P}_\alpha \dot{P}_\beta}(\mathbf{r}_1, \mathbf{r}_2, f) = \omega^2 S^{int}_{P_\alpha P_\beta}(\mathbf{r}_1, \mathbf{r}_2, f)$$

$$= [\sigma_{\alpha\beta}(\mathbf{r}_1, \mathbf{r}_2, \omega) + \sigma^*_{\beta\alpha}(\mathbf{r}_2 \mathbf{r}_1 \omega)] hf \coth\left(\frac{hf}{2k_B T}\right). \qquad (2.3.12)$$

In the absence of magnetic field, i.e., $\mathbf{B} = 0$, this spectral density is proportional to the real part of the conductivity. In the quasi-classical limit, $hf \ll 2k_B T$,

$$S^{int}_{\dot{P}_\alpha \dot{P}_\beta}(\mathbf{r}_1, \mathbf{r}_2, f) = 4k_B T \sigma'_{\alpha\beta}(\mathbf{r}_1 \mathbf{r}_2 \omega). \qquad (2.3.13)$$

As the spectral densities of the Langevin sources are known, one can find, in principle, the spectral densities of the fields $\delta \mathbf{E}$, $\delta \mathbf{H}$, and the total random currents. One has to solve the Maxwell equations,

$$\mathbf{curl}\, \delta \mathbf{H} - \frac{1}{c}\frac{\partial \delta \mathbf{D}}{\partial t} = \frac{4\pi}{c}\delta \mathbf{j}^{int}, \quad \mathbf{curl}\, \delta \mathbf{E} + \frac{1}{c}\frac{\partial \delta \mathbf{B}}{\partial t} = 0, \qquad (2.3.14)$$

and find the fluctuations of the fields in terms of the Langevin sources $\delta \mathbf{j}^{int}$. Then the correlation functions and spectral densities of the fields have to be composed and expressed in terms of the spectral densities of the Langevin sources found above.

The spectral densities of the Langevin sources (2.3.8),(2.3.12), and (2.3.13) are valid only in equilibrium systems. Of course, these sources can be found, in principle, for nonequilibrium systems. However, in such systems the spectral densities are not directly related to the linear response. It means that in nonequilibrium systems the linear response and spectral density of noise yield independent information on the kinetics of the system and complement each other.

3
Fluctuations in nonequilibrium gases

One of the most important parts of the physics of fluctuations in solids is the physics of fluctuations in solid-state plasma, i.e., in a gas of charge carriers, electrons and/or holes. The spectral density of electric fluctuations in an equilibrium conductor (zero bias voltage and current) is given by the Nyquist fluctuation–dissipation relation (Sec. 2.2) in terms of the dissipative part of the complex impedance $Z(f)$. It means that the problem of calculation and measurement of noise in an equilibrium state is reduced to the problem of, respectively, calculation and measurement of the complex resistance. This problem is usually considered to be simpler. In an equilibrium system, the spectral density of noise does not contain any information other than that contained in $Z(f)$. However, the fluctuation–dissipation relation holds only in the equilibrium state. For conductors that are in nonequilibrium states, for instance, for conductors with hot electrons in strong electric fields, there is no general relation between the spectral density of noise, on one hand, and any characteristics of the response, frequency dependent or static, of the current or voltage, on the other hand. Thus, the calculation of current noise can not be reduced to the calculation of mean current (e.g., its dependence on voltage) and is an individual problem.

This chapter is devoted mainly to fluctuations in a gas of hot charge carriers (for brevity we call them 'hot electrons'). As is well known, electrons become hot in strongly biased conductors. A gas of hot electrons is a strongly nonequilibrium system. In particular, the hot electrons are not in equilibrium with the phonon thermal bath of the conductor. The term 'gas' is used in this chapter for brevity to denote a perfect gas, i.e., a system of particles, the energy of mutual interaction of which is small as compared with their kinetic energy. The long-range potential created by the electrons and charged impurities is taken into account through the Poisson equation. The short-range interaction between the electrons can often be neglected. In this case, only the scattering of electrons by phonons and impurities should be taken into account. However, at not too small electron densities and at low temperatures the short-range interaction between the electrons becomes significant. Its effect may be considered as electron–electron $(e-e)$ collisions. This gas is similar to some extent to a common, Boltzmann, gas in which the particles are interacting via a short-range potential. Therefore, the methods developed and equations derived for fluctuations in a gas of hot electrons can be applied as well to fluctuations in common gases, both equilibrium and nonequilibrium.

The problem of fluctuations in nonequilibrium gases, being fundamental, attracted much attention. Leontovich (1935) was the first who pointed out that 'in principle, it is possible to treat the fluctuations around stationary but thermodynamically nonequilibrium states to which the common theory is not applicable'. He calculated the one-time correlations in model random processes: 'monomolecular' (in each random event only one 'molecule' changes its state) and 'bimolecular' (two 'molecules' simultaneously change their states). He noticed that in the 'bimolecular' processes a correlation between 'molecules' arises (Sec. 3.6). Hashitsume (1956) introduced the notions of fluctuations of the distribution function around its equilibrium value and of the stochastic Boltzmann equation for these fluctuations. Kadomtsev (1957) was the first to write the Boltzmann equation with Langevin sources for fluctuations of the distribution function. Later it was called the Boltzmann–Langevin equation. A fundamental contribution to the theory of fluctuations about a stationary but nonequilibrium state was made by Lax (1960, 1966b) who developed a complete theory of noise in systems in which particles undergo one-particle ('monomolecular') random transitions. The development of the basic and general methods by Kogan & Shulman (1969) (Langevin approach) and, independently, by Gantsevich, Gurevich & Katilius (1969) (approach based on equations for the correlation functions that can be associated with the Markov approach) completed, in some sense, a definite period in the history of this field (see Bareikis, Katilius & Miliušytė, 1989 for a more detailed survey of the history of the problem of fluctuations in nonequilibrium gases).

3.1 Some basic concepts of hot-electrons' physics

The effect of a voltage applied across a conductor is not limited only to the creation of a current, i.e., a directional flow of charge carriers (for definiteness, electrons). The latter gain energy from the current source and dissipate it to some thermal bath, usually to the thermal lattice vibrations (phonons). As the accelerated electrons do not lose their energy immediately, they become 'heated', that is, their mean energy is increased. Unlike the current, which at small voltages U (or electric fields F) is linear in U, the degree of electron heating is at the same voltages proportional to U^2 (or F^2). It means that the heating becomes significant in higher electric fields.

The heating of the charge carriers results in many effects. First of all, in strong electric fields the current–voltage characteristic (CVC) deviates from Ohm's law. This deviation may be so dramatic that negative differential (small-signal) resistance appears in part of the CVC (Sec. 3.9). Impact ionization of impurities or generation of electron-hole pairs by hot electrons may result in a breakdown of the conductor. In general, all transport coefficients of the charge carriers change their values. Hot electrons can be the source of microwave radiation and electroluminescence.

The notion of hot electrons can be defined more precisely in terms of the distribution of electrons in momentum $\hbar\mathbf{p}$, i.e., in terms of the function $f(\mathbf{p})$. This function is the mean occupancy number of electrons in a state with momentum $\hbar\mathbf{p}$, or, equivalently, $f(\mathbf{p})d\mathbf{p}/(2\pi)^3$ is the number of electrons (per unit volume and definite spin quantum number) in a region $d\mathbf{p} = dp_x dp_y dp_z$ around the momentum $\hbar\mathbf{p}$. In a uniform conductor with isotropic energy spectrum of the charge carriers, $f(\mathbf{p})$ depends, in fact, on two variables only: on the energy $E = \hbar^2 p^2/2m$ and $\cos\theta$, where m is the effective mass and θ is the angle between $\mathbf{p}$ and the electric field $\mathbf{F}$. The function $f(\mathbf{p})$ can be expanded in a series of Legendre polynomials $P_l(\cos\theta)$:

$$f(\mathbf{p}) = \sum_{l=0}^{\infty} f_l(E)P_l(\cos\theta). \tag{3.1.1}$$

The functions $f_l(E)$ are expansion coefficients. In the absence of current only $f_0(E) \neq 0$, all other $f_l(E) = 0$.

As $P_1(\cos\theta) = \cos\theta$ and the Legendre polynomials are orthogonal, the current is directly expressed in terms of $f_1(E)$. The higher-order terms in the r.h.s. of Eq. (3.1.1) ($l = 2, 3, \ldots$) represent more complex dependence of the distribution function on the momentum direction. As $P_0(\cos\theta) = 1$, the function $f_0(E)$ is the electron distribution in energy. In the absence of current this function is the equilibrium Fermi distribution of electrons,

$$f^{eq}(E) = \left[\exp\left(\frac{E - E_F}{k_B T}\right) + 1\right]^{-1}, \tag{3.1.2}$$

where E_F is the chemical potential of electrons (Fermi level), T_0 is the lattice (phonon) temperature. In an electric field the function $f_0(E)$ deviates from $f^{eq}(E)$. The electron gas in which this deviation is strong is called a hot

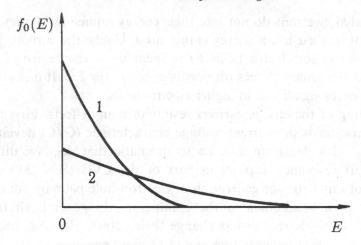

Fig. 3.1 Equilibrium energy distribution function (curve 1) and the energy distribution function of hot electrons (curve 2). E is the electron energy.

electron gas. Obviously, the electric field increases the number of high-energy electrons at the expense of the low-energy ones (Fig. 3.1).

If the electron gas is nonuniform and nonstationary, the distribution function depends also on the radius-vector $\mathbf{r}$ and time t, i.e., $f = f(\mathbf{p}\mathbf{r}t)$. This function can be used only under quasi-classical conditions. It means that the variation of $f(\mathbf{p}\mathbf{r}t)$ on the scale of the de-Broglie electron wavelength must be very small. In other words, if L is the characteristic length of the distribution function spatial variation, the inequality $pL \gg 1$ must be satisfied. The variation of the distribution function in time must be smooth enough. If ΔE is the characteristic energy of the distribution function's variation as a function of particle's energy E, the frequencies of the distribution function variations must be much smaller than $\Delta E / h$, where h is the Planck's constant.

The distribution function obeys the Boltzmann kinetic equation:

$$\frac{\partial f}{\partial t} + \mathbf{v}\frac{\partial f}{\partial \mathbf{r}} + \frac{e\mathbf{F}}{\hbar}\frac{\partial f}{\partial \mathbf{p}} - S(\mathbf{p}\mathbf{r}t) = 0. \qquad (3.1.3)$$

The second and third terms in the l.h.s. are the rates, taken with opposite signs, of the distribution function's variations due to its gradient in space and to the electric field (force), respectively. $S(\mathbf{p}\mathbf{r}t)$ is the rate of variation due to the particles' scattering and collisions.

In the case of elastic scattering of electrons by impurities and inelastic scattering by phonons, the collision integral $S(\mathbf{p}\mathbf{r}t)$ can be written, respectively, in the forms:

$$S^{(\text{imp})}(\mathbf{p}rt) = \sum_{\mathbf{p}'} W(\mathbf{p}'\mathbf{p})[f(\mathbf{p}') - f(\mathbf{p})],$$

$$S^{(\text{ph})}(\mathbf{p}rt) = \sum_{\mathbf{p}'} \{W(\mathbf{p}'\mathbf{p})f(\mathbf{p}')[1 - f(\mathbf{p})] - W(\mathbf{p}\mathbf{p}')f(\mathbf{p})[1 - f(\mathbf{p}')]\}. \tag{3.1.4}$$

Here $W(\mathbf{p}'\mathbf{p})$ is the probability, per unit time, of a scattering event as a result of which an electron changes its initial momentum $\hbar\mathbf{p}'$ to $\hbar\mathbf{p}$. The first and the second terms in the r.h.s. of both Eqs. (3.1.4) are the flows of particles coming into the state $\mathbf{p}$ and leaving this state, respectively. The factors $1 - f(\mathbf{p})$ take into account the Pauli principle: the transition into an occupied state is impossible.

The contribution of the interparticle collisions to $S(\mathbf{p}rt)$ in a nondegenerate gas equals:

$$S^{(\text{ee})}(\mathbf{p}rt) = \sum_{\mathbf{p}'\mathbf{k}\mathbf{k}'} \{W(\mathbf{k}\mathbf{k}'|\mathbf{p}\mathbf{p}')f(\mathbf{k})f(\mathbf{k}') - W(\mathbf{p}\mathbf{p}'|\mathbf{k}\mathbf{k}')f(\mathbf{p})f(\mathbf{p}')\}. \tag{3.1.5}$$

In the r.h.s., $W(\mathbf{p}\mathbf{p}'|\mathbf{k}\mathbf{k}')$ is the probability (per unit time) of such a collision of two particles with initial momenta $\hbar\mathbf{p}$ and $\hbar\mathbf{p}'$ after which they acquire momenta $\hbar\mathbf{k}$ and $\hbar\mathbf{k}'$. These probabilities can be expressed in terms of collision cross-sections.

The Boltzmann kinetics of gases is based on a fundamental assumption: the time τ_p between two successive scattering events (free-flight time) must be many times longer than the duration of an individual scattering or collision process, τ_c. In other words, the collisions should not 'overlap'. In the case of collisions of classical particles with short-range interaction of radius a, the time $\tau_c \simeq a/v$, where v is the particle's velocity. The time $\tau_p \simeq 1/na^2v$, where n is the density of particles, a^2 is the collision cross-section. The ratio $\tau_c/\tau_p \simeq na^3$. The quantity na^3 is called the 'gas parameter'. It is small ($\ll 1$) when the mean distance between the particles is many times greater than their dimension, i.e., when the gas is not dense. Similar conditions must be fulfilled when other scattering processes are dominant.

The properties of hot electrons depend not only on their mobility, or, alternatively, on the time $\tau_1 \simeq \tau_p$ of decay of an initially directed flow of electrons. They depend also on the time of the electrons' energy dissipation, or relaxation, τ_0. The mechanisms of momentum and energy scattering may be, in general, different. The dominant mechanism of momentum scattering is often, at low temperatures, the scattering by impurities. At the same time, the electrons' energy is dissipated via emission and absorption of acoustic and/or optical phonons. The physics of hot electrons takes the simplest form when the electrons are scattered only by acoustic phonons. Such conditions are met at crystal temperatures T_0 much lower than the Debye temperature Θ_D (the energy $k_B\Theta_D$ is of the order of optical phonon energies). In this case the scattering by phonons is almost elastic ('quasi-elastic') because the change of electron energy, which is equal to the energy of the emitted or absorbed phonon, $\hbar\Omega$, is much smaller than the electron energy $E(\mathbf{p})$. In fact, an electron with a momentum $\hbar p$ can change its momentum at most by $-2\hbar p$. The energy of a phonon with wave vector $2p$ equals $\hbar\Omega(2p) = 2\hbar sp$, where s is the sound speed. The ratio $\hbar\Omega/E(p) \simeq s/v$, where v is the electron velocity. Usually the ratio $s/v \ll 1$, i.e., the inelasticity of scattering by acoustic phonons is small. In other words, unlike the momentum, which is changed drastically, $|\Delta p|/p \sim 1$, the relative change of electron energy is small. Moreover, the electrons not only lose energy by phonon emission, but gain it by phonon absorption. The probabilities of phonon emission and absorption are proportional to $N_q + 1$ and N_q, respectively, where

$$N_q = \left[\exp(\hbar\Omega(\mathbf{q})/k_B T_0) - 1\right]^{-1} \tag{3.1.6}$$

is the number of phonons in the crystal vibration mode with wave vector $\mathbf{q}$. The relative excess of emission processes over the absorption ones is $1/(2N_q + 1)$. At small $\hbar\Omega \ll k_B T_0$, this ratio is of the order of $\sim \hbar\Omega/k_B T_0 \sim s/v \ll 1$. It means that the energy relaxation time, τ_0, is greater than τ_1^{phon} by a factor $\sim v^2/s^2 \gg 1$. The ratio τ_0/τ_1 becomes even greater if the momentum is scattered by impurities and the real $\tau_1 \ll \tau_1^{\text{phon}}$. It reaches $\sim 10^5$ in n-InSb at $T_0 \approx 6$ K.

Due to the small inelasticity of acoustic phonon scattering, the deviation of the electron distribution in momentum space from an isotropic one is small even when the electrons become hot. In fact, the time necessary for a significant change of the electrons' energies is sufficient for an almost complete randomization of their momenta. The functions $f_l(E)$ with $l = 2, 3, \ldots$ are therefore negligible, and the only functions in the r.h.s. of Eq. (3.1.1) that should be retained are $f_0(E)$ and $f_1(E)$. This approximation was introduced and substantiated by Pidduck (1916) and Davydov (1935, 1936, 1937).

When the scattering is strongly inelastic, a significant heating of electrons is accompanied by a strong anisotropy of their distribution function.

The collision integral $S(\mathbf{p})$ is a sum of $S_0(E)$, independent of θ, and S_1, which is proportional to $\cos\theta$. According to the definition of τ_1,

$$S_1(E) = -\frac{f_1(E)}{\tau_1}\cos\theta. \tag{3.1.7}$$

The equation for $S_0(E)$ is derived by considering the rate of change of the number of electrons in a thin spherical shell in momentum space with energies from E to $E + dE$. This rate is equal to $N(E)S_0(E)dE$, where $N(E)$ is the density of electron states. At the same time, this rate is equal to the change of the flow of electrons through the surface $E(\mathbf{p}) = E$ when E is varied by dE. If this flow per unit volume is denoted as $J(E)$ then, obviously, $S_0(E) = (dJ/dE)/N(E)$. By definition,

$$J(E) = \frac{1}{V} 2 \sum_{\mathbf{p}} \sum_{\mathbf{p}'} [W_{em}(\mathbf{p}'\mathbf{p})f_0(E(\mathbf{p}')) - W_{abs}(\mathbf{p}\mathbf{p}')f_0(E(\mathbf{p}))]$$
$$\times \Theta[E - E(\mathbf{p})]\Theta[E(\mathbf{p}') - E]. \tag{3.1.8}$$

Here W_{em} and W_{abs} are the probabilities of emission and absorption of a phonon per unit time, respectively, and $\Theta(x)$ is the Heaviside step function (Eq. (1.6.22)). The ratio

$$W_{em}(\mathbf{p}'\mathbf{p})/W_{abs}(\mathbf{p}\mathbf{p}') = \exp[\hbar\Omega(\mathbf{q})/k_B T_0].$$

If the inelasticity of scattering is small, the difference $f_0(E) - f_0(E + \hbar\Omega)$ can be approximated by $-(df_0/dE)\hbar\Omega$. Equation (3.1.8) can be represented in the form:

$$J(E) = EN(E)\left[f_0(E) + k_B T_0 \frac{df_0}{dE}\right] \frac{1}{\tau_0(E)}, \tag{3.1.9}$$

where

$$\frac{1}{\tau_0(E)} = \frac{1}{E} \int_0^E dE(\mathbf{p}) \sum_{\mathbf{p}'} \Theta(E(\mathbf{p}') - E)W_{abs}(\mathbf{p}\mathbf{p}')\hbar\Omega(\mathbf{p}' - \mathbf{p})/k_B T_0 \tag{3.1.10}$$

is the inverse energy relaxation time.

The kinetic equation (3.1.3) yields two equations for terms independent of and proportional to $\cos\theta$. Taking into account Eq. (3.1.9) one obtains:

$$f_1(E) - eFv\tau_1(-df_0/dE) = 0,$$
$$f_0 + k_B T^*(E)\frac{df_0}{dE} = 0. \tag{3.1.11}$$

Here the 'temperature' $T^*(E)$ of the electrons was introduced:

$$k_B T^*(E) = k_B T_0 + \frac{2e^2 F^2 \tau_1 \tau_0}{3m}. \tag{3.1.12}$$

Integration of the second Eq. (3.1.11) yields the Davydov distribution in energy (Davydov, 1937):

$$f_0(E) = n \exp\left[-\int_0^E \frac{dE'}{k_B T^*(E')}\right]$$

$$\times \left\{\int_0^\infty dE' \, N(E') \exp\left[-\int_0^{E'} \frac{dE''}{k_B T^*(E'')}\right]\right\}^{-1}, \qquad (3.1.13)$$

where n is the electron density.

At low temperatures T_0 and high electron densities n, the frequency of electron–electron collisions τ_{ee}^{-1} becomes greater than the inverse energy relaxation time, τ_0^{-1}. Due to effective exchange of energy between the electrons, their distribution function, within an accuracy $\sim \tau_{ee}/\tau_0$, can be approximated (if the inelasticity of electron energy dissipation scattering events is small) by the equilibrium Fermi or Boltzmann function with a temperature T. The latter is called electron temperature. Of course, $T > T_0$. The value of T is determined from the power balance equation:

$$\sigma(T)F^2 = P(T). \qquad (3.1.14)$$

The l.h.s. is the specific Joule power gained by electrons from the electric field F, $\sigma(T)$ is the conductivity. The r.h.s. of Eq. (3.1.14) is the power lost by electrons in electron-phonon scattering per unit volume. It can be represented as

$$P(T) = nc_e k_B \frac{T - T_0}{\tau_e(T)}. \qquad (3.1.15)$$

Here $\tau_e(T)$ has the meaning of the energy, or electron temperature, relaxation time, and c_e is the dimensionless heat capacity of electrons per electron ($c_e \approx 3/2$). From the last two equations one easily obtains the augmentation of electron temperature:

$$k_B(T - T_0) = \frac{\sigma F^2 \tau_e(T)}{nc_e} = \frac{e\mu F^2 \tau_e(T)}{c_e}. \qquad (3.1.16)$$

In strong electric fields and/or high crystal temperatures, the electron energies become comparable with the energies of optical phonons, and the approximation of small inelasticity of scattering does not hold. In many electron devices the electrons are moving in nonuniform and even nonstationary fields. In these cases the Boltzmann equation can not be solved analytically, only numerical calculations yield useful results. The Monte Carlo method of electron kinetics simulation is widely used. A detailed review of this method and of the results obtained in this field has been published by Jacoboni & Reggiani (1983).

3.2 Simple model of hot-electrons' noise in semiconductors

The distribution function $f(\mathbf{p}\mathbf{r}t)$ of the charge carriers discussed in the previous section is the *mean* number of these carriers in a state with momentum $\mathbf{p}$ at the point $\mathbf{r}$ and at instant t. Due to scattering by impurities, phonons, boundaries, etc., the number of charge carriers in each state fluctuates around this mean value. These fluctuations, $\delta f(\mathbf{p}\mathbf{r}t)$, manifest themselves in current and/or voltage noise and in scattering of light by the gas of charge carriers. As was shown in Sec. 3.1, in semiconductors with high enough density of free charge carriers and at low temperatures, when the electron–electron scattering time τ_{ee} is smaller than the time of energy relaxation τ_0 and the inelasticity of scattering is also small, the carriers' distribution in energy $f_0(E)$ is close to an equilibrium one with an electron, or hole, temperature T, which in high electric fields is higher than the lattice (phonon) temperature T_0. Under the same conditions and with the same accuracy, the charge carriers' fluctuations with frequencies $f \ll \tau_{ee}^{-1} \leq \tau_p^{-1}$ can be reduced to fluctuations of two quantities only: (1) the fluctuations of the current $\delta I(t)$, that is, the fluctuations of the directional part of the electron distribution, and (2) the fluctuations $\delta T(t)$ of the electron temperature which determine the fluctuations of the distribution in energy. The Langevin sources of these fluctuations are: (1) the fluctuation current source $\delta I^{\text{int}}(t)$ (intrinsic) which is generated by the scattering of electron momenta, and (2) the source of electron gas energy fluctuations $\delta J_e^{\text{int}}(t)$ which are generated by the random exchange of energy between the electron gas and the phonon thermal bath. This simple model of fluctuations in a hot electron gas allows one to demonstrate the main features of noise in high electric fields.

The equations for the fluctuations can be written using simple considerations (Kogan & Shulman, 1967). They can be derived quite rigorously, on the grounds of the general theory presented in the next section, by properly taking into account the frequent interelectron collisions. The equations for $\delta I(t)$, $\delta T(t)$, and the voltage fluctuations $\delta U(t)$ read:

$$\delta I(t) = R^{-1}\delta U + U(dR^{-1}/dT)\delta T(t) + \delta I^{\text{int}}(t),$$
$$Vnc_e\frac{d\delta T(t)}{dt} = I\delta U(t) + U\delta I(t) - V\frac{dP}{dT}\delta T(t) + \delta J_e^{\text{int}}(t). \tag{3.2.1}$$

Here $R(T)$ is the resistance of the sample, V its volume, n and c_e are the electron density and the heat capacity of electrons per one electron, respectively, and $P(T)$ is the specific power dissipated by the electrons into the phonon bath. According to the first equation, the current fluctuation is induced not only by the fluctuation of the voltage but, at a fixed nonzero voltage, by the fluctuation of the electron temperature if, of course, the conductivity depends on the electron temperature. The second equation is the equation of energy balance.

Eliminating $\delta T(t)$ from Eq. (3.2.1) one obtains an equation for the Fourier component of the current:

$$\delta I(\omega) = Z^{-1}(f)\delta U(\omega) + \delta \tilde{I}^{\text{int}}(\omega). \qquad (3.2.2)$$

Here $\delta \tilde{I}^{\text{int}}(t)$ is the total Langevin current source (see below).

The equation for the small-signal impedance follows from the same Eq. (3.2.1):

$$Z(f) = R_d \frac{1 + i\omega\tau_0\tfrac{1}{2}(\sigma_d/\sigma + 1)}{1 + i\omega\tau_0\tfrac{1}{2}(\sigma/\sigma_d + 1)}. \qquad (3.2.3)$$

Here R_d and σ_d are the dc small-signal resistance and conductivity, respectively,

$$\sigma_d = \sigma(T)\frac{P' + \sigma'F^2}{P' - \sigma'F^2}, \qquad (3.2.4)$$

F is the electric field, $\tau_0 = nc_e/P'$ is the energy relaxation time, the prime here and throughout this section denotes the derivative with respect to the electron temperature, and $\sigma = \sigma(T) = j(F)/F$ is the dc conductivity.

The total current Langevin source $\delta \tilde{I}^{\text{int}}(\omega)$ in Eq. (3.2.2) depends on both δI^{int} and δJ_e^{int}:

$$\delta \tilde{I}^{\text{int}}(\omega) = \left[\frac{U(dR^{-1}/dT)}{VP'(1 - i\omega\tau_0)}\delta J_e^{\text{int}} + \delta I^{\text{int}}\right]\left[1 - \frac{\sigma'F^2/P'}{1 - i\omega\tau_0}\right]^{-1}. \qquad (3.2.5)$$

The current fluctuations or, in other words, the fluctuations of the electron drift velocity $\delta v_d(t)$, and the electron temperature fluctuations, that is, fluctuations of the distribution in energy, are coupled in high electric fields if the conductivity σ depends on the electron temperature T. In the absence of a mean current, the current and the electron temperature are fluctuating independently, i.e., are uncorrelated. The mechanism of coupling of $\delta I(t)$ and $\delta T(t)$ is obvious: the fluctuation of current at nonzero voltage results in a fluctuation of the Joule power which, in its turn, produces an electron temperature fluctuation δT. This fluctuation changes the conductivity and, consequently, the current.

This coupling manifests itself in an observable effect suggested first by Price (1965). As the electric field is increased and the charge carriers become hot, a frequency dependence (dispersion) of the spectrum of current and/or voltage noise appears at frequencies of the order of the inverse energy relaxation time τ_0^{-1} apart from the dispersion at inverse momentum relaxation time τ_1^{-1}.

According to Eq. (3.2.5) the spectral density of noise is determined by the spectral densities of Langevin sources δI^{int} and δJ_e^{int}. At frequencies $\omega \ll \tau_1^{-1}$ the former is given by the Nyquist equation: $S_I^{\text{int}} = 4k_B T R^{-1}$. The energy flux fluctuation can be represented in terms of the fluctuations of the rates of absorption δJ_q^+ and emission δJ_q^- of phonons with wave vector $\mathbf{q}$ and frequency $\Omega(\mathbf{q})$:

$$\delta J_e^{\text{int}}(t) = \sum_{\mathbf{q}} \hbar \Omega(\mathbf{q}) [\delta J_q^+(t) - \delta J_q^-(t)]. \tag{3.2.6}$$

Since different scattering events are uncorrelated, the correlation functions of the Langevin fluxes are (1) nonzero only for fluxes corresponding to identical random events, and (2) proportional to the corresponding mean fluxes:

$$\langle \delta J_{\mathbf{q}_1}^{\pm}(t_1) \delta J_{\mathbf{q}_2}^{\pm}(t_2) \rangle = \delta(t_1 - t_2) \delta_{\mathbf{q}_1 \mathbf{q}_2} J_{\mathbf{q}_1}^{\pm},$$

$$\langle \delta J_{\mathbf{q}_1}^+(t_1) \delta J_{\mathbf{q}_2}^-(t_2) \rangle = 0, \tag{3.2.7}$$

where $J_{\mathbf{q}}^{\pm}$ is the mean rate of absorption (emission) of phonons with wave vector $\mathbf{q}$. Therefore

$$\langle \delta J_e^{\text{int}}(t_1) \delta J_e^{\text{int}}(t_2) \rangle = \delta(t_1 - t_2) \sum_{\mathbf{q}} [\hbar \Omega(\mathbf{q})]^2 (J_{\mathbf{q}}^+ + J_{\mathbf{q}}^-),$$

$$S_{J_e}(f) = 2 \sum_{\mathbf{q}} [\hbar \Omega(\mathbf{q})]^2 (J_{\mathbf{q}}^+ + J_{\mathbf{q}}^-), \tag{3.2.8}$$

Under the conditions at which the model considered is valid (high electron density and, consequently, high concentrations of impurities, low temperatures) the electron momentum is scattered mainly by impurities, and the energy by phonons. Hence, the sources δI^{int} and δJ_e^{int} are determined by different mechanisms and are uncorrelated. Therefore the spectral density of current fluctuations parallel to the mean current equals (at fixed voltage, $\delta U = 0$):

$$S_{\tilde{I}_{\parallel}}(f) = S_I^{\text{int}} + (1 + \omega^2 \tau_U^2)^{-1} \times$$
$$\times \left\{ \frac{(\sigma')^2 F^2}{4k_B T \sigma(T)(P' - \sigma' F^2)^2} \frac{1}{V} S_{J_e} + \left[\frac{(P')^2}{(P' - \sigma' F^2)^2} - 1 \right] \right\} S_I^{\text{int}}. \tag{3.2.9}$$

Here

$$\tau_U^{-1} = \frac{P' - \sigma' F^2}{nc_e} \tag{3.2.10}$$

is the inverse relaxation time of the electron temperature at constant voltage (in nonlinear systems the relaxation time depends on the regime of the relaxation process).

Using the relationship between the total current fluctuation and the uniform part of the current density fluctuation, $\delta \tilde{I} = A \delta \tilde{j}$, where A is the cross-section area of the conductor, one easily obtains from Eq. (3.2.9) the spectral density of the current density fluctuations parallel to the mean current and electric field:

$$
S_{\tilde{j}\|}(f) = \frac{4k_B T \sigma(T)}{V} \Big\{ 1 + (1 + \omega^2 \tau_U^2)^{-1} \times
$$

$$
\times \Big\{ \frac{(\sigma')^2 F^2}{4k_B T \sigma(T)(P' - \sigma' F^2)^2} \frac{1}{V} S_{J_e} + \Big[\frac{(P')^2}{(P' - \sigma' F^2)^2} - 1 \Big] \Big\} \Big\}.
$$

(3.2.11)

At small electric fields F, the part of this noise which is linear in F^2 can be represented as resistance noise: the semiconductor's resistance fluctuates due to the exchange of energy between electrons and phonons and to coupling between current and electron temperature fluctuations:

$$
\frac{S_R(f)}{R^2} = \frac{1}{V(1 + \omega^2 \tau_U^2)} \Big[\Big(\frac{\sigma'}{\sigma P'} \Big)^2 \frac{1}{V} S_{J_e} + 8k_B T \Big(\frac{\sigma'}{\sigma P'} \Big) \Big].
$$

(3.2.12)

The transverse current fluctuations perpendicular to the mean electric field are not coupled to the fluctuations of the electron temperature because in the linear approximation such fluctuations do not affect the Joule power. Therefore

$$
S_{\tilde{j}\perp}(f) = \frac{4k_B T \sigma(T)}{V} = \lim_{\omega \tau_U \gg 1} S_{\tilde{j}\|}(f).
$$

(3.2.13)

An analysis (Kogan & Shulman, 1967) shows that the sign of the difference of two plateaus of the spectral density $S_{\tilde{I}\|}(f)$, the low-frequency ($\omega \tau_U \ll 1$) and the high-frequency ($\omega \tau_U \gg 1$) one or, equivalently, the sign of the difference $S_{\tilde{j}\|}(f) - S_{\tilde{j}\perp}(f)$, coincides with the sign of σ' (Fig. 3.2). It means that this sign is determined by the type of nonlinearity of the current–voltage characteristic (CVC). If the CVC is superlinear, i.e., $\sigma' > 0$, the longitudinal current noise is higher than the transverse one, and the low-frequency longitudinal current noise is higher than the high-frequency one. The relations are opposite in the case of sublinear CVC. At the same time the sign of the difference between the low-frequency and high-frequency plateaus of the spectral density $S_U(f)$ of voltage fluctuations (at fixed current) is opposite to the sign of σ'.

The noise temperature, that is the electrical noise power transmitted from the sample with small-signal impedance $Z(f)$ to a matched load with an impedance $Z^*(f)$ in a unit frequency band, in general, equals:

$$
T_N = \frac{S_{\tilde{I}}(f)}{4k_B \operatorname{Re} Z^{-1}(f)}.
$$

(3.2.14)

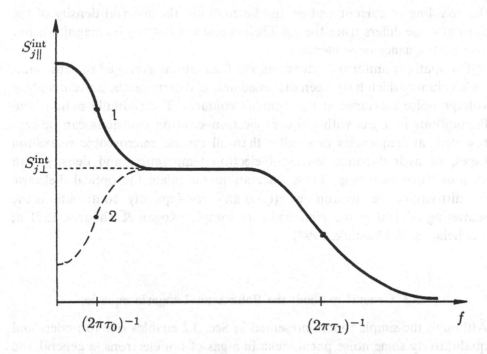

Fig. 3.2 The spectra of current fluctuations at fixed voltage and different types of deviations from Ohm's current–voltage characteristic (CVC). $S_{j\parallel}^{int}$ and $S_{j\perp}^{int}$ are the spectral densities of random currents parallel and perpendicular to the electric field, τ_0 and τ_1 are the energy and momentum relaxation times, respectively. 1: superlinear CVC, 2: sublinear CVC.

Substituting Eqs. (3.2.3),(3.2.9), and (3.2.13) into the last equation, one obtains for the longitudinal and transverse noise temperatures:

$$T_{N\parallel}(f) = T\left\{1 + \frac{(\sigma')^2 F^2}{n^2 c_e^2}\left[F^2 + \frac{S_{J_e}}{4VT\sigma}\right]\left[(\tau_I \tau_U)^{-1} + \omega^2\right]^{-1}\right\}, \tag{3.2.15}$$

$$T_{N\perp} = T,$$

where

$$\tau_I^{-1} = (P' + \sigma' F^2)/nc_e \tag{3.2.16}$$

is the inverse relaxation time of the electron temperature at constant current.

At high frequencies $T_N \to T$. However, at low frequencies $\omega \ll \sqrt{\tau_I \tau_U}$, the longitudinal noise temperature is higher than T at any mechanism of nonlinearity. The transverse noise temperature is at all frequencies $\omega \ll \tau_{ee}^{-1}$ equal to the electron temperature.

Examination of the simple model of hot electron noise shows that this noise can not be viewed merely as an enhanced Nyquist–Johnson noise. Owing to the coupling of current and energy fluctuations, the spectral density of the current noise differs from the equilibrium one not only by its magnitude but also by frequency dependence.

The spatially uniform fluctuations, i.e., fluctuations averaged over the sample's volume, which have been discussed above, determine the current and/or voltage noise measured at the sample's contacts. The spatially nonuniform fluctuations in a gas with frequent electron–electron collisions can be represented, at frequencies ω smaller than all inverse microscopic relaxation times, as hydrodynamic waves of electron temperature and density with various wave vectors $\mathbf{q}$. These fluctuations modulate the optical dielectric permittivity of the electron gas $\epsilon(\mathbf{q}, \omega)$ and, consequently, contribute to the scattering of light in the semiconductor sample (Kogan & Shadrin, 1971 a; Kochelap & Zakhleniuk, 1994).

3.3 General method: the Boltzmann–Langevin equation

Although the simple model presented in Sec. 3.2 enables one to understand qualitatively some noise phenomena in a gas of hot electrons, a general and quantitative theory of fluctuations in gases is needed. This theory has to have the same accuracy and the same range of applicability as the theory of the mean kinetic properties of gases. In other words, the equations for the fluctuations in gases (see below) must have, under quasi-classical conditions, the same range of applicability as the Boltzmann equation (Sec. 3.1).

The theory of fluctuations presented below uses the Langevin approach (Sec. 1.10). This means that its basic equation is the kinetic equation for the fluctuation $\delta f(\mathbf{pr}t)$ of the distribution function, i.e., for the random quantity itself, not for the required correlation function which is a nonrandom characteristic of the random process. As any application of the Langevin approach, it is based on the assumption that the relaxation times of the fluctuations are greater by many times than the correlation time of the Langevin sources. In the case under consideration, the former are the momentum relaxation time, τ_1, and other, even longer, relaxation times, the latter is the duration of an individual collision or scattering process, τ_c. As was mentioned in Sec. 3.1, the Boltzmann equation is based on the same assumption.

According to the Langevin approach, the equation for $\delta f(\mathbf{pr}t)$ is an inhomogeneous one. Its l.h.s. is the linearized l.h.s. part of the Boltzmann equation (3.1.3), the r.h.s. is the Langevin source, which in the problem under consideration is a random flux of particles into the given cell $(\mathbf{p}, \mathbf{r})$ of the phase space of the particles:

$$\left\{\frac{\partial}{\partial t} + \mathbf{v}(\mathbf{p})\frac{\partial}{\partial \mathbf{r}} + e\mathbf{F}(\mathbf{r}t)\frac{\partial}{\partial \hbar \mathbf{p}}\right\}\delta f(\mathbf{p}\mathbf{r}t) - S'_\mathbf{p}\{\delta f\} + \frac{\partial f}{\partial \hbar \mathbf{p}}e\delta\mathbf{F} = \delta J(\mathbf{p}\mathbf{r}t). \quad (3.3.1)$$

This equation is called the Boltzmann–Langevin equation (BLE).

The terms in the l.h.s. of the BLE following the time-derivative are the random fluxes (taken each with the opposite sign) of particles into the given phase-space cell $(\mathbf{p}\mathbf{r})$ which are produced by the fluctuations of the distribution function and may be, and are, expressed in terms of $\delta f(\mathbf{p}\mathbf{r}t)$. Particularly, $S'_\mathbf{p}\{\delta f\}$ is that part of the random flux of particles caused by scattering events which is a result of the occupancy number changes $\delta f(\mathbf{p}\mathbf{r}t)$. In the case of scattering by impurities and phonons, the linearization of (3.1.4) yields:

$$S'_\mathbf{p}\{\delta f\} = \sum_{\mathbf{p}'}\left\{\left[W_{\mathbf{p}'\mathbf{p}}(1 - f(\mathbf{p})) + W_{\mathbf{p}\mathbf{p}'}f(\mathbf{p})\right]\right.$$
$$\left. - \delta_{\mathbf{p}\mathbf{p}'}\sum_{\mathbf{p}''}\left[W_{\mathbf{p}\mathbf{p}''}(1 - f(\mathbf{p}'')) + W_{\mathbf{p}''\mathbf{p}}f(\mathbf{p}'')\right]\right\}\delta f(\mathbf{p}') \equiv \sum_{\mathbf{p}'}K(\mathbf{p}\mathbf{p}')\delta f(\mathbf{p}'). \quad (3.3.2)$$

The random flux of particles $\delta J(\mathbf{p}\mathbf{r}t)$ in the r.h.s. of the BLE is also generated by the particles' scattering, but it is basically different from $S'_\mathbf{p}\{\delta f\}$. It is nonzero even if at some instant the fluctuations $\delta f(\mathbf{p}\mathbf{r}t)$, and, consequently, $S'_\mathbf{p}\{\delta f\}$ are occasionally zero at all $\mathbf{p}$. Its correlation time is the duration of an individual scattering event, τ_c. It cannot be expressed as a functional of $\delta f(\mathbf{p}\mathbf{r}t)$. It plays the role of an intrinsic, or Langevin, source of fluctuations.

The required correlation function of the fluctuations $\langle\delta f(\mathbf{p}_1\mathbf{r}_1t_1)\delta f(\mathbf{p}_2\mathbf{r}_2t_2)\rangle$ can be found by solving the BLE (3.3.1) provided the correlation function of the Langevin sources $\delta J(\mathbf{p}\mathbf{r}t)$ is known. It can be derived quite rigorously from first principles (see, for instance, Kogan, 1991) but it is more instructive to use fundamental and simple physical considerations, namely, the absence of correlations between different scattering events (Kogan & Shulman, 1969). The same property of scattering processes in rare gases underlies, in fact, the Boltzmann kinetic equation (3.1.3).

Let us find the correlation function of the Langevin fluxes $\delta J(\mathbf{p}\mathbf{r}t)$ first for the case of scattering by impurities and phonons, i.e., when the interparticle collisions are rare. The flux into the state $\mathbf{p}$ is equal to the difference of fluxes corresponding to transitions from all $\mathbf{p}'$ into the given $\mathbf{p}$ and from $\mathbf{p}$ to all $\mathbf{p}'$:

$$\delta J(\mathbf{p}\mathbf{r}t) = \sum_{\mathbf{p}'}[\delta J(\mathbf{p}'\mathbf{p}, \mathbf{r}t) - \delta J(\mathbf{p}\mathbf{p}'; \mathbf{r}t)] \equiv \delta J^+(\mathbf{p}\mathbf{r}t) - \delta J^-(\mathbf{p}\mathbf{r}t), \quad (3.3.3)$$

where $\delta J(\mathbf{p}\mathbf{p}'; \mathbf{r}t)$ corresponds to transitions from $\mathbf{p}$ to $\mathbf{p}'$. Therefore the correlation function equals:

$$\langle \delta J(\mathbf{p}_1\mathbf{r}_1t_1)\delta J(\mathbf{p}_2\mathbf{r}_2t_2)\rangle$$
$$= \langle \delta J^+(\mathbf{p}_1\mathbf{r}_1t_1)\delta J^+(\mathbf{p}_2\mathbf{r}_2t_2)\rangle + \langle \delta J^-(\mathbf{p}_1\mathbf{r}_1t_1)\delta J^-(\mathbf{p}_2\mathbf{r}_2t_2)\rangle \qquad (3.3.4)$$
$$- \langle \delta J^+(\mathbf{p}_1\mathbf{r}_1t_1)\delta J^-(\mathbf{p}_2\mathbf{r}_2t_2)\rangle - \langle \delta J^-(\mathbf{p}_1\mathbf{r}_1t_1)\delta J^+(\mathbf{p}_2\mathbf{r}_2t_2)\rangle.$$

The elementary fluxes $\delta J(\mathbf{p}_1\mathbf{p}_1')$ and $\delta J(\mathbf{p}_2\mathbf{p}_2')$ are correlated only if they correspond to one and the same scattering event, i.e., their initial states and, separately, their final states coincide: $\mathbf{p}_1 = \mathbf{p}_2$, $\mathbf{p}_1' = \mathbf{p}_2'$. Within the quasi-classical approximation, any scattering event is confined to one point $\mathbf{r}$ and one instant t, i.e., the correlation function is proportional to $\delta(\mathbf{r}_1-\mathbf{r}_2)\delta(t_1-t_2)$. Since the scattering events are random and uncorrelated, they are similar to a shot noise, or Poissonian random process (Sec. 1.5). The correlation function of the fluxes is proportional to the mean flux, i.e., (Kogan & Shulman, 1969):

$$\langle \delta J(\mathbf{p}_1\mathbf{r}_1t_1)\delta J(\mathbf{p}_2\mathbf{r}_2t_2)\rangle$$
$$= V\delta(\mathbf{r}_1 - \mathbf{r}_2)\delta(t_1 - t_2)\{\delta_{\mathbf{p}_1\mathbf{p}_2}[J^+(\mathbf{p}_1\mathbf{r}_1t_1) + J^-(\mathbf{p}_1\mathbf{r}_1t_1)] \qquad (3.3.5)$$
$$- J(\mathbf{p}_2\mathbf{p}_1;\mathbf{r}_1t_1) - J(\mathbf{p}_1\mathbf{p}_2;\mathbf{r}_1t_1)\}.$$

Here $J^+(\mathbf{p}\mathbf{r}t)$ and $J^-(\mathbf{p}\mathbf{r}t)$ are the mean fluxes of particles into and from the state $\mathbf{p}$, respectively, V is the normalization volume for the states with definite momentum.

Equation (3.3.5) has a simple meaning. The Langevin fluxes $\delta J^+(\mathbf{p}_1)$ and $\delta J^+(\mathbf{p}_2)$ are correlated only if $\mathbf{p}_1 = \mathbf{p}_2$, and their correlation function is proportional to the mean flux $J^+(\mathbf{p}_1)$. The term with $J^-(\mathbf{p}_1)$ has a similar meaning. The fluxes $\delta J^+(\mathbf{p}_1)$ and $\delta J^-(\mathbf{p}_2)$ are correlated only if $\mathbf{p}_1 \neq \mathbf{p}_2$, and only those scattering events contribute to their correlation, in which a particle is transferred from $\mathbf{p}_2$ to $\mathbf{p}_1$. The contribution of these processes to the correlation function (3.3.5), $-J(\mathbf{p}_2\mathbf{p}_1)$, is negative because the corresponding changes of the occupancy numbers of the states $\mathbf{p}_1$ and $\mathbf{p}_2$ have opposite signs. Both terms, $-J(\mathbf{p}_1\mathbf{p}_2)$ and $-J(\mathbf{p}_2\mathbf{p}_1)$, arise from the conservation of particles in each scattering event, not only on average: the appearance of a particle in $\mathbf{p}_1$ is accompanied by its disappearance in some $\mathbf{p}_2$.

The same approach can be used to derive the correlation function of the Langevin sources in a gas with pair collisions between the particles (see Fig. 3.3). It can be applied also to the case of collisions involving a number of particles greater than two, and to inelastic collisions. The fluctuation of the net flux of particles into the state with momentum $\mathbf{p}$ equals (the variables $\mathbf{r}$ and t are omitted for brevity):

$$\delta J(\mathbf{p}) = \delta J^+(\mathbf{p}) - \delta J^-(\mathbf{p}) = \sum_{\mathbf{p}'}[\delta J^+(\mathbf{p}\mathbf{p}') - \delta J^-(\mathbf{p}\mathbf{p}')]. \qquad (3.3.6)$$

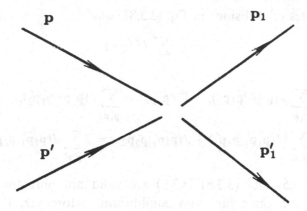

Fig. 3.3 Schematic diagram of a collision between two particles. $\mathbf{p}$ and $\mathbf{p}'$ are the momenta of the particles before the collision, $\mathbf{p}_1$ and $\mathbf{p}'_1$ after the collision.

Here $\delta J^{\pm}(\mathbf{p})$ are the random fluxes of particles into and from the state $\mathbf{p}$, respectively, $\delta J^{+}(\mathbf{pp}')$ is the random rate of such collisions in which the final momenta of particles are $\mathbf{p}$ and $\mathbf{p}'$, $\delta J^{-}(\mathbf{pp}')$ is the random rate of collisions of particles having initial momenta $\mathbf{p}$ and $\mathbf{p}'$. The fluctuations $\delta J^{\pm}(\mathbf{pp}')$ can be represented in the form of sums over initial and final momenta, respectively:

$$\delta J^{+}(\mathbf{pp}') = \sum_{\mathbf{kk}'} \delta J(\mathbf{kk}'|\mathbf{pp}'), \quad \delta J^{-}(\mathbf{pp}') = \sum_{\mathbf{kk}'} \delta J(\mathbf{pp}'|\mathbf{kk}'). \tag{3.3.7}$$

The random Langevin rate of collisions of particles with initial momenta $\mathbf{k}$ and $\mathbf{k}'$ and final momenta $\mathbf{p}$ and $\mathbf{p}'$ was denoted here by $\delta J(\mathbf{kk}'|\mathbf{pp}')$.

Since each collision is correlated only with itself, the correlation function of the Langevin sources equals (Kogan & Shulman, 1969):

$$\langle \delta J(\mathbf{p}_1 \mathbf{r}_1 t_1) \delta J(\mathbf{p}_2 \mathbf{r}_2 t_2) \rangle = V \delta(\mathbf{r}_1 - \mathbf{r}_2) \delta(t_1 - t_2) \{ \delta_{\mathbf{p}_1 \mathbf{p}_2} [J^{+}(\mathbf{p}_1) + J^{-}(\mathbf{p}_1)]$$
$$+ J^{+}(\mathbf{p}_1 \mathbf{p}_2) + J^{-}(\mathbf{p}_1 \mathbf{p}_2) - J(\mathbf{p}_1 | \mathbf{p}_2) - J(\mathbf{p}_2 | \mathbf{p}_1) \}. \tag{3.3.8}$$

The delta-functions take into account that the dimensions of the region in space and the duration of an elementary collision event are negligible as compared with the relaxation length and relaxation time of the fluctuations, respectively. The last two terms in Eq. (3.3.8) originate from the correlation of $\delta J^{+}(\mathbf{p}_1)$ and $\delta J^{-}(\mathbf{p}_2)$. This correlation is due to those collisions in which among the colliding particles there is one with momentum $\mathbf{p}_1$ (or $\mathbf{p}_2$), and among the flying away particles there is one with momentum $\mathbf{p}_2$ (or $\mathbf{p}_1$). The mean numbers of such collisions are denoted by $J(\mathbf{p}_1|\mathbf{p}_2)$ and $J(\mathbf{p}_2|\mathbf{p}_1)$, respectively.

The mean rates of collisions in Eq. (3.3.8) equal:

$$J^{\pm}(\mathbf{p}) = \sum_{\mathbf{p}'} J^{\pm}(\mathbf{p}\mathbf{p}'),$$

$$J^{+}(\mathbf{p}_1\mathbf{p}_2) = \sum_{\mathbf{p}_1'\mathbf{p}_2'} J(\mathbf{p}_1'\mathbf{p}_2'|\mathbf{p}_1\mathbf{p}_2), \quad J^{-}(\mathbf{p}_1\mathbf{p}_2) = \sum_{\mathbf{p}_1'\mathbf{p}_2'} J(\mathbf{p}_1\mathbf{p}_2|\mathbf{p}_1'\mathbf{p}_2'),$$

$$J(\mathbf{p}_1|\mathbf{p}_2) = \sum_{\mathbf{p}_1'\mathbf{p}_2'} [J(\mathbf{p}_1\mathbf{p}_1'|\mathbf{p}_2\mathbf{p}_2') + J(\mathbf{p}_1\mathbf{p}_1'|\mathbf{p}_2'\mathbf{p}_2)] = 2\sum_{\mathbf{p}_1'\mathbf{p}_2'} J(\mathbf{p}_1\mathbf{p}_1'|\mathbf{p}_2\mathbf{p}_2'). \tag{3.3.9}$$

Equations (3.3.5) and (3.3.8)–(3.3.9) are valid not only for equilibrium gases but also for gases far from equilibrium. Moreover, they hold for nonuniform and nonstationary gases provided their variations in space and time are quasi-classical (Sec. 3.1). Thus, the BLE (3.3.1) with Langevin sources given by Eqs. (3.3.5) and (3.3.8)–(3.3.9) are valid within the same conditions as the Boltzmann equation itself.

The vector $\mathbf{p}$ may include not only the momentum but also other quantum numbers of a particle, for instance, the spin quantum number. For brevity, these quantum numbers are not written down explicitly.

If the number of particles is conserved in each scattering event (we consider only such systems) the condition $\sum_{\mathbf{p}} \delta J(\mathbf{p}\mathbf{r}t) = 0$ has to be satisfied. One can easily check that the correlation functions (3.3.5) and (3.3.8) satisfy this condition:

$$\sum_{\mathbf{p}_1} \langle \delta J(\mathbf{p}_1\mathbf{r}_1t_1)\delta J(\mathbf{p}_2\mathbf{r}_2t_2)\rangle = \sum_{\mathbf{p}_2} \langle \delta J(\mathbf{p}_1\mathbf{r}_1t_1)\delta J(\mathbf{p}_2\mathbf{r}_2t_2)\rangle = 0. \tag{3.3.10}$$

The equation for the Langevin sources for interparticle collisions must satisfy also the conditions that follow from the conservation of total momentum and energy. One can easily verify that Eq. (3.3.8) satisfies these conditions.

Equations (3.3.1) and (3.3.8)–(3.3.9) apply as well to molecular (Boltzmann) gases. The fundamental problem of fluctuations in these gases attracted much attention. Kadomtsev (1957) was the first to write the Boltzmann equation with Langevin sources for the fluctuations of the distribution function. He derived the correlation function of the Langevin sources (unfortunately, due to an error, it is correct only when applied to equilibrium gases). Bixon & Zwanzig (1969); Fox & Uhlenbeck (1970); Logan & Kac (1976) derived the equation for the sources in the Boltzmann–Langevin equation for an equilibrium gas using different methods. The derived equation coincides with Eq. (3.3.8) after its simplification by taking into account the detailed balance that holds in equilibrium systems. The equation for the correlation function of a gas with interparticle collisions has been derived also on the basis of Bogolubov's hierarchy of equations (Kogan, 1972). For a review of this field see Ernst & Cohen (1981).

The Langevin approach presented above may be applied to all systems in which the individual scattering or transition events are well separated, not 'overlapping'.

An approach to the theory of quasi-classical fluctuations in gases, which is different from the Langevin one but equivalent to it, has been developed by Gantsevich, Gurevich & Katilius (1969). It has been detailed by the same authors (1979) in their review article (see also Bareikis, Katilius & Miliušytė, 1989, and Bareikis, Katilius, Pozhela *et al.*, 1992). This alternative approach is based on equations for the 'one-sided' correlation function $\psi^+ = \Theta(t_1 - t_2)\langle \delta f(\mathbf{p}_1\mathbf{r}_1t_1)\delta f(\mathbf{p}_2\mathbf{r}_2t_2)\rangle$, where $\Theta(t)$ is the Heaviside step function, and can be related to the Markov approach (Sec. 1.6). This correlation function satisfies a kinetic equation as the function of, say, the first group of variables. The initial value of the correlation function at $t_1 - t_2 = 0$, i.e., the one-time correlation function ('variance'), is easily found in the absence of interparticle collisions using simple statistical considerations (Sec. 3.6). However, this problem is more complicated in such nonequilibrium gases in which the frequency of interparticle collisions is not negligible. Then the one-time correlation function contains an additional term which owes its origin to the collisions between particles (it disappears in the equilibrium state). This term can be calculated by solving a different, inhomogeneous, kinetic equation (Sec. 3.6). The equivalence of the two approaches, the Langevin approach presented above, and the Markov approach of equations for the correlation functions, which seem quite different, was proved (Kogan & Shulman, 1970).

The theory presented in this section is a quasi-classical one. It can be applied only when the spatial and temporal variations of the distribution function and of those of its fluctuations which determine the measured noise are smooth enough. When the conditions for the quasi-classical approach are not satisfied, one has to use a more general, quantum approach. In quantum mechanics a particle cannot have simultaneously definite coordinate $\mathbf{r}$ and momentum $\mathbf{p}$. Therefore, the distribution function which depends on $\mathbf{p}$ and $\mathbf{r}$ cannot be used as the characteristic of a non-quasi-classical, quantum physical system. It is well known that such systems are described in terms of quantum Green functions. The fluctuations of physical quantities can be expressed, correspondingly, in terms of the fluctuations of these quantum Green functions. The equations for the correlation functions of Green functions' fluctuations have been derived by Kogan (1991). The equations are inhomogeneous and can be viewed as a quantum generalization of the Boltzmann–Langevin equation. The r.h.s. has the meaning of the correlation function of Langevin sources corresponding to scattering. Of course, in the quasi-classical limit this r.h.s. turns into the correlation function of Langevin sources given by Eqs. (3.3.5) and (3.3.8)–(3.3.9). In the absence of scattering the equation becomes homogeneous and its solution yields the correlation function of fluctuations in quantum ballistic systems (Sec. 5.2).

3.4 Current fluctuations and noise temperature

It is shown in this section how to calculate the correlation function and spectral density of noise starting from the Boltzmann–Langevin equation (3.3.1).

If the correlation function of the Langevin sources is known, one can obtain, in principle, the required correlation function of the fluctuations of observable physical quantities. For this purpose one must solve the Boltzmann–Langevin equation (3.3.1). It is reasonable to rewrite it in the form:

$$\frac{\partial \delta f(\mathbf{prt})}{\partial t} + \hat{L}\delta f = -\frac{\partial f}{\partial \hbar \mathbf{p}} e\delta \mathbf{F}(\mathbf{rt}) + \delta J(\mathbf{prt}). \tag{3.4.1}$$

The linear operator $\hat{L}$ is defined by the equation:

$$\hat{L}(\mathbf{pr};\mathbf{p'r'};t)\delta f(\mathbf{p'r'}t) = \left[\mathbf{v}\frac{\partial}{\partial \mathbf{r}} + e\mathbf{F}(\mathbf{rt})\frac{1}{\hbar}\frac{\partial}{\partial \mathbf{p}}\right]\delta f(\mathbf{prt}) - S'_{\mathbf{p}}\{\delta f\}. \tag{3.4.2}$$

The solution to the linear inhomogeneous Eq. (3.4.1) can be expressed in terms of the Green's function $G(\mathbf{prt}|\mathbf{p'r't'})$ of the linearized Boltzmann kinetic equation. It satisfies the equation:

$$\frac{\partial}{\partial t}G(\mathbf{prt}|\mathbf{p'r't'}) + \frac{1}{V}\sum_{\mathbf{p''}}\int d\mathbf{r''}\hat{L}(\mathbf{pr};\mathbf{p''r''};t)G(\mathbf{p''r''}t|\mathbf{p'r't'})$$
$$= \delta_{\mathbf{pp'}}V\delta(\mathbf{r}-\mathbf{r'})\delta(t-t'), \tag{3.4.3}$$

and the same boundary conditions as does the fluctuation $\delta f(\mathbf{prt})$.

The solution to Eq. (3.4.1) is a sum of two terms:

$$\delta f(\mathbf{prt}) = \delta f^F(\mathbf{prt}) + \delta f^{\text{int}}(\mathbf{prt}). \tag{3.4.4}$$

Here

$$\delta f^F(\mathbf{prt}) = \frac{1}{V}\int d\mathbf{r'}dt'\sum_{\mathbf{p'}}G(\mathbf{prt}|\mathbf{p'r't'})\left[-\frac{\partial f(\mathbf{p'r't'})}{\partial \hbar \mathbf{p'}}e\delta \mathbf{F}(\mathbf{r't'})\right] \tag{3.4.5}$$

is the response of the electron gas to the fluctuation of the electric field, and

$$\delta f^{\text{int}}(\mathbf{p}\mathbf{r}t) = \frac{1}{V} \int d\mathbf{r}' dt' \sum_{\mathbf{p}'} G(\mathbf{p}\mathbf{r}t|\mathbf{p}'\mathbf{r}'t') \delta J(\mathbf{p}'\mathbf{r}'t') \tag{3.4.6}$$

may be called the intrinsic fluctuation of the distribution function because it is a response to the Langevin (intrinsic) source in the Boltzmann–Langevin equation at fixed electric field.

Consequently, the current density fluctuation is also a sum of two terms:

$$\delta \mathbf{j}(\mathbf{r}t) = 2 \int \frac{d\mathbf{p}}{(2\pi)^3} e\mathbf{v}(\mathbf{p}) \delta f(\mathbf{p}\mathbf{r}t) = \delta \mathbf{j}^F(\mathbf{r}t) + \delta \mathbf{j}^{\text{int}}(\mathbf{r}t). \tag{3.4.7}$$

The first part is the response to the electric field fluctuation $\delta \mathbf{F}(\mathbf{r}t)$:

$$\delta j_\alpha^F(\mathbf{r}t) = \int d\mathbf{r}' dt' \sigma_{\alpha\beta}(\mathbf{r}t;\mathbf{r}'t') \delta F_\beta(\mathbf{r}'t'), \tag{3.4.8}$$

where

$$\sigma_{\alpha\beta}(\mathbf{r}t;\mathbf{r}'t') = 2 \int \frac{d\mathbf{p}}{(2\pi)^3} \frac{d\mathbf{p}'}{(2\pi)^3} e^2 v_\alpha(\mathbf{p}) G(\mathbf{p}\mathbf{r}t|\mathbf{p}'\mathbf{r}'t') \left[-\frac{\partial f(\mathbf{p}'\mathbf{r}'t')}{\partial \hbar p_\beta'} \right] \tag{3.4.9}$$

is the generalized small-signal conductivity of the electron gas. In homogeneous and stationary media it depends on the differences $\mathbf{r} - \mathbf{r}'$ and $t - t'$, and its Fourier transform as a function of these differences is just the conductivity, which depends on the wave vector $\mathbf{k}$ and frequency ω (see below).

The second part of the current fluctuation (3.4.7), being the response to the intrinsic Langevin source in the Boltzmann–Langevin equation, may be called the intrinsic current fluctuation. It equals:

$$\delta \mathbf{j}^{\text{int}}(\mathbf{r}t) = 2 \int \frac{d\mathbf{p}}{(2\pi)^3} e\mathbf{v}(\mathbf{p}) \delta f^{\text{int}}(\mathbf{p}\mathbf{r}t). \tag{3.4.10}$$

Equation (3.4.7) for the total current fluctuation has to be substituted into the equation for the electric field fluctuation, that is, into the Maxwell equation (Sec. 2.3) or, if the currents are quasi-stationary, into the Kirchhoff equation. These equations become linear inhomogeneous equations in which the inhomogeneity is the intrinsic (Langevin) current $\delta \mathbf{j}^{\text{int}}$. Just this current plays the role of the Langevin source of the current and/or voltage fluctuations in quasi-stationary circuits and of electromagnetic fluctuations (Secs. 2.2 and 2.3). Using Eqs. (3.4.6) and (3.4.10), this current can be expressed in terms of the Langevin source $\delta J(\mathbf{p}\mathbf{r}t)$ in the Boltzmann–Langevin equation. Because the correlation function of these sources is known even for nonequilibrium gases (Sec. 3.3), the correlation function of the intrinsic currents can also be calculated for gases, including those which are far from equilibrium.

The desired equation for the correlation function of the intrinsic fluctuations,

$$\langle \delta f^{\text{int}}(\mathbf{p}_1 \mathbf{r}_1 t_1) \delta f^{\text{int}}(\mathbf{p}_2 \mathbf{r}_2 t_2) \rangle$$

$$= \int d\mathbf{r}_1' dt_1' \sum_{\mathbf{p}_1'} \int d\mathbf{r}_2' dt_2' \sum_{\mathbf{p}_2'} G(\mathbf{p}_1 \mathbf{r}_1 t_1 | \mathbf{p}_1' \mathbf{r}_1' t_1') G(\mathbf{p}_2 \mathbf{r}_2 t_2 | \mathbf{p}_2' \mathbf{r}_2' t_2')$$

$$\times \langle \delta J(\mathbf{p}_1' \mathbf{r}_1' t_1') \delta J(\mathbf{p}_2' \mathbf{r}_2' t_2') \rangle,$$

$$(3.4.11)$$

can be significantly simplified in the case of a nondegenerate gas of particles with negligible rate of interparticle collisions, that is, under conditions that are often met in semiconductors. To perform this we need to examine some properties of the Green's function $G(\mathbf{p}\mathbf{r}t|\mathbf{p}'\mathbf{r}'t')$.

It has a simple meaning. It is the change of the occupancy number of the state with momentum $\hbar\mathbf{p}$ at the point $\mathbf{r}$ and instant t, as the result of the introduction at a preceding instant $t' \leq t$ of a particle into a state with momentum $\mathbf{p}'$ at the point $\mathbf{r}'$. It follows from Eq. (3.4.3) that its initial value

$$\lim_{t \to t'+0} G(\mathbf{p}\mathbf{r}t|\mathbf{p}'\mathbf{r}'t') = \delta_{\mathbf{p}\mathbf{p}'} V \delta(\mathbf{r} - \mathbf{r}'). \qquad (3.4.12)$$

After a sufficiently long time, that is, at $(t - t') \to \infty$, the effect of the introduction of an extra particle is reduced to such a change of the particles' distribution function, independent of the initial $\mathbf{p}', \mathbf{r}'$, and t', which corresponds to an increase of the total number of particles N in the gas by unity (at a fixed electric field):

$$\lim_{(t-t') \to \infty} G(\mathbf{p}\mathbf{r}t|\mathbf{p}'\mathbf{r}'t') = \left(\frac{\partial f(\mathbf{p}\mathbf{r}t)}{\partial N} \right)_F. \qquad (3.4.13)$$

It means that it is not the Green's function G which decays to zero at $(t - t') \to \infty$, but the function

$$g(\mathbf{p}\mathbf{r}t|\mathbf{p}'\mathbf{r}'t') = G(\mathbf{p}\mathbf{r}t|\mathbf{p}'\mathbf{r}'t') - \Theta(t - t') \left(\frac{\partial f(\mathbf{p}\mathbf{r}t)}{\partial N} \right)_F, \qquad (3.4.14)$$

where $\Theta(t)$ is the Heaviside step function (compare the introduction of g with the introduction of p in Sec. 1.6). In a nondegenerate gas in the absence of interparticle interaction $\partial f / \partial N = f/N$.

Let us substitute Eq. (3.3.5) for the correlation function of Langevin sources, in the absence of collisions between particles, into Eq. (3.4.11). Taking into account the equation for G as a function of the second set of variables (primed) and Eq. (3.3.2) for the operator K,

$$-\left[\frac{\partial}{\partial t'} + \mathbf{v}(\mathbf{p}')\frac{\partial}{\partial \mathbf{r}'} + e\mathbf{F}(\mathbf{r}'t')\frac{\partial}{\partial \mathbf{p}'}\right]G(\mathbf{p}\mathbf{r}t|\mathbf{p}'\mathbf{r}'t')$$

$$-\sum_{\mathbf{p}''} G(\mathbf{p}\mathbf{r}t|\mathbf{p}''\mathbf{r}'t')K(\mathbf{p}''\mathbf{p}') = V\delta_{\mathbf{p}\mathbf{p}'}\delta(\mathbf{r} - \mathbf{r}')\delta(t - t'),$$

(3.4.15)

one obtains:

$$\langle \delta f^{\text{int}}(\mathbf{p}_1\mathbf{r}_1t_1)\delta f^{\text{int}}(\mathbf{p}_2\mathbf{r}_2t_2)\rangle = G(\mathbf{p}_1\mathbf{r}_1t_1|\mathbf{p}_2\mathbf{r}_2t_2)f(\mathbf{p}_2\mathbf{r}_2t_2)$$

$$+ G(\mathbf{p}_2\mathbf{r}_2t_2|\mathbf{p}_1\mathbf{r}_1t_1)f(\mathbf{p}_1\mathbf{r}_1t_1) - f(\mathbf{p}_2\mathbf{r}_2t_2)f(\mathbf{p}_1\mathbf{r}_1t_1)/N.$$

(3.4.16)

It follows from this equation and from the definition (3.4.14) that in the absence of $e - e$ collisions

$$\langle \delta f^{\text{int}}(\mathbf{p}_1\mathbf{r}_1t_1)\delta f^{\text{int}}(\mathbf{p}_2\mathbf{r}_2t_2)\rangle$$

$$= g(\mathbf{p}_1\mathbf{r}_1t_1|\mathbf{p}_2\mathbf{r}_2t_2)f(\mathbf{p}_2\mathbf{r}_2t_2) + g(\mathbf{p}_2\mathbf{r}_2t_2|\mathbf{p}_1\mathbf{r}_1t_1)f(\mathbf{p}_1\mathbf{r}_1t_1).$$

(3.4.17)

This equation holds for any nondegenerate electron gas with rare $e - e$ collisions, even for nonuniform and nonstationary electron gases. It means that the noise is fully determined by two quantities: (1) the mean distribution function $f(\mathbf{p}\mathbf{r}t)$, and (2) the solution to the inhomogeneous linearized Boltzmann kinetic equation $g(\mathbf{p}_1\mathbf{r}_1t_1|\mathbf{p}_2\mathbf{r}_2t_2)$.

Simple but lengthy calculations yield the following equation for the correlation function of intrinsic fluctuations in a gas in which the effect of interparticle collisions can not be neglected:

$$\langle \delta f^{\text{int}}(\mathbf{p}_1\mathbf{r}_1t_1)\delta f^{\text{int}}(\mathbf{p}_2\mathbf{r}_2t_2)\rangle$$

$$= g(\mathbf{p}_1\mathbf{r}_1t_1|\mathbf{p}_2\mathbf{r}_2t_2)f(\mathbf{p}_2\mathbf{r}_2t_2) + g(\mathbf{p}_2\mathbf{r}_2t_2|\mathbf{p}_1\mathbf{r}_1t_1)f(\mathbf{p}_1\mathbf{r}_1t_1)$$

$$+ \Theta(t_1 - t_2)\frac{\partial f(\mathbf{p}_1\mathbf{r}_1t_1)}{\partial N}f(\mathbf{p}_2\mathbf{r}_2t_2) + \Theta(t_2 - t_1)\frac{\partial f(\mathbf{p}_2\mathbf{r}_2t_2)}{\partial N}f(\mathbf{p}_1\mathbf{r}_1t_1)$$

$$- N\frac{\partial f(\mathbf{p}_1\mathbf{r}_1t_1)}{\partial N}\frac{\partial f(\mathbf{p}_2\mathbf{r}_2t_2)}{\partial N} + \Phi(\mathbf{p}_1\mathbf{r}_1t_1; \mathbf{p}_2\mathbf{r}_2t_2).$$

(3.4.18)

The last term equals:

$$\Phi(\mathbf{p}_1\mathbf{r}_1t_1; \mathbf{p}_2\mathbf{r}_2t_2) = \frac{1}{V}\int d\mathbf{r}'dt' \sum_{\mathbf{p}_1'\mathbf{p}_2'} G(\mathbf{p}_1\mathbf{r}_1t_1|\mathbf{p}_1'\mathbf{r}'t')G(\mathbf{p}_2\mathbf{r}_2t_2|\mathbf{p}_2'\mathbf{r}'t')$$

$$\times [J^+(\mathbf{p}_1'\mathbf{p}_2'\mathbf{r}'t') - J^-(\mathbf{p}_1'\mathbf{p}_2'\mathbf{r}'t')].$$

(3.4.19)

Here $J^{\pm}(\mathbf{p}\mathbf{p}')$ are the mean particle fluxes introduced in Sec. 3.3. In equilibrium this part of the correlation function is zero: owing to the detailed balance that takes place in equilibrium the fluxes J^+ and J^- in the r.h.s. are equal. One can see that in a more general case, when the interparticle collisions are not rare, the correlation function is expressed in terms of the same quantities as in the absence of such collisions (Eq. (3.4.17)) but the equation is more complex.

According to Eqs. (3.4.10) and (3.4.17), the correlation function of intrinsic current density fluctuations in a nondegenerate gas with rare $e - e$ collisions equals:

$$\langle \delta j_\alpha^{\text{int}}(\mathbf{r}_1 t_1) \delta j_\beta^{\text{int}}(\mathbf{r}_2 t_2) \rangle = \frac{e^2}{V^2} 2 \sum_{\mathbf{p}_1 \mathbf{p}_2} v_\alpha(\mathbf{p}_1) v_\beta(\mathbf{p}_2)$$

$$\times \{ g(\mathbf{p}_1 \mathbf{r}_1 t_1 | \mathbf{p}_2 \mathbf{r}_2 t_2) f(\mathbf{p}_2 \mathbf{r}_2 t_2) + g(\mathbf{p}_2 \mathbf{r}_2 t_2 | \mathbf{p}_1 \mathbf{r}_1 t_1) f(\mathbf{p}_1 \mathbf{r}_1 t_1) \}. \tag{3.4.20}$$

The factor 2 results from explicit summation over spin quantum numbers.

Both the current density correlation function (3.4.20) and the small-signal conductivity (3.4.9) are expressed in terms of two quantities: the mean distribution function $f(\mathbf{prt})$ and the function $g(\mathbf{p}_1 \mathbf{r}_1 t_1 | \mathbf{p}_2 \mathbf{r}_2 t_2)$ (in Eq. (3.4.9) the Green's function G can be substituted by g because the difference between these functions gives no contribution). The Nyquist fluctuation–dissipation relation between the current spectral density and the real part of the conductivity is easily seen to hold only in the equilibrium state ($\mathbf{F} = 0$) when

$$-\frac{1}{\hbar} \frac{\partial f(\mathbf{pr})}{\partial p_\beta} = \frac{v_\beta(\mathbf{p}) f(\mathbf{pr})}{k_B T_0}. \tag{3.4.21}$$

At nonzero but small F ('warm' electrons) the part of the current noise linear in F^2 or I^2 may be viewed as a resistance noise, similar to the noise which, in the simple model presented in Sec. 3.2, was given by Eq. (3.2.12).

In many experiments, the quantity that is directly measured is the noise temperature T_N. It is defined (Secs. 2.2 and 3.2) as the noise power transmitted from the sample to a matched load in a unit frequency band. Equation (3.2.14) for T_N can be expressed in terms of the spectral density of current density fluctuations averaged over the sample's volume V, $S_{j_\alpha j_\alpha}(f)$, and the real part of the small-signal conductivity $\text{Re}\,\sigma_{\alpha\alpha}(\omega)$:

$$T_{N\alpha} = \frac{V S_{j_\alpha j_\alpha}(f)}{4 k_B \text{Re}\,\sigma_{\alpha\alpha}(\omega)}. \tag{3.4.22}$$

The noise temperature depends not only on the magnitude of the electric field, but also on the direction α along which it is measured: parallel or normal to the field $\mathbf{F}$, and, even in cubic crystals, on the crystallographic orientation of the sample.

As is well known, the kinetic equations, both homogeneous (for $f(\mathbf{prt})$) and nonhomogeneous (for the Green's function), can be solved analytically only in few cases. The best known is the case of uniform electric field and small inelasticity of electrons' scattering (Sec. 3.1). The mean distribution function for this case is given by Eqs. (3.1.11–13). The Green's function $g(\mathbf{pp'}; \mathbf{k}, \omega)$ was calculated for the same case by Kogan & Shulman (1969) and Shulman & Kogan (1969) ($\mathbf{k}$ and ω are the wave vector and frequency in the Fourier expansion of g as a function of $\mathbf{r} - \mathbf{r'}$ and $t - t'$, respectively). In the case of $\mathbf{k} = 0$, $\omega\tau_0 \ll 1$ it equals:

$$g(\mathbf{pp'}; 0, 0) = g_0(E, \mathbf{p'}) + g_1(\mathbf{pp'}),$$

$$g_0(E, \mathbf{p'}) = \frac{1}{V} f_0(E) \int_0^\infty d\epsilon \frac{\tau_0(\epsilon)}{N(\epsilon)\epsilon k_B T^*(\epsilon) f_0(\epsilon)} [\Theta(E - \epsilon) - v(\epsilon)]$$

$$\times [\Theta[E(\mathbf{p'}) - \epsilon] - v(\epsilon) + e\mathbf{F}\mathbf{v}(\mathbf{p'})\tau_1(E(\mathbf{p'}))\delta(\epsilon - E(\mathbf{p'}))], \qquad (3.4.23)$$

$$g_1(\mathbf{pp'}; 0, 0) = \tau_1(E)\left[\frac{1}{2}(\delta_{\mathbf{p'p}} - \delta_{\mathbf{p'},-\mathbf{p}}) - \frac{f_1(E)\cos\theta_{\mathbf{p}}}{nV}\right.$$

$$\left. - e\mathbf{F}\mathbf{v}(\mathbf{p})\frac{d}{dE}g_0(E, \mathbf{p'})\right].$$

Here $E, \tau_0, \tau_1, \theta_{\mathbf{p}}, T^*(E), N(E), V$ have the same meaning as in Sec. 3.1, $\Theta(x)$ is the Heaviside step function (Eq. (1.6.22)), and

$$v(E) = \frac{2}{n}\int_E^\infty d\epsilon \, N(\epsilon)f_0(\epsilon). \qquad (3.4.24)$$

The function $g_0(E, \mathbf{p'}; 0, \omega)$ is related to the relaxation of the electron's energy. At $\omega \gg \tau_0^{-1}$ the real part of this function falls off as $(\omega\tau_0)^{-2}$. Due to the integration over momenta in Eq. (3.4.20) only the asymmetric part g_1 of g contributes directly to the current density fluctuations with $\mathbf{k} = 0$. The function g_0 enters through the last term in the r.h.s. of the Eq. (3.4.23) for g_1. In the case of transverse current fluctuations ($\delta\mathbf{j} \perp \mathbf{E}$) the contribution of this term is zero. It is also negligible in the case of longitudinal current fluctuations at high frequencies, $f \gg \tau_0^{-1}$. Therefore the spectral density of transverse current fluctuations at frequencies $f \ll (2\pi\tau_1)^{-1}$ equals:

$$S_{j\perp}^{\text{int}}(f) = S_{j\parallel}(f \gg \tau_0^{-1}) = 4ne^2 D_\perp/V, \qquad (3.4.25)$$

where

$$D_\perp = \frac{2}{3n}\int_0^\infty dE \, N(E)v^2\tau_1(E)f_0(E). \qquad (3.4.26)$$

The longitudinal current fluctuations are affected by the fluctuations of the electron gas energy at frequencies f that are smaller than or of the order of τ_0^{-1}. They are accounted for by the function g_0 (the last term in Eq. (3.4.23)). The sign of the difference between the low-frequency and high-frequency values of the spectral density depends on the mechanism of momentum scattering. This sign can be determined by integration of the difference over the frequency. It was shown that (Kogan & Shulman, 1969):

$$\int_0^\infty df \left[S_{j\parallel}^{\text{int}}(f) - 4ne^2 D_\perp / V \right] = \frac{ne^4 F^2}{m^2 V} \left\{ \langle (\tau_1 - \langle \tau_1 \rangle)^2 \rangle + \frac{3}{5} \left\langle \frac{E d\tau_1^2}{dE} \right\rangle \right\}. \quad (3.4.27)$$

Here $\langle \ldots \rangle$ denotes averaging with the function $(2/3)E(-df_0/dE)$.

If $d\tau_1/dE > 0$ the expression in the r.h.s. of Eq. (3.4.27) is obviously positive. If τ_1 does not depend on energy this expression is zero. When $d\tau_1/dE < 0$ an analysis shows that this expression is negative for realistic scattering mechanisms (the last term in the r.h.s. dominates). It means that the sign of the frequency dependence of the spectral density of longitudinal current fluctuations at $f \sim \tau_0^{-1}$ depends on the deviation of the current–voltage characteristic (CVC) from the Ohmic one: in the case of superlinear CVC the low-frequency plateau is higher than the high-frequency one, in the case of sublinear CVC just the opposite takes place (see Fig. 3.2 and compare with Sec. 3.2). However, these conclusions apply only to electron gases with small inelasticity of scattering.

The absolute value of the difference of the two plateaus can be estimated by dividing the r.h.s. of Eq. (3.4.27) by the width of the frequency band, $\sim \tau_0^{-1}$, that contributes to the integral in Eq. (3.4.27). The relative height is

$$\frac{|S_{j\parallel}^{\text{int}}(f) - 4ne^2 D_\perp / V|}{4ne^2 D_\perp / V} \simeq \frac{T^* - T_0}{T_0}. \quad (3.4.28)$$

Here T^* is of the order of the electron temperature (Eq. (3.1.12)). Thus, the relative change of $S_{j\parallel}(f)$ from low frequencies to the high ones is of the order of the relative 'heating' of the electron gas.

In the case of small inelasticity of electron scattering, explicit equations for $S_{j\parallel}^{\text{int}}(0)$ and $S_{j\perp}^{\text{int}}(0)$ have been found (Gurevich & Katilius, 1965; Kogan & Shulman, 1969). As these spectral densities are related to the electrons' diffusion coefficients, $D_\parallel$ and $D_\perp$, the corresponding expressions are given in Sec. 3.5.

All analytical results presented above for hot electrons have been obtained for systems with quasi-elastic scattering only and, hence, an almost isotropic distribution function. A different regime which can be treated, to some extent, analytically is the so called 'streaming regime'. It is, in some sense, just the opposite of the regime of quasi-elastic scattering. Each charge carrier is accelerated by the electric field almost ballistically up to the optical phonon energy $\hbar\Omega$. When its energy exceeds this threshold energy, it emits an optical phonon (extremely inelastic scattering) and returns to the region of small energies. Then the process is repeated. The distribution function is 'needle-shaped'. The noise in this regime is discussed in Sec. 3.8.

A significant part of the theoretical results for the mean quantities and for the noise of hot electrons in various semiconductors was obtained by numerical methods. The most widely used is the Monte Carlo method of computer simulation of the charge carriers' random motion in strong electric fields, even spatially nonuniform. A detailed review of this field was published by Jacoboni & Reggiani (1983). Many results are presented also in the review by Bareikis, Katilius, Pozhela *et al.* (1992). Direct numerical solutions of the integro-differential kinetic equations for $f(\mathbf{p}\mathbf{r}t)$ and for the Green's function $g(\mathbf{p}\mathbf{r}t; \mathbf{p}'\mathbf{r}'t')$ are also possible.

A brief account of experiments on noise in high electric fields is presented in Sec. 3.5.

3.5 Current fluctuations and diffusion in a gas of hot electrons

In this section the relationship between the zero-frequency spectral density of current noise and the diffusion coefficient of hot electrons is discussed. We shall find the conditions for which this simple relationship holds.

In equilibrium systems the spectral density of intrinsic current fluctuations and the conductivity are connected by the fluctuation–dissipation relationship (Eq. (2.1.31)):

$$S_{j_\alpha j_\beta}^{\text{int}}(f) = 2\langle \delta j_\alpha^{\text{int}}(t_1)\delta j_\beta^{\text{int}}(t_2)\rangle_\omega = \frac{4k_B T_0}{V}\frac{\sigma_{\alpha\beta}(\omega) + \sigma_{\beta\alpha}^*(\omega)}{2}$$

$$= \frac{4k_B T_0}{V}\text{Re}\,\sigma_{\alpha\beta}(\omega) \qquad (\mathbf{B}\to 0). \tag{3.5.1}$$

Here T_0 is the temperature of the equilibrium system, V is the volume over which the fluctuations are averaged, $\mathbf{B}$ is the magnetic field.

Apart from Eq. (3.5.1), another relationship holds in equilibrium systems, specifically, the Einstein relationship between the tensor of the particles' diffusion coefficients, which determines the spreading of a packet of these particles, and the conductivity tensor:

$$D_{\alpha\beta} = \frac{1}{e^2}\frac{\partial E_F}{\partial n}\sigma_{\alpha\beta}(0), \tag{3.5.2}$$

where E_F is the chemical potential of the particles (Fermi energy), n is their density. Comparing Eqs. (3.5.1) and (3.5.2) one obtains an equation which may be called the fluctuation–diffusion relationship:

$$S_{j_\alpha j_\beta}^{\text{int}}(0) = \frac{4k_B T_0}{V}\frac{e^2}{\partial E_F/\partial n}\frac{D_{\alpha\beta} + D_{\beta\alpha}}{2}. \tag{3.5.3}$$

In the case of a nondegenerate gas and in the absence of interparticle interaction the derivative $\partial E_F/\partial n = k_B T_0/n$. Then Eq. (3.5.3) becomes:

$$S_{j_\alpha j_\beta}^{int}(0) = \frac{4e^2 n}{V} \frac{D_{\alpha\beta} + D_{\beta\alpha}}{2}. \tag{3.5.4}$$

A similar relationship was derived above for Brownian particles (Eq. (1.9.16)). It is worth noticing that the temperature does not enter explicitly in the r.h.s. of Eq. (3.5.4).

Interestingly enough, despite the fact that relationships (3.5.1) and (3.5.2) do not hold in nonequilibrium systems, for instance, for hot electrons, the relationship (3.5.4) is valid even in high electric fields (Price, 1965) if the electron–electron interaction is negligible, that is, if the correlation function of the intrinsic fluctuations of the occupancy numbers can be approximated by Eq. (3.4.17). This statement is proved below.

Let us examine the evolution in time and space of a packet of electrons which is assumed to be a small (linear) perturbation of an otherwise uniform and stationary hot electron gas. Assume also that the characteristic wave vectors k of the packet are many times smaller than all inverse electron relaxation lengths, and its characteristic frequencies are smaller than the inverse relaxation times. The equation for the diffusion coefficients which relate the current fluctuation and the gradient of the electron density can be represented in terms of the function $g(\mathbf{p}\mathbf{r}t|\mathbf{p}'\mathbf{r}'t')$ (Eq. (3.4.14)), which is, in fact, the inverse operator of the linearized kinetic equation (Gantsevich, Gurevich & Katilius, 1969):

$$D_{\alpha\beta} = \frac{2}{V} \sum_{\mathbf{p}\mathbf{p}'} v_\alpha(\mathbf{p}) g(\mathbf{p}\mathbf{p}'; \mathbf{k} = 0, \omega = 0) v_\beta(\mathbf{p}') \frac{\partial f(\mathbf{p}')}{\partial N}. \tag{3.5.5}$$

Here $g(\mathbf{p}\mathbf{p}'; \mathbf{k}, \omega)$ is the Fourier transform of g as a function of $\mathbf{r} - \mathbf{r}'$ and $t - t'$, the factor 2 accounts for the electron spin.

When the electron gas is nondegenerate and the electron–electron collisions are rare, $\partial f(\mathbf{p})/\partial N = f(\mathbf{p})/nV$. According to Eq. (3.4.20), under the same conditions the spectral density of intrinsic current fluctuations at zero frequency equals:

$$S_{j_\alpha j_\beta}^{int}(0) = 2\langle \delta j_\alpha^{int}(\mathbf{r}_1 t_1) \delta j_\beta^{int}(\mathbf{r}_2 t_2)\rangle_{\mathbf{k}=0,\omega=0}$$

$$= 2\frac{e^2}{V} 2 \sum_{\mathbf{p}\mathbf{p}'} [v_\alpha(\mathbf{p})v_\beta(\mathbf{p}') + v_\beta(\mathbf{p})v_\alpha(\mathbf{p}')] g(\mathbf{p}\mathbf{p}'; \mathbf{k} = 0, \omega = 0) f(\mathbf{p}').$$

$$\tag{3.5.6}$$

Comparing Eqs. (3.5.5) and (3.5.6) one can see that Price's fluctuation–diffusion relation (3.5.4) holds under the conditions stated above.

The equation for the diffusion coefficients can be derived also by calculating the conductivity of the electron gas $\sigma_{\alpha\beta}(\mathbf{k}\omega)$. It yields the attenuation, quadratic in the components of the wave vector $\mathbf{k}$, of a smooth perturbation (Shulman & Kogan, 1969). Just this attenuation determines the tensor of diffusion coefficients $D_{\alpha\beta}$.

In strong electric fields the small-signal conductivity $\sigma'_{\alpha\beta}$, the spectral density of current fluctuations $S^{\text{int}}_{j_\alpha j_\beta}$, and the diffusion coefficient $D_{\alpha\beta}$ are all anisotropic even in such systems (isotropic medium, cubic crystals) in which these quantities are isotropic in small electric fields. In the simplest case of an isotropic dependence of the charge carriers' energy on momentum, as in many $A^{III}B^V$ and $A^{II}B^{VI}$ semiconductors, the corresponding second rank tensors are determined by two elements only, longitudinal and transverse with respect to the dc electric field. In particular, there are two diffusion coefficients: the longitudinal $D_\parallel$ and transverse $D_\perp$.

The diffusion coefficients of the hot electrons can be calculated using Eqs. (3.5.5), (3.1.3), (3.4.3), and (3.4.14). Explicit expressions can be derived for small inelasticity of electron scattering (Eq. (3.4.23)). Then

$$D_\perp = \frac{1}{n}\int_0^\infty dE\, N(E)\frac{v^2\tau_1(E)}{3}f_0(E),$$

$$D_\parallel = D_\perp + \mu^2 F^2 \tau_0^{(1)}. \tag{3.5.7}$$

Here $N(E)$ is the density of electron states with spin states included, $f_0(E)$ is the distribution of electrons in energies E at a given electric field F, $\tau_1(E)$ is the momentum relaxation time,

$$\tau_0^{(1)} = \frac{1}{n}\int_0^\infty dE\, N(E)E\tau_0(E)\left(-\frac{df_0}{dE}\right)\left[\left(\frac{\mu(E)/\mu - n(E)/n}{N(E)Ef_0(E)/n}\right)^2 - \left(\frac{2e\tau_1}{3m\mu}\right)^2\right],$$

$$\mu(E) = \frac{e}{n}\int_E^\infty dE'\, N(E')\frac{v^2\tau_1(E')}{3}\left(-\frac{df_0}{dE'}\right),$$

$$n(E) = \int_E^\infty dE'\, N(E')f_0(E'),$$

$\mu \equiv \mu(0)$ is the effective electron mobility.

One can easily find that the sign of the difference $D_\parallel - D_\perp$, like the sign of the current spectral densities (Sec. 3.4), is related to the sign of the deviation of the CVC from Ohm's law. If τ_1 does not depend on energy E, that is, the CVC is linear even in high electric fields, the quantity $\tau_0^{(1)} = 0$, and the diffusion coefficient is isotropic. If the CVC is superlinear, the quantity

$\tau_0^{(1)} > 0$ and $D_\parallel > D_\perp$. The relative anisotropy of the diffusion coefficient is of the order of the electron gas 'heating' (Eq. (3.4.28)).

The diffusion coefficient can be expressed in terms of the noise temperature T_N and small-signal dc conductivity using Eqs. (3.4.22) and the fluctuation–diffusion relationship Eq. (3.5.4):

$$D_{\alpha\alpha} = k_B T_{N\alpha} \frac{\sigma_{\alpha\alpha}}{e^2 n}. \tag{3.5.9}$$

If the energies of the charge carriers are isotropic functions of their momenta, the tensors $\sigma_{\alpha\beta}$ and $D_{\alpha\beta}$ are diagonal in the frame, one axis of which is parallel to the field **F**. The small-signal conductivities equal:

$$\sigma_\perp = en\mu_\perp(F); \quad \sigma_\parallel = \frac{dj}{dF} = en\frac{dv_d}{dF}. \tag{3.5.10}$$

Here $\mu_\perp(F)$ is the field-dependent transverse mobility, $v_d = \mu(F)F$ is the drift velocity. Thus,

$$D_\perp = \frac{1}{e}k_B T_{N\perp}(F)\mu_\perp(F),$$
$$D_\parallel = \frac{1}{e}k_B T_{N\parallel}\frac{dv_d}{dF}. \tag{3.5.11}$$

Using measured values of $j = j(F)$, charge carriers' density n, and the noise temperatures, one is able to find the diffusion coefficients. In the case of small inelasticity of scattering, when the electron distribution function is almost isotropic even in high electric fields, $\mu_\perp(F) = \mu(F) = j(F)/enF$.

Erlbach & Gunn (1962) were the first to measure the hot electrons' noise. They measured the voltage fluctuations, in n-Ge at room temperature, in the direction normal to the applied electric field. The transverse noise temperature $T_{N\perp}$ increased with the electric field by $\simeq 230$ K at $F = 1.7$ kV/cm. At small fields $T_{N\perp} \propto F^2$, but at higher fields it grows with F almost linearly. Hart (1970) measured the noise temperature $T_{N\parallel}$ of current fluctuations in n-Ge at 77 K in the direction of the field for three different orientations of the sample's axis: [111], [110], and [100]. The frequencies were small as compared with both intervalley and intravalley electron inverse relaxation times (see Sec. 3.7). When the field **F** $\parallel$ [100], the intervalley noise is predicted to vanish due to symmetry (Sec. 3.7) and only the noise due to intravalley random electron motion should be observed. This last noise is much smaller than the noise measured in the two other directions. It means that at low frequencies the intervalley noise dominates.

Bareikis, Pozhela & Matulionene (1969) measured the field dependence of the noise temperatures $T_{N\parallel}$ and $T_{N\perp}$ in p-Ge at temperatures 10, 77 and 300 K. The noise measurements have been performed at frequency $f = 9.8$ GHz.

The field dependence of the mean current and noise temperatures have been also measured by Nougier & Rolland (1973) at temperatures between 77 and 300 K at $f = 0.5$ GHz, which is smaller than the inverse relaxation times of the charge carriers. Using the relationships between the diffusion coefficients of the charge carriers, on one hand, and noise temperatures and small-signal mobilities on the other hand, (Eq. (3.5.11)), Nougier & Roland obtained the field dependence of both diffusion coefficients. The diffusion coefficient $D_\parallel$ was found to be smaller than $D_\perp$. Since the CVC of p-Ge is sublinear, this result is in agreement with the theory.

3.6 One-time correlation in nonequilibrium gases

The one-time correlation function of the occupancy numbers' fluctuations in an equilibrium electron gas is a statistical, not a kinetic, quantity: it does not depend on any relaxation time. For a perfect gas it can be easily calculated. According to the Pauli principle, the number of electrons in a definite state with momentum $\mathbf{p}$ may be either $n(\mathbf{p}) = 0$ or 1. The one-time correlation equals $\langle n^2(\mathbf{p}) \rangle - \langle n(\mathbf{p}) \rangle^2$ (Eq. (1.2.1)). As at both possible values of the occupancy number $[n(\mathbf{p})]^2 = n(\mathbf{p})$, the correlation is proportional to $f(\mathbf{p})[1 - f(\mathbf{p})]$, where $f(\mathbf{p})$ is the mean occupancy number. In a perfect gas, in which the interaction between particles is negligible, the occupancies of different one-electron states are uncorrelated (see below). Hence the one-time correlation function equals:

$$\langle \delta f(\mathbf{p}_1, t)\delta f(\mathbf{p}_2 t) \rangle = \delta_{\mathbf{p}_1\mathbf{p}_2} f(\mathbf{p}_1)[1 - f(\mathbf{p}_1)] = \delta_{\mathbf{p}_1\mathbf{p}_2} k_B T \frac{\partial f(\mathbf{p}_1)}{\partial E_F}, \qquad (3.6.1)$$

where E_F is the Fermi level (chemical potential) of the particles. The factor $1 - f(\mathbf{p})$, which takes into account the Pauli principle, is of quantum origin. It means that the electron states with energies deep under the Fermi level, being permanently occupied by electrons, do not contribute to the noise.

The fluctuation of the total number of particles in the system under consideration is $\delta N = \sum_{\mathbf{p}} \delta f(\mathbf{p})$. Summation of both parts of Eq. (3.6.1) over $\mathbf{p}_1$ or $\mathbf{p}_2$ does not yield zero, that is, $\delta N \neq 0$. It means that the correlation function (3.6.1) refers to systems in which the total number of particles is not fixed. Each such system is a part of a much greater system with which it exchanges particles. Instead, the chemical potential is fixed by the greater system.

In a more general case that includes nonequilibrium gases, the one-time correlation function can be found as the limit of the two-time correlation function

$$\psi(\mathbf{p}_1\mathbf{r}_1 t|\mathbf{p}_2\mathbf{r}_2 t) = \lim_{t_1 \to t, t_2 \to t} \langle \delta f(\mathbf{p}_1\mathbf{r}_1 t_1) \delta f(\mathbf{p}_2\mathbf{r}_2 t_2) \rangle. \qquad (3.6.2)$$

This two-time correlation function satisfies the Boltzmann–Langevin equation (3.3.1). The latter can be applied both to systems with nonzero flow of particles through the boundaries (the total number of particles is not conserved) and to systems with a fixed total number N of particles (no flow of particles through the boundaries due, for instance, to charge conservation, i.e., to Coulomb forces). In the latter case the correlation functions, both two-time and one-time, must satisfy a simple condition: integration over $\mathbf{r}_1$ and $\mathbf{p}_1$ or $\mathbf{r}_2$ and $\mathbf{p}_2$ must yield zero. This condition becomes simpler in the case of an uniform system. The correlation function averaged over coordinates, $\psi(\mathbf{p}_1 t_1|\mathbf{p}_2 t_2)$, must yield zero after summation (or integration) over all $\mathbf{p}_1$ or over all $\mathbf{p}_2$.

The last condition is, of course, satisfied in the Langevin approach in the absence of flows through the boundaries. The Langevin sources in the Boltzmann–Langevin equation (3.3.1) correspond to such collisions after each of which the particles remain in the system. Thus, $\sum_{\mathbf{p}} \delta J_{\mathbf{p}}(t) = 0$. Consequently, the Boltzmann–Langevin equation after summation of each of its terms over all momenta yields:

$$\frac{\partial \sum_{\mathbf{p}} \delta f(\mathbf{p})}{\partial t} = \frac{\partial \delta N}{\partial t} = 0. \qquad (3.6.3)$$

The correlation between particles resulting from the restriction $N = \text{const.}$ adds a term in the r.h.s. of Eq. (3.6.1). In the case of an equilibrium system it can be easily found by taking into account that it has to be symmetric in both momenta, and to compensate exactly the first term in the r.h.s. of Eq. (3.6.1) after summation over either of the two momenta:

$$\langle \delta f(\mathbf{p}_1, t) \delta f(\mathbf{p}_2 t) \rangle = k_B T \left(\frac{\partial N}{\partial E_F} \right) \left[\delta_{\mathbf{p}_1 \mathbf{p}_2} \frac{\partial f(\mathbf{p}_1)}{\partial N} - \frac{\partial f(\mathbf{p}_1)}{\partial N} \frac{\partial f(\mathbf{p}_2)}{\partial N} \right]. \qquad (3.6.4)$$

We consider below only a nondegenerate gas. In this case Eq. (3.6.4), generalized to fluctuations varying in space, takes the form:

$$\langle \delta f(\mathbf{p}_1\mathbf{r}_1, t) \delta f(\mathbf{p}_2\mathbf{r}_2 t) \rangle = V\delta(\mathbf{r}_1 - \mathbf{r}_2)\delta_{\mathbf{p}_1\mathbf{p}_2} f(\mathbf{p}_1\mathbf{r}_1 t) - f(\mathbf{p}_1\mathbf{r}_1 t) f(\mathbf{p}_2\mathbf{r}_2 t)/N. \qquad (3.6.5)$$

Equation (3.6.5) holds as well in nonequilibrium gases, however, only when the interparticle collisions are rare. In fact, according to Eq. (3.4.12) and (3.4.14), the common value of the two terms in the r.h.s. of (3.4.17) at $|t_1 - t_2| \to 0$ coincides with the r.h.s. of Eq. (3.6.5).

The correlation function of a nonequilibrium gas with interparticle colli-
sions contains an additional term $\Phi(\mathbf{p}_1\mathbf{r}_1t_1|\mathbf{p}_2\mathbf{r}_2t_2)$ (Eqs. (3.4.18−19)). The
one-time correlation function can be obtained from Eq. (3.4.18) in the limit
$t_1 \to t_2$ (from either side). One has to take into account the value of the
function g at equal instants of time (Eqs. (3.4.12) and (3.4.14)):

$$\langle \delta f(\mathbf{p}_1\mathbf{r}_1t)\delta f(\mathbf{p}_2\mathbf{r}_2t)\rangle = V\delta(\mathbf{r}_1 - \mathbf{r}_2)\delta_{\mathbf{p}_1\mathbf{p}_2}f(\mathbf{p}_1\mathbf{r}_1t)$$
$$- N\frac{\partial f(\mathbf{p}_1\mathbf{r}_1t)}{\partial N}\frac{\partial f(\mathbf{p}_2\mathbf{r}_2t)}{\partial N} + \Phi(\mathbf{p}_1\mathbf{r}_1t|\mathbf{p}_2\mathbf{r}_2t) \tag{3.6.6}$$

The r.h.s. of Eq. (3.6.6) differs from the r.h.s. of Eq. (3.6.5): (1) each
function $f(\mathbf{p}\mathbf{r}t)$ in the second term is substituted by $N\partial f(\mathbf{p}\mathbf{r}t)/\partial N$, where N
is the total number of particles in the gas; (2) the term $\Phi(\mathbf{p}_1\mathbf{r}_1t|\mathbf{p}_2\mathbf{r}_2t)$ (Eq.
(3.4.19)), which represents an additional correlation induced by interparticle
collisions, appears. According to Eqs. (3.4.3) and (3.4.19), at $t_1 \neq t_2$ it
satisfies the following equation:

$$\left(\frac{\partial}{\partial t_1} + \frac{\partial}{\partial t_2}\right)\Phi(\mathbf{p}_1\mathbf{r}_1t_1|\mathbf{p}_2\mathbf{r}_2t_2)$$
$$+ \hat{L}(\mathbf{p}_1\mathbf{r}_1t_1|\mathbf{p}'\mathbf{r}'t_1)\Phi(\mathbf{p}'\mathbf{r}'t_1|\mathbf{p}_2\mathbf{r}_2t_2) + \hat{L}(\mathbf{p}_2\mathbf{r}_2t_2|\mathbf{p}'\mathbf{r}'t_2)\Phi(\mathbf{p}_1\mathbf{r}_1t_1|\mathbf{p}'\mathbf{r}'t_2)$$
$$= \sum_{\mathbf{p}'}\{G(\mathbf{p}_2\mathbf{r}_2t_2|\mathbf{p}'\mathbf{r}_1t_1)[J^+(\mathbf{p}_1\mathbf{p}'\mathbf{r}_1t_1) - J^-(\mathbf{p}_1\mathbf{p}'\mathbf{r}_1t_1)] \tag{3.6.7}$$
$$+ G(\mathbf{p}_1\mathbf{r}_1t_1|\mathbf{p}'\mathbf{r}_2t_2)[J^+(\mathbf{p}_2\mathbf{p}'\mathbf{r}_2t_2) - J^-(\mathbf{p}_2\mathbf{p}'\mathbf{r}_2t_2)]\}.$$

Operator $\hat{L}$ is defined by Eq. (3.4.2), summation and integration over primed
variables is implied. The first term in the r.h.s. is nonzero at $t_2 > t_1$,
the second is nonzero in the opposite case. At $|t_1 - t_2| \to 0$ both terms
tend to one and the same value (Eq. (3.4.12)). The equation for the
one-time additional correlation becomes (Gantsevich, Gurevich & Katilius,
1969):

$$\frac{\partial}{\partial t}\Phi(\mathbf{p}_1\mathbf{r}_1t|\mathbf{p}_2\mathbf{r}_2t) + \hat{L}(\mathbf{p}_1\mathbf{r}_1t|\mathbf{p}'\mathbf{r}'t)\Phi(\mathbf{p}'\mathbf{r}'t|\mathbf{p}_2\mathbf{r}_2t) + \hat{L}(\mathbf{p}_2\mathbf{r}_2t|\mathbf{p}'\mathbf{r}'t)$$
$$\times \Phi(\mathbf{p}_1\mathbf{r}_1t|\mathbf{p}'\mathbf{r}'t) = V\delta(\mathbf{r}_1 - \mathbf{r}_2)[J^+(\mathbf{p}_1\mathbf{p}_2\mathbf{r}_1t) - J^-(\mathbf{p}_1\mathbf{p}_2\mathbf{r}_1t)]. \tag{3.6.8}$$

According to Eqs. (3.3.9) and (3.6.8), the source of the additional correlation
is such collisions after which two particles *simultaneously* appear in the states
with momenta $\mathbf{p}_1$ and $\mathbf{p}_2$ and collisions in which two particles *simultaneously*
leave these states.

In an equilibrium gas the detailed balance holds: the mean rate of each
kind of transition is equal to the mean rate of transitions in the opposite
direction. However, in a nonequilibrium gas there is no detailed balance,
even in a stationary state, and hence the difference $J^+(\mathbf{p}_1\mathbf{p}_2) - J^-(\mathbf{p}_1\mathbf{p}_2) \neq 0$.

In a stationary state this difference is counterbalanced, for instance, by the action of the field which makes the gas a nonequilibrium one or by 'ring' processes in which other states, different from the two considered ones, participate. It is worthy of note that the additional one-time correlation in a nonequilibrium gas arises even though the individual collisions between particles are uncorrelated (Sec. 3.3).

An interesting problem is the role of the additional correlation in the case of very frequent electron–electron collisions when the energy distribution function is close to a Maxwellian one with some electron temperature T (Secs. 3.1 and 3.2). On the one hand, just the electron–electron collisions are the source of the additional correlation. On the other hand, the Maxwellian function turns the r.h.s of Eq. (3.6.8) to zero, and, therefore, to $\Phi(\mathbf{p}_1\mathbf{p}_2) = 0$. Shulman (1970) examined this problem taking into account the corrections to the Maxwellian function. These corrections are nonzero only for a nonequilibrium gas and are of the order of τ_{ee}/τ_0 (Sec. 3.1). The main conclusion is that the additional correlation $\Phi(\mathbf{p}_1\mathbf{p}_2) \neq 0$ and must be taken into account. The relationship (3.5.4) between the zero-frequency spectral density of the intrinsic currents and the longitudinal diffusion coefficient $D_\parallel$ does not hold.

The problem of the one-time additional correlation is expounded in detail by Gantsevich, Gurevich & Katilius (1979), and by Bareikis, Katilius & Miliušytė (1989).

3.7 Intervalley noise in multivalley semiconductors

The conduction band of some semiconductors (Ge, Si) has several minima which are symmetrically located in the Brillouin zone. In the absence of uniaxial strain of the crystal the energies of these minima are equal. The region of the Brillouin zone, i.e., the region of the quasi-momentum space around each conduction band minimum is called a valley. The surfaces of constant electron energies in the valleys are ellipsoids of revolution, and the electron effective mass is anisotropic: the longitudinal mass m_l, in the direction of the ellipsoid's longest axis, is greater than the transverse one, m_t. The mobility of charge carriers is well known to depend on their effective mass. Hence, the electron mobility in a given valley is anisotropic, i.e., is a tensor $\mu_{\alpha\beta}^{(v)}$, where α and β are tensor indices, v is the number of the given valley. It varies from 1 to the total number of valleys v. This tensor is diagonal in the reference frame with one of the axes parallel to the ellipsoid's axis of revolution. One of the diagonal components of this diagonal tensor is the longitudinal mobility μ_l, two others are equal to the transverse mobility

μ_t. The measured total conductivity is a sum of contributions of electrons in all valleys:

$$\sigma_{\alpha\beta} = e \sum_{v=1}^{v} \mu_{\alpha\beta}^{(v)} n_v. \tag{3.7.1}$$

Here n_v is the density of conduction electrons in the v-th valley. The total density of conduction electrons is obviously $n = \sum_{v=1}^{v} n_v$.

In the absence of elastic strain and strong electric field (the electrons are not hot) all densities $n_v = n/v$ are equal and, owing to symmetrical location of the valleys, the total conductivity is isotropic, i.e., proportional to $\delta_{\alpha\beta}$. The effective mobility is $\mu = (2\mu_t + \mu_l)/3$.

The conduction electrons randomly switch between the valleys. Each such transition, say, from the valley v to the valley v', is accompanied by emission or absorption of a phonon ('intervalley phonon') with a wave vector $\mathbf{q} = \mathbf{k}_{v'} - \mathbf{k}_v$, where $\hbar\mathbf{k}_v$ is the quasi-momentum of the v-th minimum of the conduction band in the Brillouin zone. Usually the absolute value of the difference $|\mathbf{k}_{v'} - \mathbf{k}_v| \sim a_0^{-1}$, where a_0 is the lattice parameter. It means that the energy of the intervalley phonons is $\hbar\Omega(\mathbf{q}) \sim k_B\Theta_D$, where Θ_D is the Debye temperature of the crystal. In general, especially at low temperatures $T < \Theta_D$, the rate of intervalley transitions is much smaller than the rate of intravalley scattering because the energies of phonons emitted or absorbed in the last processes are much smaller than $k_B\Theta_D$. An electron changes its momentum and energy many times within a valley between two successive intervalley transitions. Therefore the rate of intervalley transitions from valley v into valley v' may be characterized by a probability $W_{vv'}$ which depends only on the numbers of the valleys, and does not depend on the momenta in the first and second valleys. If the valleys are equivalent, this probability may depend only on the initial valley and not on v', the final valley. It may be denoted by the number of the initial valley only, W_v. If the electrons are not hot, all W_v are equal.

Under a strong electric field the conduction electrons become hot. The heating is, however, different in different valleys. The maximum heating is in the valley in which the direction of light electron effective mass coincides with or is close to the direction of the electric field $\mathbf{F}$ in the specimen. The probability W_v of intervalley transitions is greater for the 'hot' valley than for a 'cold' valley. Hence, the density of electrons in the 'cold' valleys increases at the expense of the electron density in 'hot' valleys. The conductivity $\sigma_{\alpha\beta}$ becomes, in general, anisotropic.

Under stationary conditions the electron densities in the valleys, n_v, satisfy the following simple equation:

$$\frac{dn_v}{dt} = \sum_{v'(\neq v)} W_{v'v} n_{v'} - n_v \sum_{v'(\neq v)} W_{vv'} = 0. \qquad (3.7.2)$$

As $W_{vv'} = W_v$ is independent of v', this equation becomes $v W_v n_v = \sum_{v'} W_{v'} n_{v'}$. It means that the mean rate of electron transitions $J_{vv'} = W_v n_v$ from a valley v to any other valley in a unit volume is the same for all valleys, and can be denoted by J. It is convenient to introduce the intervalley relaxation time $\tau_v = 1/v W_v$. The solution to Eq. (3.7.2) then reads:

$$n_v = n \frac{\tau_v}{\sum_{v'} \tau_{v'}}, \quad J = \frac{n}{v \sum_{v'} \tau_{v'}}. \qquad (3.7.3)$$

As the intervalley transitions are random, the number of electrons in the specimen belonging to a given valley fluctuates around its mean value, $N_v = n_v V$, where V is the volume of the specimen. The electron mobility in each valley is anisotropic. Hence, the electrons in different valleys have different mobilities at given directions of the electric field and current in the specimen. In other words, an intervalley transition changes the effective mobility of an electron, and the fluctuations of the numbers of electrons in valleys, $\delta N_v(t)$, produce fluctuations of the conductivity, $\delta\sigma_{\alpha\beta}(t)$. At a given field $\mathbf{F}$ and a constant total density of conduction electrons n, the fluctuations of the current density equal:

$$\delta j_\alpha(t) = \frac{e}{V} \sum_v \mu_{\alpha\beta}^{(v)} F_\beta \delta N_v(t). \qquad (3.7.4)$$

The resulting noise is called intervalley noise. It was predicted by Price (1960).

In the Langevin approach (Sec. 1.10) the kinetic equation for the fluctuations $\delta N_v(t)$ reads:

$$\frac{\partial \delta N_v}{\partial t} + \delta N_v \sum_{v'(\neq v)} W_{vv'} - \sum_{v'(\neq v)} W_{v'v} \delta N_{v'} = \delta J_v(t). \qquad (3.7.5)$$

The second and third terms in the l.h.s. represent that part of the fluctuations of the rates of transitions to the given valley v and from this valley to all other valleys v', which results from the fluctuations of the populations of the valleys, δN_v. The Langevin source in the r.h.s., $\delta J_v(t)$, results from the randomness of uncorrelated intervalley transitions. It can not be expressed in terms of $\delta N_v(t)$. Since the total number of electrons is conserved, $\sum_v \delta N_v(t) = 0$ and $\sum_v \delta J_v(t) = 0$.

The solution to Eq. (3.7.5) for the Fourier amplitudes of the fluctuations is easily obtained:

$$\delta N_v(\omega) = (-i\omega + \tau_v^{-1})^{-1} \left[\delta J_v(\omega) - (\theta_1 + i\omega\theta_2)^{-1} \sum_{v'} \frac{\delta J_{v'}(\omega)}{-i\omega + \tau_{v'}^{-1}} \right], \quad (3.7.6)$$

where

$$\theta_1(\omega) = \sum_{v=1}^{v} \frac{\tau_v}{1 + \omega^2\tau_v^2}, \quad \theta_2(\omega) = \sum_{v=1}^{v} \frac{\tau_v^2}{1 + \omega^2\tau_v^2}. \quad (3.7.7)$$

The Langevin source is a sum of sources corresponding to all kinds of intervalley transitions:

$$\delta J_v(t) = \sum_{v'(\neq v)} [\delta J_{v'v}(t) - \delta J_{vv'}(t)]. \quad (3.7.8)$$

Because different intervalley transitions are uncorrelated, each Langevin source is correlated only with itself. It is a Poisson-like random process similar to shot noise. According to Eqs. (1.5.6) and (1.5.8) its correlation function equals:

$$\langle \delta J_{v_1v_1'}(t_1)\delta J_{v_2v_2'}(t_2) \rangle = \delta_{v_1v_2}\delta_{v_1'v_2'}J_{v_1v_2}\delta(t_1 - t_2), \quad (3.7.9)$$

where $J_{v_1v_2}$ is the mean rate of transitions from the v_1-th valley to the v_2-th valley in the entire sample of volume V. This correlation function is nonzero only if both initial and both final states coincide, i.e., $v_1 = v_2$ and $v_1' = v_2'$. The delta-function $\delta(t_1 - t_2)$ in the r.h.s. is a result of neglection of intravalley relaxation times as compared with the intervalley relaxation times τ_v. Consequently, Eq. (3.7.9) is supposed to be used only for noise frequencies that are lower than the inverse intravalley relaxation times.

One can easily derive from Eqs. (3.7.8) and (3.7.9) that in the case of equivalent valleys, when the mean rates of transition are independent of the numbers v and v' of the valleys,

$$\langle \delta J_{v_1}(t_1)\delta J_{v_2}(t_2) \rangle = 2JV[v\delta_{v_1v_2} - 1]\delta(t_1 - t_2). \quad (3.7.10)$$

From Eqs. (3.7.3), (3.7.4), (3.7.6), and (3.7.10) one obtains the spectral density of current fluctuations:

$$S_{j\alpha\beta}(f) = \frac{4e^2n}{V\sum_{v'}\tau_{v'}} \sum_{\alpha'\beta'v_1v_2} \mu_{\alpha\alpha'}\mu_{\beta\beta'}F_{\alpha'}F_{\beta'}$$

$$\times \left\{ \frac{\delta_{v_1v_2}}{\omega^2 + \tau_{v_1}^{-2}} + \frac{(\tau_{v_1}^{-1}\tau_{v_2}^{-1} - \omega^2)\theta_2(\omega) - (\tau_{v_1}^{-1} + \tau_{v_2}^{-1})\theta_1(\omega)}{[\omega^2\theta_2^2(\omega) + \theta_1^2(\omega)](\omega^2 + \tau_{v_1}^{-2})(\omega^2 + \tau_{v_2}^{-2})} \right\}. \quad (3.7.11)$$

At small electric fields when the electrons are not hot, all $\tau_v = \tau$ are equal and the mean densities of electrons in all valleys are also equal. Equation (3.7.11) becomes much simpler:

$$S_{j_\alpha j_\beta}(f) = \frac{4n\tau e^2}{V(1 + \omega^2 \tau^2)} \sum_{\alpha' \beta'} \Big[\frac{1}{v} \sum_v \mu_{\alpha\alpha'}^{(v)} \mu_{\beta\beta'}^{(v)} - \mu^2 \delta_{\alpha\alpha'} \delta_{\beta\beta'} \Big] F_{\alpha'} F_{\beta'}. \qquad (3.7.12)$$

For example, let us examine n-Si. Six conduction band minima lie on cubic axes of [100] type (two minima on each). If the electric field and the current are parallel with one of these axes, say [001], all tensor indices in Eq. (3.7.12) are z. For the valley on the [001] axis the mobility is μ_l, for the two other valleys the mobility in the same direction is μ_t. The quantity in rectangular brackets equals:

$$\frac{1}{v} \sum_v \mu_{zz}^{(v)} \mu_{zz}^{(v)} - \mu^2 = \frac{2}{9}(\mu_t - \mu_l)^2. \qquad (3.7.13)$$

However, if the electric field is directed along [111], with respect to which the valleys are symmetrical, the r.h.s. of Eq. (3.7.11) and (3.7.12) turns into zero and the intervalley noise disappears. The same happens in n-Ge when the electric field is directed along [100] because in n-Ge the conduction band minima lie on the [111] axes.

The intervalley noise can be distinguished from other noise sources: it appears only under nonzero bias like any other resistance noise, and it depends on the orientation of the current relative to the crystallographic axes. Of course, it can be measured only at frequencies that are lower than or of the order of inverse intervalley scattering times.

As was mentioned in Sec. 3.5, Hart (1970) found that in n-Ge at 77 K the noise temperature $T_{N\parallel}$ is much smaller when the field direction is parallel to [100] (symmetrical with respect to the conduction band valleys) than when it is parallel to [111] or [110]. The frequencies in his experiment were lower than both intervalley and intravalley electron inverse relaxation times. It means that at these frequencies the intervalley noise is dominant when the field is directed along crystallographic axes different from [100]. If F ∥ [100] only the noise due to intravalley random electron motion should be observed.

3.8 Noise of hot electrons emitting optical phonons in the streaming regime

The emission of an optical phonon by a charge carrier, the energy of which exceeds the optical phonon energy $\hbar\Omega$, is known to be a very fast process. In n-GaAs, for instance, the time for emission of an optical phonon is of the order of $2 \cdot 10^{-13}$ s. The probability of optical phonon emission by an electron steeply increases above the threshold $\hbar\Omega$. It means that if the charge carrier, being accelerated by the electric field, reaches the energy $\hbar\Omega$, it emits an optical phonon in a short time and descends to small energies. In fact, the energy in excess of $\hbar\Omega$ acquired by the charge carrier in the 'active region' $(E > \hbar\Omega)$ is $\sim 2(\tau_{em}/\tau_F)\hbar\Omega$, where τ_{em} is the time spent by the electron in the 'active region', $\tau_F = \sqrt{2m\hbar\Omega}/eF$ is the time of acceleration of a charge carrier from small energies $E \ll \hbar\Omega$ to the energy $E_0 = \hbar\Omega$ by the electric field F, and m is the effective mass. Owing to the small ratio τ_{em}/τ_F, this energy is usually a small part of $\hbar\Omega$. In high-purity semiconductors at low temperatures the relaxation time for scattering by acoustic phonons and impurities, τ_p, may be much greater than τ_F. Under these conditions the charge carriers are accelerated from low energies up to $\hbar\Omega$ almost ballistically. In the extreme case of $\tau_F/\tau_p \to 0$ the momentum distribution function of the charge carriers is 'needle-shaped' (Shockley, 1951; Pinson & Bray, 1964; Vosilyus & Levinson, 1966; see also the review by Komiyama, Kurosawa & Masumi, 1985): the momenta of all charge carriers are directed parallel to the accelerating force $e\mathbf{F}$ (such motion of electrons is also called the 'streaming regime'). Each charge carrier moves periodically: it is accelerated from a small energy to $\hbar\Omega$, then almost immediately returns to the initial state, and so on. However, the phases of this periodic motion of different charge carriers, or, equivalently, their 'starting' times t_l (l numbers the charge carriers) are random and uncorrelated. The velocity of the l-th charge carrier $v(t - t_l)$ is a periodic 'saw-like' function of time t with a mean value $\bar{v} = eF\tau_F/2m$. Between t_l and $t_l + \tau_F$ it is equal to $(eF/m)(t - t_l)$. Its Fourier expansion equals:

$$v(t) = \bar{v} + \frac{\bar{v}}{\pi} \sum_{k=-\infty(k\neq0)}^{k=+\infty} \frac{1}{ik} e^{-ik\omega_0 t}. \qquad (3.8.1)$$

Here $\omega_0 = 2\pi/\tau_F$ is the characteristic angular frequency.

The current in the sample is expressed in terms of a sum over the velocities of all electrons in the specimen:

$$I(t) = Aj(t) = A\frac{e}{V} \sum_l v(t - t_l). \qquad (3.8.2)$$

Here A is the sample's cross-sectional area, V is its volume, $j(t)$ is the current density. The mean value of the current is $\bar{I} = AeN\bar{v}/V = eAn\bar{v}$, where N is the total number of charge carriers in the sample, n is their density. The fluctuation of the current equals:

$$\delta I(t) \equiv I(t) - \bar{I} = \frac{eA\bar{v}}{\pi V} \sum_l \sum_{\substack{k=-\infty \\ (k \neq 0)}}^{k=+\infty} \frac{1}{ik} e^{-ik\omega_0(t-t_l)}. \tag{3.8.3}$$

The current correlation function, i.e., the mean product of fluctuations at two different times t_1 and t_2, is expressed by Eq. (3.8.3) in terms of a double sum over the charge carriers. Due to the randomness of the times t_l (or phases $\omega_0 t_l$) and absence of their correlation, this double sum is reduced to a single sum, and, at last, to a Kronecker symbol:

$$\sum_l \sum_{l'} e^{-i\omega_0(kt_l + k't_{l'})} = \sum_l e^{-i(k+k')\omega_0 t_l} = N\delta_{k',-k}. \tag{3.8.4}$$

Hence, the correlation function equals:

$$\psi_{I\parallel}(t_1 - t_2) \equiv \langle \delta I(t_1)\delta I(t_2) \rangle = \frac{2nA^2 e^2 \bar{v}^2}{\pi^2 V} \sum_{k=1}^{+\infty} \frac{\cos[k\omega_0(t_1 - t_2)]}{k^2}. \tag{3.8.5}$$

According to the Wiener–Khintchine theorem (1.3.8), the spectral density of longitudinal current fluctuations equals (Levinson & Matulis, 1968):

$$S_{I\parallel}(f) = \frac{2A^2 e^2 n\bar{v}^2}{\pi^2 V} \sum_{\substack{k=-\infty \\ (k \neq 0)}}^{k=+\infty} \frac{1}{k^2} \delta(f - kf_0). \tag{3.8.6}$$

It is a sum of infinitely narrow peaks at the main frequency $f_0 = 1/\tau_F = \omega_0/2\pi$ and its harmonics.

Within the simple model presented above, all processes that randomize the periodic motion of each charge carrier have been neglected as weak ones. Consequently, the phase of these oscillations is preserved infinitely. The model in its simplest version lacks a finite correlation time. This flaw of the model becomes obvious after inspection of the correlation function (3.8.5): it is a periodical function of time with a period τ_F and it does not decay to zero at $|t_1 - t_2| \to \infty$. In fact, the sum in the r.h.s. of Eq. (3.8.5) is exactly expressed in terms of the second Bernoulli polynomial. In the interval $0 \leq t \leq \tau_F$

$$\psi_{I\parallel}(t) = \frac{nA^2 e^2 \bar{v}^2}{3V} \left[1 - 6\frac{t}{\tau_F} + 6\left(\frac{t}{\tau_F}\right)^2 \right], \tag{3.8.7}$$

and is periodically continued beyond the interval $0 \leq t \leq \tau_F$ (Fig. 3.4, dashed curve).

The processes that randomize the periodical motion of charge carriers are: the scattering of the charge carriers in the 'passive region' ($E < \hbar\Omega$) and different penetration of the charge carriers into the 'active region' ($E > \hbar\Omega$)

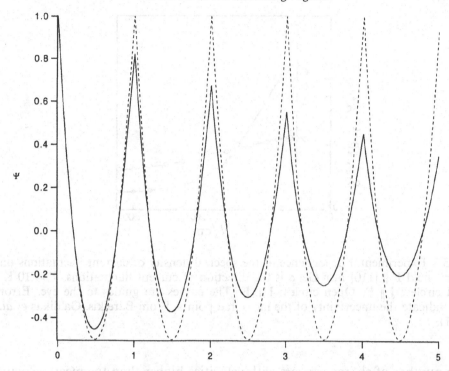

Fig. 3.4 Correlation function of electrons emitting optical phonons in a streaming regime. $\Psi = \psi_I(t)/\psi_I(0)$, τ_F is the time of acceleration of an electron up to the energy of an optical phonon. Dashed curve: no randomization of the periodical motion of each individual electron, solid curve: exponential decay of the correlation function oscillations due to their randomization. Decay parameter $\gamma = 1/5\tau_F$.

at each emission of an optical phonon. These relaxation processes broaden the peaks of the spectral density given by Eq. (3.8.6) and give rise to a low background between the peaks (Levinson & Matulis, 1968). The line-widths of the peaks due to the first process mentioned above are the same for all peaks. Those due to the second process are proportional to k^2, where k is the number of the harmonic. The amplitude of the oscillations of the correlation function given by Eq. (3.8.7) decays exponentially if the first process is more important, and according to a more complicated law if the second one is the most important. In Fig. 3.4 this decay is shown schematically by the solid curve.

The noise generated by charge carriers when the emission of optical phonons is the dominant scattering mechanism is interesting for the shape of its correlation function: it takes negative values in definite intervals of $t_1 - t_2$. It means that the fluctuations separated by these time-intervals are anticorrelated. This phenomenon is easily understood. Suppose that at a time

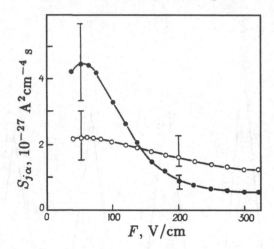

Fig. 3.5 Experimental dependence of the spectral density of current fluctuations on electric field $\mathbf{F} \parallel [110]$ in p-Ge. α is the direction of current fluctuations. $T = 10$ K. Solid circles: $\mathbf{j} \parallel \mathbf{F}$. Open circles: $\mathbf{j} \perp \mathbf{F}$. The curves are guides to the eye. Error bars indicate the uncertainty of the measured points. From Bareikis, Galdikas *et al.* (1981).

t the number of charge carriers with velocities higher than the mean velocity $\bar{v}$ exceeds the number of charge carriers with lower velocities, i.e., the current fluctuation is positive. However, after a time $\sim \tau_F/2$, that is, after the major part of charge carriers reach the energy $\hbar\Omega$ and emit optical phonons, they become the slowest, and the current fluctuation becomes negative. Hence, the product $\delta I(t)\delta I(t + \tau_F/2)$ is negative.

These features of the noise in the streaming regime have been verified by Monte Carlo simulations, and by experiments (see the review by Bareikis, Katilius, Pozhela *et al.*, 1992). As the inverse scattering time by impurities and acoustic phonons is of the order of ~ 10 GHz, the streaming regime is reached only when $f_0 = 1/\tau_F > 10$ GHz. The measurements of the noise spectrum at such high frequencies are technically very difficult. To some extent these measurements can be substituted by measurements of the dependence of the spectral density of noise on the electric field F at fixed high frequency f. Varying the electric field, and, hence, the transit time τ_F (see above) one can achieve the resonant condition $\tau_F^{-1} = f$. The spectral density of longitudinal current fluctuations, $S_{I\parallel}$, as a function of the field F in the vicinity of this 'resonance' is expected to have a peak. This peak is actually seen in Fig. 3.5. The spectral density of the longitudinal current noise falls off at higher fields and this noise becomes even smaller than the transverse current noise.

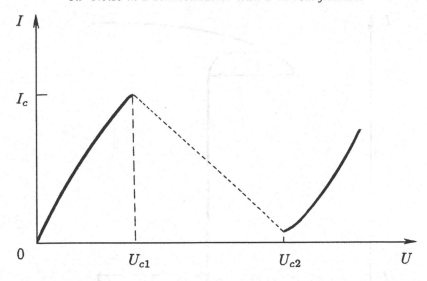

Fig. 3.6 *N*-type current–voltage characteristic (CVC). The states corresponding to the part with negative differential resistance between critical voltages U_{c1} and U_{c2} (dashed curve) are unstable in sufficiently long samples.

3.9 Noise in a semiconductor with a postbreakdown stable current filament

The effect of electron nonequilibrium 'heating' in high electric fields is in some semiconductors so strong that the current–voltage characteristic (CVC) becomes *N*-type (Fig. 3.6) or *S*-type (Fig. 3.7). In the first case the voltage U is a multivalued function of the current I, in the second the current I, is a multivalued function of U. The main feature of these CVC is the presence of a branch with negative differential resistance (NDR) $dU/dI < 0$. If the semiconductor sample with *N*-type CVC is sufficiently long and if the voltage U across the sample is fixed (the impedance of the outer circuit is small) and corresponds to the NDR region, the uniform distribution of the electric field along the sample is unstable. The stable distribution is nonuniform. It has the form of an electric domain (Gunn domain), which is a solitary wave of the electric field. The effect of periodical creation of domains at the cathode, their motion along the sample, disappearance at the anode, and of the resulting current oscillations is called the Gunn effect. It is used in high-frequency oscillators.

The *S*-type CVC is observed in many semiconductors: *n*-InSb, Ge at low temperatures, in amorphous chalcogenide semiconductors (in this case the effect is called 'switching'). It is viewed as a result of some breakdown. *S*-type CVC is observed also in diode structures (*n-p-n*, and others), where it is associated with double injection of electrons and holes. This type of

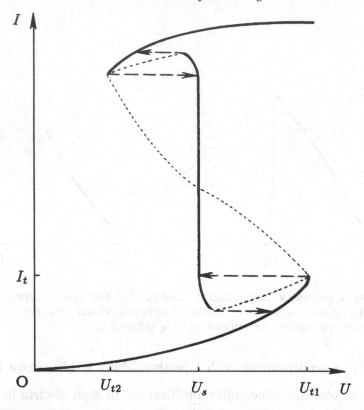

Fig. 3.7 S-type CVC. The states corresponding to negative differential resistance and uniform distribution of current density (between U_{t2} and U_{t1}) and the states with a current filament but with positive differential resistance are unstable (dashed curves). The branch at voltages around U_s corresponds to a current filament in the sample (solid curve). Arrows indicate voltage discontinuities forming two hysteresis loops.

CVC is observed as well in the positive column of electric arc discharge, and in many other systems. If the transverse dimensions of the sample are sufficiently large as compared with a definite microscopic length l, which depends on the mechanism of the S-type CVC, the uniform distribution of current density over the sample's cross-section is unstable at those values of current that correspond to the NDR. If the load impedance is sufficiently high (current fixed), the stable current distribution is a current filament. Inside it the current density is higher than in the surrounding part of the specimen. The CVC is shown schematically in Fig. 3.7 by solid lines. It differs from that CVC which would be measured if the current distribution were uniform (central dashed curve). When the current and voltage reach the values I_t and U_{t1}, respectively, the voltage switches to a lower value U_s.

As the current I is increased above I_t, the voltage remains almost constant, and the corresponding branch of the CVC is almost vertical. That is why the voltages U_t and U_s are called the threshold breakdown voltage and the voltage sustaining the breakdown, respectively. At even greater currents the voltage drops once more and the CVC switches to the upper branch with positive differential resistance. At decreasing current the voltage augments usually at lower values of the current (than it dropped), and therefore two hysteresis loops are observed (for a review see Volkov & Kogan, 1968). Inhomogeneities of the sample and of the near-electrode regions may distort this picture. In what follows we consider only a reversible breakdown.

The small-signal resistance in the presence of a stable current filament is negative (of course, it differs from the NDR of the unstable CVC at uniform current density). It means that the frequency-dependent Re $Z(f)$ and the spectral density of voltage noise $S_U(f)$ differ even qualitatively, because the former is negative up to some frequency while the latter is always positive.

Very often the measured CVC after breakdown is vertical or almost vertical. It means that the radius r_f of the current filament which has been created in the sample is many times greater than the width l_w of the filament's wall, i.e., the thickness of the transition region between the inner part of the filament and the surrounding part of the sample (Kogan, 1968; Volkov & Kogan, 1968). The length l_w is determined, depending on the mechanism of the NDR, by the diffusion of charge carriers, diffusion of heat or of the charge carriers' energy, etc. In a 'thick' filament, to which the vertical branch of the CVC corresponds, the radius $r_f \gg l_w$, the wall is almost planar, and the increase of the current is accompanied only by the increase of the filament's radius r_f, i.e., by the displacement of the wall. The electric field $F_s = U_s/L_s$ remains almost constant (L_s is the length of the specimen).

In some conductors with S-type CVC in the vicinity of the voltage switch ('jump') or after this switch, i.e., in the state with a current filament, strong chaotic oscillations are observed. These oscillations can be associated with an instability in the circuit caused by the fact that the differential resistance of the sample, even after its transition to the state with a current filament, is negative (Volkov & Kogan, 1968). However, there are many systems in which such large amplitude spontaneous oscillations do not arise. In this section the main features of noise in such stable systems are elucidated. The main interest lies in the fact that these systems are extremely nonequilibrium and nonlinear.

It is reasonable to use a simple model of a system with S-type CVC. It is an electron gas, the conductivity $\sigma(T)$ and the energy dissipation time

$\tau_e(T)$ of which increase so steeply with the electron temperature T that an overheating instability occurs at some electric field F_t. Let us consider an axially symmetric current filament in a sample with length L and radius r_s. The thickness of its wall in this model is $l_w \sim \sqrt{D\tau_e}$, where D is the heat diffusion coefficient. The voltage fluctuation is

$$\delta U(t) = \int_0^{L_s} dz \delta F_z(\rho, \phi, z, t). \qquad (3.9.1)$$

Here δF_z is the fluctuation of the longitudinal component of the field, ρ, ϕ, z are the cylindrical coordinates. The integral in the r.h.s. of Eq. (3.9.1) is independent of coordinates by virtue of $\mathbf{curl\,F} = 0$, near-electrode effects can be neglected in a long sample. As the current is fixed, its fluctuation $\delta I = 0$.

The current density fluctuation is a sum of several terms: (1) the current induced by voltage fluctuation δU, (2) the current proportional to the conductivity fluctuation $\delta\sigma = (\partial\sigma/\partial T)\delta T$ caused by the electron temperature fluctuation δT, (3) the Langevin current source δj^{int} which is caused by the randomness of electron scattering. The condition of zero total current fluctuation,

$$\int_0^{r_s} d\rho\rho \left[\sigma(T)\frac{1}{L}\delta U + F\frac{d\sigma}{dT}\delta T + \delta j_z^{\mathrm{int}} \right] = 0, \qquad (3.9.2)$$

relates the voltage fluctuation δU to the fluctuations $\delta T(\rho, t)$ and $\delta j_z^{\mathrm{int}}(\rho, t)$ averaged over the polar angle ϕ.

The fluctuation of the electron temperature satisfies the energy conservation equation:

$$c_e\frac{\partial\delta T}{\partial t} + \mathrm{div}\,\delta\mathbf{q} = \delta(\mathbf{jE}) - \delta P. \qquad (3.9.3)$$

Here c_e is the specific heat of the electron gas,

$$\delta\mathbf{q} = -\kappa\vec{\nabla}\delta T + \delta\mathbf{q}^{\mathrm{int}} \qquad (3.9.4)$$

is the fluctuation of the energy flow in the electron gas, $\kappa = c_e D$ is the thermal conductivity, δq^{int} is the Langevin energy source which is due to the randomness of the electrons' scattering, $\delta(\mathbf{jE})$ is the fluctuation of the Joule power, and $\delta P = (\partial P/\partial T)\delta T + \delta P^{\mathrm{int}}$ is the fluctuation of the specific power transferred by the electrons to the lattice phonons.

The equation for the Fourier transform $\delta T(\rho, \omega)$ follows from Eqs. (3.9.3)–(3.9.4). It may be represented in the form:

$$(\hat{H}(\rho) - i\omega)\delta T(\rho, \omega) = f(\rho, \omega). \qquad (3.9.5)$$

Here

$$\hat{H}(\rho) = -D\frac{1}{\rho}\frac{d}{d\rho}\left(\rho\frac{d}{d\rho}\right) + V(\rho), \quad V(\rho) = \frac{dP}{dT} - F^2\frac{d\sigma}{dT}, \tag{3.9.6}$$

$f(\rho, \omega)$ is a linear function of δU and all Langevin sources introduced above, D for simplicity is assumed to be constant. The sign of the function $V(\rho)$ coincides with the sign of the small-signal conductivity σ_d of a homogeneous electron gas with the given electron temperature. Therefore $V(\rho) > 0$ outside the current filament, where the electron gas is in the stable phase corresponding to the lower branch of the CVC, and deeply inside the filament, where the electron gas is in a stable phase corresponding to the upper branch of the CVC. However, in the intermediate region, i.e., in the filament's wall, $T(\rho)$ passes through the value which corresponds to the NDR branch of the CVC ($\sigma_d < 0$). Therefore in the wall ($\rho \sim r_f$) the 'potential' $V(\rho) < 0$. As a whole, $V(\rho)$ has a form of a potential well of width $\sim l_w$.

The Hermitian operator $\hat{H}$ is, formally, the Hamiltonian of a quantum particle with mass $2/D$ moving in the potential $V(\rho)$. Let $\psi_n(\rho)$ and ϵ_n ($n = 0, 1, \ldots$) be the eigenfunctions and eigenvalues of $\hat{H}$, respectively. Each eigenfunction ψ_n corresponds to a definite perturbation of the electron temperature in the semiconductor with the current filament. For instance, ψ_0 corresponds to the shift of the filament's wall, i.e., to the change of its radius, or in other words, to pulsations of the current filament. It is important that in the case of a large radius of the filament, $r_f \gg l_w$, the lowest eigenvalue $\epsilon_0 < 0$, and its absolute value is small: $|\epsilon_0| \sim D/r_f^2$ (see below). The rest of the eigenvalues $\epsilon_n \geq \sim \tau_e^{-1}$ (Kogan, 1968). Therefore

$$|\epsilon_0|/\tau_e^{-1} \sim (l_w/r_f)^2 \ll 1. \tag{3.9.7}$$

The eigenvalue ϵ_0 and the eigenfunction $\psi_0(\rho)$ can be found by differentiating the equation for the stationary distribution of the electron temperature in the semiconductor with the current filament (Eq. (3.1.14)):

$$-\frac{D}{\rho}\frac{d}{d\rho}\left(\rho\frac{d}{d\rho}\right)T + P(T) - \sigma(T)F^2 = 0. \tag{3.9.8}$$

Differentiating all terms with respect to ρ, one finds:

$$\left(\hat{H}(\rho) + \frac{D}{\rho^2}\right)\frac{dT}{d\rho} = 0. \tag{3.9.9}$$

The derivative $dT/d\rho$ does not change its sign anywhere and is close to zero everywhere except the filament's wall, where it has a negative peak. The eigenfunction $\psi_0(\rho)$, being the solution to the equation

$$\left(\hat{H} - \epsilon_0\right)\psi_0 = 0, \tag{3.9.10}$$

has the same form. Equating $D/\rho^2 \approx D/r_f^2$ in the most significant region of Eq. (3.9.9) and comparing it with Eq. (3.9.10), one finds that ψ_0 differs from $dT/d\rho$ only by a normalizing factor, and $\epsilon_0 = -D/r_f^2$.

Using the smallness of $|\epsilon_0|$ as compared with other eigenvalues ϵ_n, one is able to separate the contributions of the filament's wall pulsations into the small-signal impedance $Z(f)$ of the sample and the spectral density of voltage noise $S_U(f)$ from the contributions of other eigenmodes, which correspond mainly to the perturbations of electron temperature inside the filament and in the outer region.

The voltage fluctuation can be expressed in terms of δT and δj_z^{int} using Eq. (3.9.2), δT can be found from Eq. (3.9.5) in terms of the eigenfunctions ψ_n and the Langevin sources whose correlation functions are known (Sec. 3.2). The spectral density of voltage noise is obtained in the form (Kogan & Shadrin, 1971b):

$$S_U(f) = |Z(f)|^2 \left\{ S_{Iw}(0)(1 + \omega^2/\epsilon_0^2)^{-1} + S_{I\text{in}} + S_{I\text{out}} \right\}. \qquad (3.9.11)$$

The expression in curly brackets is the total spectral density of the Langevin current. It consists of three terms. The first is associated with the random pulsations of the filament's radius. The second and third are associated with the Langevin currents inside ($S_{I\text{in}}$) and outside ($S_{I\text{out}}$) the filament, both regions being considered as conductors with fixed dimensions. In Eq. (3.9.11) $Z(f)$ is the small-signal impedance of the sample. The admittance also consists of three terms:

$$Z^{-1}(f) = Z_w^{-1}(0)(1 + i\omega/|\epsilon_0|)^{-1} + Z_{\text{in}}^{-1} + Z_{\text{out}}^{-1}. \qquad (3.9.12)$$

The first term in the r.h.s. is associated with the motion of the filament's wall under the small signal, the second and third are the complex conductivities of the regions inside and outside the filament, respectively, considered as hot-electron conductors with nonfluctuating dimensions. The equivalent circuit is shown in Fig. 3.8. The impedance of the filament's wall is a sum of negative $Z_w(0) < 0$ and inductance $L = c^2|Z_w(0)/\epsilon_0|$.

The spectral densities $S_{I\text{in}}$ and $S_{I\text{out}}$, as well as the impedances Z_{in} and Z_{out}, depend on frequency f only at frequencies of the order of or greater than τ^{-1}, where τ is the bulk relaxation time. In the model under consideration it is the energy relaxation time. At $\omega = 2\pi f \ll \tau^{-1}$ these quantities are constants, and the spectral density may be represented in the form:

$$S_U(f) = S_U(0)\left(1 + \omega^2/\omega_N^2\right)\left(1 + \omega^2/\omega_Z^2\right)^{-1}, \qquad (3.9.13)$$

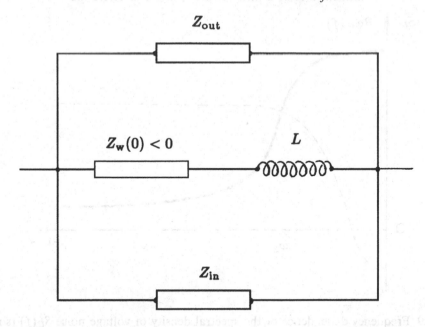

Fig. 3.8 Equivalent circuit of a semiconductor with a stable current filament. Z_{in} and Z_{out} are the small-signal impedances of the regions inside and outside the filament, Z_w is the impedance of the filament's wall. The latter consists of a negative resistance $Z_w(0) < 0$ and an inductance L.

where

$$\omega_N = [S_{Iw}/(S_{Iin} + S_{Iout})]^{\frac{1}{2}}|\epsilon_0|,$$
$$\omega_Z = [|Z_w^{-1}(0)|/(Z_{in}^{-1} + Z_{out}^{-1})]|\epsilon_0|. \tag{3.9.14}$$

These frequencies can be estimated as follows: $\omega_N \sim (l_w/r_f)^{\frac{1}{2}}\tau^{-1}$, $\omega_Z \sim (l_w/r_f)\tau^{-1}$, that is, $|\epsilon_0| \ll \omega_Z \ll \omega_N \ll \tau^{-1}$. The zero-frequency small-signal resistance is negative and its absolute value is small:

$$Z_w(0) = -\frac{L_s}{\pi r_f^2}(\sigma_{in} - \sigma_{out})^{-1}\frac{l_w}{r_f}. \tag{3.9.15}$$

Consequently, the CVC of the sample with a 'thick' filament is close to a vertical line.

The ratio

$$\frac{S_{Iw}}{S_{Iin} + S_{Iout}} \sim \left(\frac{r_f}{l_w}\right)^3 \gg 1. \tag{3.9.16}$$

Having in mind these estimates, one can conclude that at low frequencies $\omega < \omega_Z$ the noise is fully determined by the random pulsations of the current filament. The noise drops at frequencies $\sim \omega_Z$ at which the contribution

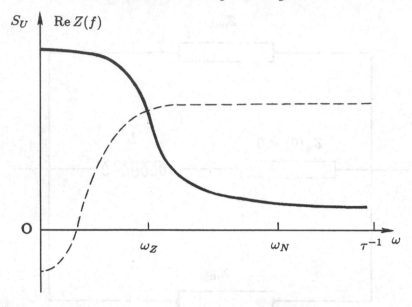

Fig. 3.9 Frequency dependence of the spectral density of voltage noise $S_U(f)$ (solid curve) and the real part of small-signal impedance Re $Z(f)$ (dashed curve) in a conductor with a current filament formed after breakdown. The characteristic relaxation time τ and the frequencies ω_Z and ω_N are defined in the text.

of filament's wall motion to Re $Z(f)$ disappears because $\omega_Z \ll \omega_N$. At $\omega \gg \omega_N$ and up to frequencies $\sim \tau^{-1}$ the noise is independent of frequency. The ratio of the noise spectral density at zero frequency to that at $\omega \gg \omega_N$ is equal to $\omega_N^2/\omega_Z^2 \gg 1$. It means that in a semiconductor with a current filament there is a considerable low-frequency voltage noise induced by the stochastic pulsations of the current filament. At $\omega \gg \omega_N$ this noise is negligible, and the noise is the same as the voltage noise across two semiconductors, connected in parallel, with hot electrons, in each of which the current density is uniformly distributed over the cross-section, and the dimensions are constant (Fig. 3.9).

The low-frequency behavior of a semiconductor with a current filament stems from the fact that in highly uniform samples the motion of the filament as a whole in the transverse direction and, at large radius $r_f \gg l_w$, the motion of the filament's wall are almost free (almost no resistance). One should expect that the qualitative conclusions on the properties of the CVC and noise spectrum found for a simple overheating model hold also in other semiconductors with different mechanisms of S-type CVC and noise.

Inhomogeneities which occur in real samples may act as barriers or local traps preventing the filament's free motion. They affect both the CVC, which deviates from an almost vertical one, and the noise spectrum at low frequencies.

High reverse bias across a Ge $p - n$ junction causes a breakdown and formation of a current filament in the depleted region of the diode. The spectral density of voltage noise $S_U(f)$ in such junctions at low frequencies is high (Bareikis, Liberis, Machulaitis & Mishkinis, 1977). The noise falls off with frequency f. It drops particularly strongly at the same frequency (~ 2 MHz) at which the small-signal impedance $Z(f)$ varies considerably. In accordance with Eq. (3.9.12), the imaginary part of the impedance is inductive. The observed qualitative features of the noise spectrum are indicative of the presence of a current filament in the diode. They are in agreement with the model theory presented above.

4

Generation–recombination noise

The states of electrons and holes in semiconductors and insulators can be divided into two classes: delocalized electron states in the conduction band (holes in the valence band) and localized states. Being in a delocalized state, the charge carrier, electron or hole, contributes to the conductivity. The electrons and holes localized in impurities or defects do not participate in the conduction (they can participate in hopping conduction but its mobility is comparatively much lower). The transition of an electron or a hole from a localized state to a delocalized one or the creation of an electron–hole pair is called generation, the inverse process is called recombination. The term 'trapping' is also used when an electron or a hole is captured by an impurity. A comprehensive review of the physics of nonradiative recombination processes is presented in the book by Abakumov, Perel' & Yassievich (1991).

Since the elementary generation and recombination processes are random, the number of charge carriers, i.e., electrons or holes in delocalized states, fluctuates around some mean value which determines the mean conductance of the specimen. The fluctuations of the charge carriers' number produce fluctuations of the resistance and, consequently, of current and/or voltage if a nonzero mean current is passing through the specimen. This noise is called generation–recombination noise (G–R noise). It is, perhaps, the most important mechanism of modulation noise, i.e., noise produced by random modulation of the resistance.

The life-time of the charge carriers in delocalized band states is usually, but not always, much longer than the time of momentum and energy relaxation. Hence, an electron shortly after its transition to a delocalized state in the conduction band 'forgets' this definite state and becomes a thermalized charge carrier. The trapping (recombination) process is also a rather complex one. It includes energy release by emission of phonons and relaxation of the trap: the lattice around the trap adjusts to the presence of an electron in the trap. These processes are usually also fast, and by the instant of a new transition to the delocalized states the trapped electron completely forgets its previous 'history'. It means that usually the generation and recombination processes are typical Markov processes (Sec. 1.6). The random variables are the numbers of electrons and holes in the bands and in the localized states.

In semiconductors with a high density of recombination centers, for example, in a material doped by shallow impurities and compensated, the life-time of the charge carriers may be, at low temperatures, of the same order as the relaxation time of energy scattering. In this case the generation–recombination processes must be considered on the same footing as other scattering processes. The random variables are then the occupancy numbers of states with various energies, that is, the distribution function $f(E)$ (see Ch. 3). This random process can also be considered by methods presented in Ch. 3 if the time between successive scattering events is much longer than the duration of the scattering event itself. However, this random process is much more complicated than that which is called G–R noise and was described above.

The G–R noise strongly depends on the properties of the semiconductor: unipolar or bipolar conduction, type of the trapping or recombination centers, the length of the specimen as compared with the lengths of drift and diffusion during the trapping or recombination time, etc. Much experimental research was devoted to G–R noise in various semiconductors and semiconductor devices. In this chapter we discuss only the simplest problems in this field. For a detailed review on the theory of G-R and diffusion noise see van Vliet & Fassett (1965) and references therein. Many useful results are also published in the Proceedings of the International Conferences on Noise in Physical Systems (see the list of references).

4.1 G–R noise in monopolar semiconductors

Let us consider first the simplest case of a semiconductor with only one type of impurity, for definiteness, donors which exchange electrons with the conduction band. Let the dimensions of the specimen be much greater than all correlation lengths, so the effect of boundaries can be neglected. Let N, N_D, N_D^0, N_D^+, and N_A be the numbers of conduction electrons, donors, neutral donors (occupied by electrons), ionized donors, and ionized, i.e., negatively charged, acceptors that partly compensate the donor impurity, respectively, in the whole specimen. The rate J^- of the electron trapping or, equivalently, donor neutralization, is proportional to the number of conduction electrons and empty (ionized) donors. It can be written as $J^- = \alpha N_D^+ N / V$. Here α is the trapping coefficient, i.e., trapping cross-section times the electron thermal velocity, V is the specimen's volume. The division by V is caused by the dependence of the trapping rate on the *concentration* of traps, not on their total number in the specimen. The rate J^+ of the conduction electron generation (donor ionization) is proportional to the number of neutral donors and can be represented as $J^+ = \beta N_D^0$. The coefficient β can be related to α using the statistical-mechanical equations for the numbers N, N_D^0 and $N_D^+ = N_D - N_D^0$ in the equilibrium state:

$$N = V N_c \exp\left(\frac{E_F - E_c}{k_B T}\right), \quad N_D^0 = N_D \left[g^{-1}\exp\left(\frac{E_D - E_F}{k_B T}\right) + 1\right]^{-1}. \quad (4.1.1)$$

Here E_F, E_D, E_c are the Fermi level, the donor level, and the energy of the conduction band bottom, respectively, N_c is the effective number of states in the conduction band per unit volume, $E_c - E_D$ is the ionization energy of a donor, and g is the degeneracy of the neutral donor ground state. Equating J^+ and J^- and using Eq. (4.1.1), one obtains:

$$\beta = \alpha n_t, \quad n_t = (N_c/g)\exp\left(-\frac{E_c - E_D}{k_B T}\right). \quad (4.1.2)$$

The quantity n_t is the density of conduction electrons when the Fermi level coincides with the donor level.

The Langevin kinetic equation for G–R fluctuations (Sec. 1.10) is obtained by linearization of the G–R kinetic equation and by adding the Langevin sources of fluctuations in the r.h.s.:

$$\frac{\partial \delta N}{\partial t} + \delta(\alpha N_D^+ N / V - \beta N_D^0) = \delta J^+(t) - \delta J^-(t). \quad (4.1.3)$$

Here $\delta(\ldots)$ means the linearization of the expression in brackets, $\delta J^+(t)$ and $\delta J^-(t)$ are the Langevin sources of fluctuations. These random fluxes correspond to the generation and recombination processes, respectively.

Equation (4.1.3) must be complemented by the equation of charge conservation, i.e., $\delta N_D^+ = -\delta N_D^0 = \delta N$. Taking this condition into account and Fourier transforming Eq. (4.1.3), one obtains:

$$\delta N(\omega) = (-i\omega + \tau^{-1})^{-1}[\delta J^+ - \delta J^-(\omega)]. \quad (4.1.4)$$

Here τ is the electron life-time, i.e., the relaxation time of G–R fluctuations. Its inverse equals:

$$\tau^{-1} = \alpha(n_D^+ + n + n_t) = \alpha(n_A + 2n + n_t). \quad (4.1.5)$$

If different elementary generation and recombination events are uncorrelated (this assumption is not always true) the sources δJ^+ and δJ^- are also uncorrelated and the correlation functions of Langevin generation sources δJ^+, and of the recombination ones, δJ^-, are proportional to the corresponding mean rates:

$$\langle \delta J^\pm(t_1)\delta J^\pm(t_2)\rangle = J^\pm \delta(t_1 - t_2). \tag{4.1.6}$$

The spectral density is proportional to the absolute value of the Fourier transform of the fluctuation squared (Sec. 1.3). Using Eqs. (1.3.8), (4.1.4) and (4.1.6) one obtains the spectral density of G–R fluctuations:

$$S_N(f) = \frac{2(J^+ + J^-)}{\omega^2 + \tau^{-2}}. \tag{4.1.7}$$

Here J^+ and J^- are the mean rates of generation and recombination:

$$J^+/V = \alpha n_t n_D^0, \quad J^-/V = \alpha n_D^+ n. \tag{4.1.8}$$

Since under stationary conditions $J^+ = J^-$, Eq. (4.1.7) may be further simplified:

$$S_N(f) = \frac{4J^-\tau^2}{1+\omega^2\tau^2} = \frac{4Vnn_D^+\tau}{n_D^+ + n + n_t}\frac{1}{1+\omega^2\tau^2}. \tag{4.1.9}$$

In the derivation of Eq. (4.1.9) the condition $\delta N_D^0 = -\delta N$ was used. Therefore this equation relates to a system with constant total number of electrons in all states, not to one with fixed Fermi level (the difference is briefly discussed in Sec. 3.6).

The conductance fluctuation equals $\delta R^{-1} = R^{-1}\delta N/N$. The spectral density of relative resistance fluctuations follows from Eq. (4.1.9):

$$\frac{S_R(f)}{R^2} = \frac{4n_D^+\tau}{Vn(n_D^+ + n + n_t)}\frac{1}{1+\omega^2\tau^2}. \tag{4.1.10}$$

The spectral density of relative fluctuations is inversely proportional to the volume V over which the fluctuations are averaged, as should be expected for any macroscopic body with dimensions much greater than the correlation length of fluctuations. At low temperatures $n_D^+ \approx n_A \gg n, n_t$. Then n_D^+ drops out from Eqs. (4.1.9) and (4.1.10)):

$$\frac{S_R(f)}{R^2} \simeq \frac{4\tau}{Vn(1+\omega^2\tau^2)}. \tag{4.1.11}$$

Measurements of the spectral density of G–R fluctuations can be used to find the life-time τ of the charge carriers. One can measure the frequency at which the spectral density is half the low-frequency value. This frequency equals $f_c = 1/2\pi\tau$. One can find τ also by measuring the low-frequency spectral density of the G–R noise, $S_R(0)/R^2$ and using Eq. (4.1.11), if the density of charge carriers n is known from Hall effect measurements.

In general, when the number M of different impurity levels which participate in the generation and recombination of charge carriers is more than one, the number of random variables equals $M + 1$. It includes the fluctuation of the conduction electrons' number, $\delta N(t)$, and the fluctuations of the number of electrons localized in the impurities, $\delta N_i(t)$, $i = 1, \ldots M$. These variables must satisfy $M + 1$ Langevin kinetic equations. The r.h.s. of each of these equations is the Langevin source corresponding to generation of charge carriers from and trapping into the corresponding level. However, the equations are not independent, due to the charge conservation condition: the sum $\delta N + \delta N_1 + \ldots + \delta N_M = 0$. This equation allows one to eliminate one of the variables. One has to solve M linear inhomogeneous kinetic Langevin equations. The necessary correlation functions of the Langevin sources can be easily found using the same principle that was used above: each kind of Langevin transition rate is correlated only with itself, and its correlation function is $\delta(t_1 - t_2)$ times the mean rate of the transitions considered. Finally, the spectral density of G–R noise is expressed in terms of these mean transition rates and M inverse relaxation times τ^{-1} which are the eigenvalues of the matrix in the l.h.s. of the system of Langevin kinetic equations. For a review of G–R noise, including G–R noise in multilevel semiconductors, see van Vliet & Fassett (1965).

The cross-section of charge carrier trapping to an impurity center depends on the charge carrier's energy. Fluctuations of the energy distribution function (Secs. 3.2–3.4) produce fluctuations of the trapping rate. Consequently, a correlation exists between the fluctuations of the charge carriers' number and their distribution in energy. If the life-time τ of the charge carriers is much greater than the energy relaxation time τ_0, this correlation is small at frequencies $f <\simeq \tau^{-1}$. The simple theory of G–R noise, that does not take into account this correlation, remains valid at these frequencies.

Perhaps, the best understood recombination process is the trapping of electrons by attracting charged (Coulomb) impurities at low temperatures. The theory presented above treated the trapping of a charge carrier to an impurity as one instantaneous transition. Trapping to attractive Coulomb, i.e., charged, impurities is more complex. First of all, it is effective only at low temperatures when $k_B T \ll E_i$, where E_i is the ionization energy of the impurity. Secondly, these impurities are shallow, E_i is usually smaller than the energy of an optical phonon in the same crystal. The trapped charge carrier can release its energy only by emission of acoustic phonons. The probability of transition directly to the ground state by emission of one acoustic phonon with energy $\hbar\Omega \simeq E_i$ is very small. Therefore, the most probable trapping process is the cascade one through excited states (Lax, 1960; Abakumov, Perel' & Yassievich, 1977). By emission of a low-energy acoustic phonon in the vicinity of the charged impurity, the charge carrier is trapped first into an excited state of this impurity. Subsequently, the charge carrier descends further along the 'staircase' of the excited levels, ultimately reaching the ground state. However, the phonons are not only emitted by the charge carrier but also absorbed (the probabilities of emission and absorption are proportional to $\bar{N} + 1$ and $\bar{N}$, respectively, where $\bar{N}$ is the mean number of phonons in one vibrational mode). The charge carrier trapped into an excited state of the impurity can be emitted back to conduction states, not reaching the ground level. Transitions of charge carriers from the band states to the bound excited states at Coulomb impurities and back, without reaching the ground state, manifest themselves in an additional resistance noise. One may expect that, at $\tau \gg \tau_0$, the characteristic cut-off frequency of this noise is many times higher then τ^{-1} and its magnitude is smaller than the magnitude given by Eq. (4.1.11).

It is interesting to compare the G–R noise with the Nyquist–Johnson one. Because $S_R/R^2 = S_U/U^2$, their ratio at low frequencies $f \ll 1/2\pi\tau$ equals:

$$\frac{S_U^{(GR)}(0)}{S_U^{(NJ)}(0)} = \frac{e\mu F^2 \tau}{k_B T}. \tag{4.1.12}$$

Here F is the electric field, μ is the electron mobility. According to the estimates given by Eqs. (3.1.12) and (3.1.16), the 'heating' of the electrons in a field F is of the order of $T - T_0 \simeq e\mu F^2/k_B\tau_0$, where τ_0 is the energy relaxation time. Therefore

$$\frac{S_U^{(GR)}(0)}{S_U^{(NJ)}(0)} = \frac{\tau}{\tau_0} \frac{T - T_0}{T}. \tag{4.1.13}$$

Usually in semiconductors $\tau \gg \tau_0$. Consequently, even in small electric fields, when the deviation of the electron energy distribution function from the equilibrium one is small, G–R noise at frequencies $f < \tau^{-1}$ may be much higher than the equilibrium Nyquist–Johnson noise.

4.2 Noise produced by recombination and diffusion

The fluctuations of the number of charge carriers in a semiconductor specimen which results in its resistance fluctuations may be caused not only by the randomness of the generation and recombination processes but also by diffusion of charge carriers into and out of that segment of the specimen which determines the resistance. In this section the interplay of these two different random processes is discussed. The problem was solved by Lax & Mengert (1960) (see also the review by van Vliet & Fassett, 1965).

In a unipolar semiconductor with only one type of charge carrier, the charge carriers' density fluctuations with spatial dimensions ('wavelength') greater than the Debye screening length are suppressed by Coulomb forces. If the conductor's length is much greater than the screening length, the total number of charge carriers and trapped charges in the sample is effectively constant due to these Coulomb forces (quasi-neutrality). In a bipolar semi-conductor electrons and holes neutralize each other. Hence the variation of the electron–hole pairs' density over lengths that are many times longer than the screening length is not suppressed by Coulomb forces.

Let us find the spectral density of noise in a finite segment of an infinitely long specimen. The electrons and holes with densities n and p, respectively, are assumed to recombine and be generated in pairs via some effective recombination centers. The continuity equations for the electrons' and holes' density fluctuations may be represented as follows:

$$\begin{aligned}
\frac{\partial \delta n(\mathbf{r}, t)}{\partial t} + \operatorname{div} \delta \mathbf{j}_n(\mathbf{r}, t) &= \delta g(\mathbf{r}, t) - \delta r(\mathbf{r}, t), \\
\frac{\partial \delta p(\mathbf{r}, t)}{\partial t} + \operatorname{div} \delta \mathbf{j}_p(\mathbf{r}, t) &= \delta g(\mathbf{r}, t) - \delta r(\mathbf{r}, t).
\end{aligned} \tag{4.2.1}$$

Here δn, $\delta \mathbf{j}_n$ and δp, $\delta \mathbf{j}_p$ are the fluctuations of number densities and flow densities of electrons and holes, respectively; δg and δr are fluctuations of the electron–hole pairs' generation and recombination rates. The r.h.s. of both equations are identical since generation (or recombination) of an electron is accompanied, in the model under discussion, by generation (or, respectively, recombination) of a hole.

According to the Langevin approach (Sec. 1.10) each flow density fluc-tuation consists of two parts. One part results from the fluctuations of the number density, its gradient, and electric field δF. The second part cannot be expressed in terms of these fluctuations and is the Langevin ('intrinsic') source of fluctuations that stems from the randomness of the charge carriers' scattering:

$$\begin{aligned}
\delta \mathbf{j}_n(\mathbf{r}, t) &= -\mu_n[\mathbf{F}\delta n(\mathbf{r}, t) + n\delta \mathbf{F}(\mathbf{r}, t)] - D_n\nabla\delta n(\mathbf{r}, t) + \delta \mathbf{j}_n^{\text{int}}(\mathbf{r}, t), \\
\delta \mathbf{j}_p(\mathbf{r}, t) &= \mu_p[\mathbf{F}\delta p(\mathbf{r}, t) + p\delta \mathbf{F}(\mathbf{r}, t)] - D_p\nabla\delta p(\mathbf{r}, t) + \delta \mathbf{j}_p^{\text{int}}(\mathbf{r}, t).
\end{aligned} \tag{4.2.2}$$

Here μ_n, μ_p and D_n, D_p are the electron and hole mobilities and diffusion coefficients, respectively. For simplicity the semiconductor is assumed to be uniform, hence the mean concentrations n, p, and field F are constants.

A very important simplification has its origin in the quasi-neutrality of the fluctuations in semiconductors with a high enough mean density of mobile charge carriers and, correspondingly, with a short enough Debye screening length L_{Deb} and small enough Maxwell dielectric relaxation time $\tau_M = \epsilon/4\pi\sigma$, where σ is the conductivity and ϵ is the dielectric permittivity of the semiconductor. When the characteristic length of the fluctuation's spatial variation is many times longer than L_{Deb} and the characteristic frequency of variation of the fluctuation in time $\omega \ll \tau_M^{-1}$, the fluctuation can be considered as quasi-neutral (the last condition means that there is enough time for the charge density fluctuation, if it appears, to be screened). Under the conditions of quasi-neutrality, the charge density in the fluctuation is negligible. When the charge density fluctuation of trapped electrons or holes is also negligible due to small concentration of traps, the quasi-neutrality results in simple equations:

$$\delta n = \delta p, \qquad \nabla\delta n = \nabla\delta p. \tag{4.2.3}$$

The fluctuation $\delta g - \delta r$ also consists of two parts:

$$\delta g(\mathbf{r}, t) - \delta r(\mathbf{r}, t) = -\frac{1}{\tau}\delta p(\mathbf{r}, t) + \delta g^{\text{int}}(\mathbf{r}, t) - \delta r^{\text{int}}(\mathbf{r}, t). \tag{4.2.4}$$

When writing the first term in the r.h.s. we assumed, for the sake of simplicity, that the semiconductor is *n*-type and the holes are the minority carriers. In this case the fluctuation of the difference of generation and recombination rates is proportional to the fluctuation of minority carriers' density, δp, and τ is their life-time. The Langevin sources in the r.h.s. stem from the randomness of the generation and recombination of the electron–hole pairs. Their correlation function is found by the same method as was used in Sec. 4.1.

Subtracting the second Eq. (4.2.2) from the first one and neglecting the time derivative of the charge density $\delta\rho = e(\delta p - \delta n)$ (quasi-neutrality), one obtains for the current density fluctuation $\delta\mathbf{j} = e(\delta\mathbf{j}_p - \delta\mathbf{j}_n)$:

$$\text{div}\,\delta\mathbf{j} = 0. \tag{4.2.5}$$

In the one-dimensional case, when the measured quantities are averaged over the specimen's cross-section and depend on only one coordinate x, Eq. (4.2.5) means that the current fluctuation is independent of this coordinate. According to Eq. (4.2.2) and the quasi-neutrality conditions Eq. (4.2.3), it equals:

$$\delta j(t) = F\delta\sigma(x,t) - \sigma\frac{d}{dx}\left[\delta\phi(x,t) + \frac{D_p - D_n}{\mu_n n + \mu_p p}\delta p(x,t)\right] + \delta j^{\text{int}}(x,t). \quad (4.2.6)$$

Here δj is the current density fluctuation component parallel to OX, $\delta\sigma = e(\mu_n\delta n + \mu_p\delta p)$ is the local conductivity fluctuation, $\sigma = e(\mu_n n + \mu_p p)$ is the mean conductivity, and ϕ is the electric potential. The Langevin current density equals:

$$\delta j^{\text{int}} = e[\delta j_p^{\text{int}}(x,t) - \delta j_n^{\text{int}}(x,t)]. \quad (4.2.7)$$

All fluctuations are assumed to be averaged over the specimen's cross-section (for simplicity we don't change their notations).

The equation for the electric field fluctuation follows from Eq. (4.2.6):

$$\delta F(x,t) - \frac{(D_p - D_n)}{\mu_n n + \mu_p p}\frac{d\delta p(x,t)}{dx} = \frac{1}{\sigma}[\delta j(t) - \delta j^{\text{int}}(x,t)] - \frac{F}{\sigma}\delta\sigma(x,t). \quad (4.2.8)$$

Using this equation one can eliminate δF from Eq. (4.2.2) for the flow density fluctuations and, ultimately, from Eq. (4.2.1):

$$\delta j_p(x,t) = \frac{\mu_p p}{\sigma}\delta j(t) + v_d\delta p(x,t) - D\frac{d\delta p(x,t)}{dx} + \delta\tilde{j}^{\text{int}}(x,t). \quad (4.2.9)$$

Here μ and D are the ambipolar drift velocity and diffusion coefficient (see van Roosbroeck, 1953, and Lax & Mengert, 1960):

$$v_d = \frac{\mu_n\mu_p(n - p)}{\mu_n n + \mu_p p}F = \mu F, \qquad D = \frac{\mu_n n D_p + \mu_p p D_n}{\mu_n n + \mu_p p}. \quad (4.2.10)$$

In the case when the electrons are not hot, i.e., the fields are not too high (Ch. 3), the Einstein relation holds (Eq. (1.9.8)), and the equation for the ambipolar diffusion coefficient becomes simpler.

The ambipolar Langevin flow density in the r.h.s. of Eq. (4.2.9) equals:

$$\delta\tilde{j}^{\text{int}} = \frac{\mu_n n\delta j_p^{\text{int}} + \mu_p p\delta j_n^{\text{int}}}{\mu_n n + \mu_p p}. \quad (4.2.11)$$

Now we are ready to solve Eq. (4.2.1) for $\delta p(x,t)$:

$$\frac{\partial\delta p}{\partial t} + \mu F\frac{\partial\delta p}{\partial x} - D\frac{\partial^2\delta p}{\partial x^2} + \frac{1}{\tau}\delta p = -\frac{\partial\delta\tilde{j}^{\text{int}}}{\partial x} + \delta g^{\text{int}} - \delta r^{\text{int}}. \quad (4.2.12)$$

The Fourier transform of the fluctuations equals:

$$\delta p(k,\omega) = \frac{\delta g^{\text{int}}(k,\omega) - \delta r^{\text{int}}(k,\omega) - ik\delta\tilde{j}^{\text{int}}(k,\omega)}{-i\omega + ikv_d + Dk^2 + \tau^{-1}}. \quad (4.2.13)$$

We consider a segment, of an infinite semiconductor, of length L_x. The fluctuation of the current at fixed voltage, $\delta I(t)$, or the voltage fluctuation at fixed current are found by averaging all terms in Eq. (4.2.8) over the length L_x of the segment, i.e., by integrating them and dividing by L_x:

$$\delta I(\omega) = \frac{1}{L_x} \int_{-L_x/2}^{L_x/2} dx \int_A dy\, dz\, [\delta j^{\text{int}}(\mathbf{r}, \omega) + F\delta\sigma(\mathbf{r}, \omega)]. \qquad (4.2.14)$$

Here A is the cross-sectional area of the specimen. In contrast with Eq. (4.2.8), the fluctuations $\delta j^{\text{int}}(\mathbf{r}, t)$ and $\delta\sigma(\mathbf{r}, t)$ are not averaged over the specimen's cross-section.

The current fluctuation given by Eq. (4.2.14) is a sum of two terms: the Langevin current source and the change of the current due to the fluctuation of the conductivity. The latter is produced both by G–R processes and by the change of the charge carriers' number due to their diffusion across the ends of the segment.

Using Eq. (4.2.13) one can represent Eq. (4.2.14) in a more explicit form:

$$
\begin{aligned}
\delta I(\omega) = e \int_A dy\, dz \int \frac{dk}{2\pi} & \frac{\sin(kL_x/2)}{kL_x/2} \\
\times \Big\{ \Big[1 + kL(k,\omega)&\frac{\mu_n n}{\mu_n n + \mu_p p} \Big] \delta j_p^{\text{int}}(k, y, z, \omega) \\
- \Big[1 - kL(k,\omega)&\frac{\mu_n n}{\mu_n n + \mu_p p} \Big] \delta j_n^{\text{int}}(k, y, z, \omega) \\
+ iL(k,\omega)[\delta g^{\text{int}}&(k, y, z, \omega) - \delta r^{\text{int}}(k, y, z, \omega)] \Big\},
\end{aligned}
\qquad (4.2.15)
$$

where

$$L(k,\omega) = \frac{(\mu_n + \mu_p)F}{\omega - kv_d + i(Dk^2 + \tau^{-1})}. \qquad (4.2.16)$$

The correlation functions of the Langevin sources have been derived in the preceding section. Owing to the absence of correlation between different generation and recombination events, the correlation functions of their rates are proportional to the mean rates per unit volume, g and r (Sec. 4.1). The correlation functions of the Langevin currents are proportional to $2nD_n$ and $2pD_p$, respectively (Eq. (3.5.4)). Hence,

$$
\begin{aligned}
\langle \delta g^{\text{int}}(\mathbf{r}_1, t_1)\delta g^{\text{int}}(\mathbf{r}_2, t_2) \rangle &= \delta(\mathbf{r}_1 - \mathbf{r}_2)\delta(t_1 - t_2)g, \\
\langle \delta r^{\text{int}}(\mathbf{r}_1, t_1)\delta r^{\text{int}}(\mathbf{r}_2, t_2) \rangle &= \delta(\mathbf{r}_1 - \mathbf{r}_2)\delta(t_1 - t_2)r, \\
\langle \delta j_n^{\text{int}}(\mathbf{r}_1, t_1)\delta j_n^{\text{int}}(\mathbf{r}_2, t_2) \rangle &= 2nD_n\delta(\mathbf{r}_1 - \mathbf{r}_2)\delta(t_1 - t_2), \\
\langle \delta j_p^{\text{int}}(\mathbf{r}_1, t_1)\delta j_p^{\text{int}}(\mathbf{r}_2, t_2) \rangle &= 2pD_p\delta(\mathbf{r}_1 - \mathbf{r}_2)\delta(t_1 - t_2), \\
\langle \delta g^{\text{int}}(\mathbf{r}_1, t_1)\delta r^{\text{int}}(\mathbf{r}_2, t_2) \rangle &= 0, \langle \delta j_p^{\text{int}}(\mathbf{r}_1, t_1)\delta j_n^{\text{int}}(\mathbf{r}_2, t_2) \rangle = 0.
\end{aligned}
\qquad (4.2.17)
$$

The equation for the spectral density of the total Langevin current follows from Eqs. (4.2.15) and (4.2.17):

$$S_I(f) = 2e^2 A \int_{-\infty}^{+\infty} \frac{dk}{2\pi} \left(\frac{\sin(kL_x/2)}{kL_x/2}\right)^2 \left\{(g+r)|L(k,\omega)|^2\right.$$

$$\left. + \left|1 + kL(k,\omega)\frac{\mu_n n}{\mu_n n + \mu_p p}\right|^2 2D_p p + \left|1 - kL(k,\omega)\frac{\mu_p p}{\mu_n n + \mu_p p}\right|^2 2D_n n\right\}.$$

$$(4.2.18)$$

At $F = 0$ Eq. (4.2.18) reduces to the Nyquist equation, i.e., $S_I^{eq}(f) = 4e^2 A(D_n n + D_p p)/L_x$.

The rate of generation–recombination events can be taken as $g = r = p/\tau$. Then, taking also $n/(n+p) \approx 1$ one obtains:

$$S_I(f) = S_I^{eq}(f) + I^2 \frac{S_R(f)}{R^2}. \qquad (4.2.19)$$

Here I is the mean current, $S_R(f)$ is the spectral density of resistance fluctuations due to generation, recombination, and diffusion of the charge carriers:

$$\frac{S_R(f)}{R^2} = 4p\frac{1}{A}\left(\frac{\mu_n + \mu_p}{\mu_n n + \mu_p p}\right)^2$$

$$\times \int_{-\infty}^{+\infty} \frac{dk}{2\pi} \left(\frac{\sin(kL_x/2)}{kL_x/2}\right)^2 \frac{\tau^{-1} + Dk^2}{(\omega - kv_d)^2 + (Dk^2 + \tau^{-1})^2}. \qquad (4.2.20)$$

The meaning of the denominator in the integrand is simple. It turns out to be zero at $\omega = v_d k - iDk^2 - i/\tau$. Substituting this equation into the Fourier exponent $\exp(-i\omega t)$, one can see that a fluctuation moves as a wave with drift velocity v_d, and decays due to diffusion and recombination as $\exp(-Dk^2 t)$ and $\exp(-t/\tau)$, respectively.

Introducing the notation $z = kL_x$ and using integration in the complex z plane one obtains the following equation for the spectral density:

$$\frac{S_R(f)}{R^2} = 4p\frac{1}{V}\left(\frac{\mu_n + \mu_p}{\mu_n n + \mu_p p}\right)^2 \left\{\frac{\tau}{1 + \omega^2 \tau^2}\right.$$

$$\left. - \tau \text{Re}\left[\frac{D\tau/L_x^2}{(1 + i\omega\tau)^2(y_1 + y_2)}[y_2^2(1 - \exp(-y_1)) + y_1^2(1 - \exp(-y_2))]\right]\right\},$$

$$(4.2.21)$$

where $V = AL_x$ is the volume between the boundaries,

$$y_{1,2} = \left[\left(\frac{v_d L_x}{2D}\right)^2 + (1 + i\omega\tau)\frac{L_x^2}{D\tau}\right]^{1/2} \pm \frac{v_d L_x}{2D} \equiv \frac{L_x}{L_{1,2}}. \qquad (4.2.22)$$

The first term in the curly brackets corresponds to the G–R noise, the second one can be called the diffusion noise. The spectral density depends on three dimensionless parameters: (1) $\omega\tau$, i.e., the frequency in units of τ^{-1}; (2) $\sqrt{D\tau}/L_x$, i.e., the ratio of the length of diffusion in time τ and the length L_x; (3) $v_d L_x/D$ which is proportional to the drift velocity.

If L_x exceeds all other characteristic lengths, the exponential functions in the diffusion term are small and this term is proportional to L_x^{-1}. It tends to zero as $L_x \to \infty$. At small L_x this diffusion term may be significant as compared with the first, G–R, term. It tends to a finite value as $L_x \to 0$. At high frequencies $\omega \gg \tau^{-1}$, the diffusion term equals $(\sqrt{D/2}/L_x\tau)\omega^{-3/2}$. It falls off at high frequencies as $f^{-3/2}$, i.e., less steeply than the G–R noise, and dominates in the total noise. This behavior of the diffusion noise at high frequencies is a general property of one-dimensional diffusion (Lax & Mengert, 1960).

The theory presented above refers to quasi-neutrality conditions. The theory of G–R noise in finite compensated semiconductors, for which $v_d\tau > L_x$ and $v_d\tau_M > L_x$, where τ_M is the Maxwell time, was developed by Suris & Fuks (1980).

5

Noise in quantum ballistic systems

5.1 Introduction

In a ballistic conductor the charge carriers are moving between the electrodes without being scattered at impurities or phonons. Obviously, the length of the conductor has to be smaller than the free path length of the charge carriers, hence it has to be smaller than, say, 1 μm. Before the 1980s the only real solid state ballistic conductors were some very pure monocrystalline metals at liquid helium temperatures and microcontacts (microbridges) between bulk metals (Fig. 5.1). The latter were used in point contact spectroscopy of metals (Yanson, 1974). The resistance of the entire point contact is determined by the transverse dimensions and length of the narrow part (constriction). The conduction is ballistic if these dimensions are smaller than the electrons' free-path length. Since in metals the de-Broglie wavelength of electrons at the Fermi surface, λ_F, is of the order of interatomic distances, the microcontact's dimensions are inevitably many times greater than λ_F, and the ballistic motion of electrons in such microcontacts is always quasi-classical.

The recent progress in semiconductor technology made it possible to fabricate ballistic devices, the dimensions of which are of the same order as the wavelength of the charge carriers. These devices are called quantum ballistic systems, or quantum point contacts. They are made by giving a definite shape to the two-dimensional electron gas (2DEG).

The electrons in a 2DEG are confined to a two-dimensional quantum well. For instance, in a semiconductor heterostructure GaAlAs–GaAs–GaAlAs electrons in the GaAs layer are confined between two GaAlAs layers which serve as barriers. A 2DEG can be confined to the inversion layer at the interface between the semiconductor (Si) and an insulator (SiO$_x$) in a metal–insulator–semiconductor structure or at the interface in a heterostructure.

130

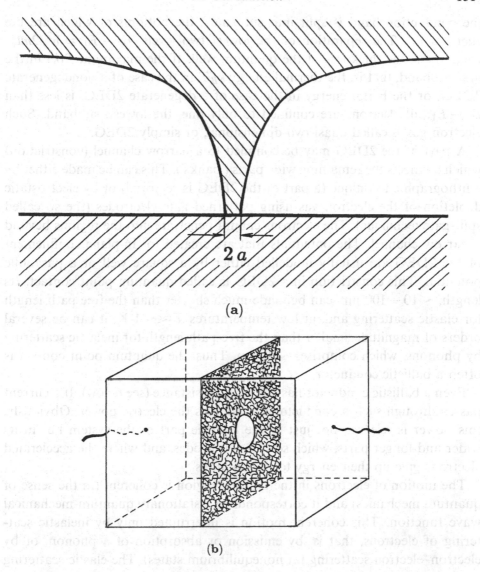

Fig. 5.1 Microcontacts between two bulk metals. (a) A contact made of a pointed wire pressed to a metal surface, (b) Two bulk metals separated by an insulating film everywhere beyond a small area where the metals are connected continuously.

The electron density n_s and the Fermi energy E_F may be continuously varied by varying the voltage V_G on a special gate electrode.

The momentum $\hbar\mathbf{k}$ and the energy $E(k) = \hbar^2 k^2/2m$ of the in-plane motion of electrons in a 2DEG are continuous (m is the effective mass). However, their transverse motion (normal to the plane of the 2DEG) is quantized, and the corresponding energy levels E_n, where $n = 1, 2, \ldots$, are discrete. Hence, the conduction band is split into subbands, the bottoms of which are the energies E_n. The separation of the lowest two levels, $E_2 - E_1 \approx \hbar^2/2mt_w^2$, where t_w is the thickness of the quantum well. If the maximum $E(k)$ in the first subband, that is, the thermal energy $k_B T$ in the case of a nondegenerate 2DEG, or the Fermi energy in the case of a degenerate 2DEG, is less than $E_2 - E_1$, all electrons are confined to only one, the lowest, subband. Such electron gas is called quasi-two-dimensional, or simply 2DEG.

A part of the 2DEG may be confined to a narrow channel (constriction) which connects the remaining wide parts ('banks'). This can be made either by a lithographic technique (a part of the 2DEG is removed) or by electrostatic depletion of the electron gas using patterned gate electrodes (the so called split-gate technique). The electrons cannot penetrate into the regions beyond a narrow channel. The potentials that are created for in-plane confinement of the 2DEG are usually modelled either by a square well or a parabolic potential. This constriction in a 2DEG is called quantum point contact. Its length, $\sim 10 - 100$ nm, can be made much shorter than the free path length for elastic scattering and, at low temperatures $T <\sim 1$ K, it can be several orders of magnitude smaller than the free-path length for inelastic scattering by phonons, which comprises $\sim 10\,\mu$m. Thus, the quantum point contact is often a ballistic conductor.

Even a ballistic conductor has a nonzero resistance (see below). If a current passes through such a conductor it dissipates the electric power. Obviously, this power is released not just in the ballistic part of the system but in its wider and longer parts, which serve as electrodes, and where the accelerated electrons give up their energy to the phonons.

The motion of electrons in the ballistic region is coherent (in the sense of quantum mechanics) and it corresponds to a stationary quantum-mechanical wave function. This coherent motion is interrupted only by inelastic scattering of electrons, that is, by emission or absorption of a phonon, or by electron–electron scattering (in nonequilibrium states). The elastic scattering by randomly located impurities does not break the coherence of the electron's motion, but makes the wave function very complex and dependent on the particular arrangement of the scattering centers. Some interesting features of the point contact are smeared out if the number of scattering centers in it is too high. The same happens if the temperature is too high. Therefore, we confine ourselves in this section mainly to contacts in which the electron gas is strongly degenerate and the electron motion is either purely ballistic (no impurities) or almost ballistic (few impurities). However, many results and conclusions apply as well to all conductors in which the inelastic electron scattering is rare and the electron wave functions are coherent throughout

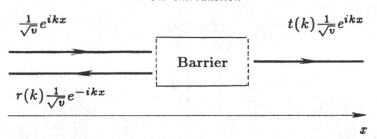

Fig. 5.2 Schematic description of the reflection from and transmission through a barrier. The incident wave is propagating from the left, v is its speed. $r(k)$ and $t(k)$ are the amplitudes of reflection and transmission, respectively.

the entire conductor, irrespective of the frequency of elastic scattering. Such conductors can be called 'quantum conductors'.

The motion of electrons in a quantum point contact between the electrodes has to be described in terms of quantum mechanics, specifically, the quantum theory of scattering. Perhaps, the most important for us in this context are the notions of the transmission and reflection amplitudes. Let electrons be incident onto an obstacle, for instance, onto a potential barrier (Fig. 5.2). In general, electrons are partly reflected by the barrier and partly transmitted through it. Far from the barrier, the wave function of the incident electrons, $\psi_i(x)$, can be taken as a plane wave, $v^{-1/2}\exp(ikx)$, where k and v are the electron wave vector and velocity, respectively. The flow density of electrons in this electron state equals:

$$j = \frac{\hbar}{2im}\left\{\psi_i^* \frac{\partial\psi(x)}{\partial x} - \frac{\partial\psi^*(x)}{\partial x}\psi(x)\right\} = 1. \qquad (5.1.1)$$

The division of the wave function by $\sqrt{v}$ just guarantees the unity flow density.

Far from the obstacle, the wave functions of the reflected and of the transmitted electrons are also plane waves with wave vectors $-k$ and k, and factors $r(k)$ and $t(k)$, respectively (Fig. 5.2). These factors are called reflection and transmission amplitudes. As the flow density (Eq. (5.1.1)) is proportional to the absolute value of the wave amplitude squared, the flow densities of the reflected and transmitted waves are $|r(k)|^2$ and $|t(k)|^2$ as compared with unity flow density of the incident wave. Thus, $|r|^2$ and $|t|^2$ have the meaning of probabilities of electron reflection and transmission. The amplitudes r and t determine also the phases of the waves. Obviously, $|t|^2 + |r|^2 = 1$ because the number of electrons is conserved. According to the fundamental Landauer formula (see below), the small-bias conductance (inverse resistance) G of a quantum conductor is proportional to $|t|^2$.

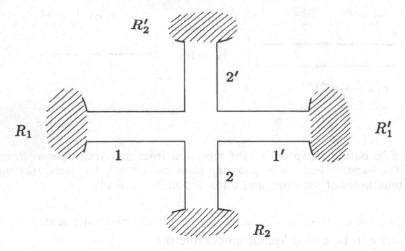

Fig. 5.3 Ballistic system with four intersecting electrodes 1, 1', 2, 2'. R_1, R'_1, R_2, R'_2 are contacts (reservoirs of electrons).

The wave function at $x \to -\infty$ is a superposition of the incident and reflected waves, at $x \to +\infty$ it has the form of a transmitted wave (Fig. 5.2):

$$\Psi(x) = \begin{cases} \frac{1}{\sqrt{v}}e^{ikx} + r(k)\frac{1}{\sqrt{v}}e^{-ikx}, & x \to -\infty; \\ t(k)\frac{1}{\sqrt{v}}e^{ikx}, & x \to +\infty. \end{cases} \qquad (5.1.2)$$

Let us turn now to a more general configuration of a quantum point contact shown in Fig. 5.3. The ballistic conductors attached to highly doped electron reservoirs can be called electrodes. Their intersection forms a constriction.

The transverse motion of electrons in each electrode is quantized due to confinement. Each state of transverse motion is called a conductive channel (the term 'channel' is taken from the quantum theory of scattering). The wave function of any wave propagating in a given electrode, being a function of two coordinates (longitudinal and transverse), is different for different channels. An electron emitted, say, from channel m of the electrode α, is partially reflected by the constriction into various channels n of the same electrode and partially transmitted through it into other electrodes $\beta \neq \alpha$.

It is reasonable to choose the quantum states, i.e., the wave functions at each definite total energy E in the following way. Each such state has the form of a wave incident onto the constriction only in one electrode, e.g., α, and in only one channel (m) of this electrode. Other parts of the same wave function in the same electrode α are the waves reflected into different channels of this electrode and propagating in the opposite direction (as compared with the incident wave). The amplitudes of the reflected waves far from the constriction can be denoted by $r_{\alpha n, \alpha m}$. In all other electrodes $\beta \neq \alpha$, this wave function has the form of transmitted waves propagating away from the constriction in various channels n of these electrodes. Their amplitudes far from the constriction are denoted by $t_{\beta n, \alpha m}$. Such a wave function, $\Psi_{\alpha m E}(\mathbf{r})$, is determined by the electrode (α) and the channel (state of transverse motion) m in this electrode from which the electrons are emitted, as well as by the total energy E of electrons. These eigenfunctions form a full orthonormal system. It is convenient to normalize them in the following way:

$$\int d\mathbf{r} \Psi^*_{\alpha m E}(\mathbf{r}) \Psi_{\beta n E'}(\mathbf{r}) = \delta_{\alpha\beta} \delta_{mn} \delta(E - E'). \tag{5.1.3}$$

The wave functions, including the reflection and transmission amplitudes, can, in principle, be calculated by solving the Schrödinger equation, for example, numerically.

The amplitudes of reflection and transmission form the scattering matrix (S-matrix). It is unitary, the unitarity being the consequence of the conservation of the number of electrons. It means that no electrons are generated or trapped on the way from one electrode to another one. Hence, the elements of the S-matrix satisfy the condition:

$$\sum_n |r_{\alpha n, \alpha m}|^2 + \sum_{\beta n} |t_{\beta n, \alpha m}|^2 = 1. \tag{5.1.4}$$

Its meaning is simple. $R_{\alpha m} = \sum_n |r_{\alpha n, \alpha m}|^2$ is the total probability of reflection for electrons emitted from the electrode α. $T_{\beta n, \alpha m} = |t_{\beta n, \alpha m}|^2$ is the probability of transmission of electrons into the channel n of the electrode β. The condition Eq. (5.1.4) means that the sum of the probabilities of all these events is unity.

Let us examine a simple model of a quantum point contact, which is shown in Fig. 5.4 (Kawabata, 1989; Kogan, 1994). The 2DEG is confined between two symmetrically located hyperbolas by an infinitely high potential along these hyperbolas. The extremums of the hyperbolas are separated by a distance w (the minimum width of the constriction), and the asymptotes' angles of incidence are equal to $\pm v_m$ $(0 < v_m \leq \pi/2)$. The potential is assumed to be constant in the absence of bias voltage or defects. Varying v_m, one obtains various forms of the point contact. At $v_m \ll 1$ it is a stripe with slowly varying width. At $v_m = \pi/2$ it consists of two half-planes separated by an insulator along a line beyond a segment of length w within which these half-planes are connected continuously.

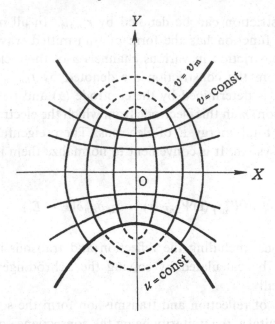

Fig. 5.4 Model of a two-dimensional quantum point contact. The current is flowing along OX. The boundaries are hyperbolas corresponding to coordinate $v = \pm v_m$. The curves $u = $ const. are ellipses, curves $v = $ const. are hyperbolas.

The appropriate orthogonal coordinates in two dimensions are the elliptic coordinates:

$$x = a \sinh u \cos v, \quad y = a \cosh u \sin v, \quad a = w/2 \sin v_m. \qquad (5.1.5)$$

The coordinate v is the transverse one, the curves $v = $ const. $(-v_m \le v \le v_m)$ are hyperbolas (Fig. 5.4). The coordinate u is the longitudinal one $(-\infty < u < +\infty)$, the curves $u = $ const. are confocal ellipses orthogonal to the hyperbolas $v = $ const.

Let E be the dimensionless energy in units of $\hbar^2/2ma^2$. The Schrödinger equation in the coordinates u, v takes the form:

$$\frac{\partial^2 \Psi}{\partial u^2} + \frac{\partial^2 \Psi}{\partial v^2} + [E - \Phi(u, v)](\sinh^2 u + \cos^2 v)\Psi = 0, \qquad (5.1.6)$$

where $\Phi(u, v)$ is the dimensionless potential energy of the bias voltage and defects. Since at the boundaries $v = \pm v_m$ the potential is infinite, the boundary condition along the edges of the quantum point contact is:

$$\Psi(u, \pm v_m) = 0. \qquad (5.1.7)$$

At $\Phi = 0$ the variables in Eq. (5.1.6) can be easily separated. The solution can be found in the form of $\Psi(u, v) = X(u)Y(v)$. The equations for X and

Y are two Schrödinger equations:

$$X'' + (E \sinh^2 u - \lambda)X(u) = 0, \qquad Y'' + (E \cos^2 v + \lambda)Y = 0. \qquad (5.1.8)$$

Here λ is the separation constant.

The equation for $Y(v)$ is, formally, a Schrödinger equation for a quantum particle in a 'potential' $-E \cos^2 v$ with boundary conditions $Y(\pm v_m) = 0$ (Eq. (5.1.7)), λ plays the role of the eigenenergy of the particle. The quantization of λ is a manifestation of the quantization of the transverse motion of electrons in the point contact. The eigenvalues λ_n have to be substituted into the equation for $X(u)$ which describes the longitudinal motion of electrons. Formally, it is an equation of motion of a quantum particle with an 'energy' $-\lambda_n$ in the 'potential' $-E \sinh^2 u$ which, at $E > 0$, is a barrier with a maximum at $u = 0$.

At each value of the dimensionless energy $E > 0$ there are an infinite number of eigenvalues λ_n. Only a finite number of them are negative. At small E and v_m there are no negative eigenvalues $\lambda_n < 0$. However, only at $\lambda_n < 0$, the 'energy' $-\lambda_n$ of the longitudinal motion (along the u coordinate) is higher than the top of the barrier $-E \sinh^2 u$, and the probability of transmission from one side of the barrier to another side (from one electrode to another one) is close to unity. On the contrary, at large $\lambda_n > 0$ the 'energy' $-\lambda_n$ is significantly lower than the maximum of the barrier, and the probability of transmission from one electrode to the opposite one by tunneling is negligible. According to the terminology of the quantum theory of scattering, the quantum states for which the probability of transmission is close to unity are called open channels, the states with a negligible probability of transmission are called closed channels.

Suppose the dimensionless energy E increases, for instance, due to an increase of the electron's Fermi energy, or, at constant energy in common units, due to an increase of the width w of the point contact and the corresponding decrease of the energy $\hbar^2/2ma^2$ (see above). The depth of the 'potential well' in the equation for $Y(v)$ increases and the eigenvalues λ_n are lowered. If an eigenvalue λ_n was initially positive it may become negative. As a result, a channel which was initially closed may become open, increasing the number of open channels by one.

In the asymptotic regions $u \to \pm\infty$, the solution to the first Eq. (5.1.8) reads

$$X(u) \to [p(u)]^{-1/2} \exp[\pm iS(u)],$$

$$p(u) = \sqrt{E \sinh^2 u - \lambda}, \qquad S(u) = \int^u du' p(u'). \qquad (5.1.9)$$

Therefore, far from the constriction the wave function of the electron emitted from the m-th channel of the left ($\alpha = L$) electrode equals (it is normalized according to Eq. (5.1.3)):

$$\Psi_{LmE}(u,v)$$

$$= \sqrt{\frac{m}{2\pi\hbar^2 p(u)}} \begin{cases} e^{iS(u)}Y_m(v) + \sum_n r_{Ln,Lm}(E)e^{-iS(u)}Y_n(v), & u \to -\infty; \\ \sum_n t_{Rn,Lm}(E)e^{iS(u)}Y_n(v), & u \to +\infty. \end{cases} \quad (5.1.10)$$

One can write a similar equation for the wave function Ψ_{RmE} of an electron incident from the right electrode. Equation (5.1.10) is, in fact, the definition of the transmission and reflection amplitudes.

According to quantum mechanics, in order to calculate the current in the electrode α and, subsequently, the fluctuations of this current, one has to write the Heisenberg operator of the longitudinal component of the current density, i.e., along the coordinate x_α. The expression for the Heisenberg current density operator is merely a generalization of the equation for the current density given in any text-book on quantum mechanics. Then one has to integrate it over the transverse coordinate y_α, i.e., over the cross-section of the electrode. In the model presented in Fig. 5.4, these longitudinal and transverse coordinates are u and v, respectively:

$$\hat{I}_\alpha(t) = \frac{e\hbar}{2im} \int dy_\alpha \left(\hat{\psi}^+(\mathbf{r}t)\frac{\partial \hat{\psi}(\mathbf{r}t)}{\partial x_\alpha} - \frac{\partial \hat{\psi}^+(\mathbf{r}t)}{\partial x_\alpha}\hat{\psi}(\mathbf{r}t) \right), \quad (5.1.11)$$

where $\hat{\psi}^+(\mathbf{r}t)$ and $\hat{\psi}(\mathbf{r}t)$ are the time-dependent Heisenberg operators of creation and annihilation of an electron at the point $\mathbf{r}$.

It is convenient to expand the second quantization field operators $\hat{\psi}(\mathbf{r}t)$ and $\hat{\psi}^+(\mathbf{r}t)$ in the full system of eigenfunctions $\Psi_{\alpha mE}(\mathbf{r})$ introduced above:

$$\hat{\psi}(\mathbf{r}t) = \sum_{\alpha m} \int dE \, \Psi_{\alpha mE}(\mathbf{r})\hat{a}_{\alpha mE}(t), \quad (5.1.12)$$

where $\hat{a}_{\alpha mE}$ is the operator of annihilation of an electron in the state $\Psi_{\alpha mE}$.

The expansion (5.1.12) for $\hat{\psi}(\mathbf{r}t)$ and the similar expansion for $\hat{\psi}^+(\mathbf{r}t)$ must be substituted into Eq. (5.1.11):

$$\hat{I}_\alpha(t) = \sum_{\beta m, \gamma n} \int dE\, dE' I_{\beta mE, \gamma nE'}(\alpha)\hat{a}^+_{\beta mE}(t)\hat{a}_{\gamma nE'}(t). \quad (5.1.13)$$

For brevity the matrix elements of the current are introduced:

$$I_{\beta mE, \gamma nE'}(\alpha) = \frac{e\hbar}{2im} \int dy_\alpha \left\{ \Psi^*_{\beta mE}(x_\alpha y_\alpha)\frac{\partial \Psi_{\gamma nE'}(x_\alpha y_\alpha)}{\partial x_\alpha} \right.$$
$$\left. - \frac{\partial \Psi^*_{\beta mE}(x_\alpha y_\alpha)}{\partial x_\alpha}\Psi_{\gamma nE'}(x_\alpha y_\alpha) \right\}, \quad (5.1.14)$$

In this section, only the expression for the mean current as a function of voltage U is derived. According to quantum statistical mechanics, the current operator $\hat{I}$ has to be averaged with the density matrix $\hat{\rho}$ of the system. The electron gas in each electrode is in the equilibrium state, and the electron distribution over the energies E_α measured from the bottom of the conduction band in the electrode α is the Fermi function $f(E_\alpha) \equiv f_\alpha(E)$. If the interaction between the electrons is neglected, the result of averaging the operators in Eq. (5.1.13) is:

$$\text{Tr}\{\hat{\rho}\hat{a}^+_{\beta mE}\hat{a}_{\gamma nE'}\} \equiv \langle a^+_{\beta mE}a_{\gamma nE'}\rangle = \delta_{\beta\gamma}\delta_{mn}\delta(E'-E)f_\beta(E). \tag{5.1.15}$$

Here Tr stands for the trace (sum of the diagonal elements) of the expression enclosed in the braces.

The mean current then equals:

$$\langle \hat{I}_\alpha \rangle = \sum_{\beta m} \int dE I_{\beta mE,\beta mE}(\alpha)f_\beta(E). \tag{5.1.16}$$

If there are only two electrodes the current is the same in both electrodes. For definiteness, we calculate the current in the right one. The wave functions of the simple model (Fig. 5.4 and Eq. (5.1.10)), taken at $u \to +\infty$, are substituted into Eq. (5.1.16). One has to take into account the Kramers degeneracy g of the electron states. In the simplest case it is equivalent to summation over two spin states ($g = 2$), i.e., the result must be doubled. The following equation for the current is obtained, which is independent of any particular model:

$$\langle I \rangle = \frac{ge}{2\pi\hbar} \int dE \{f_L(E) \sum_{mn} |t_{Rn,Lm}(E)|^2$$
$$- f_R(E) \sum_m (1 - \sum_n |r_{Rn,Rm}(E)|^2)\}. \tag{5.1.17}$$

Here $f_L(E)$ and $f_R(E)$ are the electron Fermi functions in the left and right electrode, respectively. At nonzero bias the electrochemical potentials (Fermi energies) in the two electrodes, μ_L and μ_R, are different. Their difference equals $\mu_L - \mu_R = eU$, where U is the voltage across the quantum conductor.

Equation (5.1.17) has a simple meaning. The current from the left electrode to the right one is at each energy E a sum of two currents: (1) the current of those electrons emitted from all channels m of the left electrode which with probability $|t_{Rn,Lm}|^2$ reach the right electrode, that is, all its channels n; (2) the reverse current of electrons emitted from the right electrode and not reflected back.

It is convenient to write the sums over products of transmission and reflection amplitudes using the definition of Hermitian conjugation of matrices. By definition, it includes not only the complex conjugation but also the transposition of the matrix. For example, $t^+_{Lm,Rn} = t^*_{Rn,Lm}$, where the upper cross and star denote the Hermitian and complex conjugation, respectively. In these notations:

$$\sum_n |t_{Rn,Lm}|^2 = (\mathbf{t}^+\mathbf{t})_{mm}, \qquad \sum_n |r_{Rn,Rm}|^2 = (\mathbf{r}^+\mathbf{r})_{mm}. \qquad (5.1.18)$$

For brevity, the matrices of the amplitudes of reflection and transmission have been denoted by $\mathbf{r}$ and $\mathbf{t}$, respectively.

The conservation of particles connects the probabilities of reflection and transmission, their sum equals unity (Eq. (5.1.4)):

$$\mathbf{r}^+\mathbf{r} + \mathbf{t}^+\mathbf{t} = \hat{1}. \qquad (5.1.19)$$

Here $\hat{1}$ is a unit matrix.

The equation for the mean current reads:

$$\langle I \rangle = \frac{ge}{2\pi\hbar} \int dE \left[f_L(E) - f_R(E)\right] \mathrm{Tr}(\mathbf{t}^+(E)\mathbf{t}(E)). \qquad (5.1.20)$$

The eigenvalues of the Hermitian matrix $\mathbf{t}^+\mathbf{t}$, which are usually denoted by T_m, have the meaning of probabilities of electron transmission from one electrode to another one, and $\mathrm{Tr}(\mathbf{t}^+\mathbf{t}) = \sum_m T_m$ is the sum of the probabilities of transmission in all channels.

At small voltages U:

$$f_L(E) - f_R(E) = f(E - \mu_L) - f(E - \mu_R) \approx eU\left(-\frac{\partial f}{\partial E}\right)$$
$$= U\frac{e}{k_B T}f(E)[1 - f(E)]. \qquad (5.1.21)$$

It follows from Eqs. (5.1.20) and (5.1.21) that at small voltages the current is proportional to the voltage (Ohm's law), and the conductance equals:

$$G \equiv R^{-1} = \frac{ge^2}{2\pi\hbar} \int dE\left(-\frac{\partial f}{\partial E}\right) \sum_m T_m(E)$$
$$= \frac{ge^2}{2\pi\hbar k_B T} \int dE f(E)[1 - f(E)] \sum_m T_m(E). \qquad (5.1.22)$$

At low temperatures $k_B T \ll E_F$, that is, when the electron gas is degenerate, the derivative $(-\partial f/\partial E)$ is a narrow peak at the electron gas Fermi energy E_F. It is close to $\delta(E - E_F)$. Thus, the conductance of a ballistic contact with a degenerate electron gas equals (spin degeneracy $g = 2$ is assumed):

$$G = \frac{e^2}{\pi\hbar} \sum_m T_m(E_F). \tag{5.1.23}$$

This fundamental equation for the conductance of a quantum ballistic contact in terms of the transmission amplitudes in the channels has been derived by Landauer (1957, 1987). It can be applied to any conductor within which the wave function remains coherent, i.e., the electrons are not scattered inelastically. The equation holds at any elastic scattering in the absence of spin–orbit interaction.

The examination of the simple model of Fig. 5.4 has shown that the probability T_m of transmission of an electron from one electrode to the opposite one T_m is either exponentially small ($\ll 1$, closed channels) or close to 1 (open channels). The transition from zero to unity takes place within a narrow range of the dimensionless energy E. Hence, the quantity $\sum_m T_m(E_F)$ in Eq. (5.1.23) is just the number of open channels at the Fermi energy.

As was shown above, the probabilities of transmission T_m depend on the dimensionless energy

$$E_F/(\hbar^2/2ma^2) = (\hbar^2 k_F^2/2m)/(\hbar^2/2ma^2) = (k_F a)^2, \tag{5.1.24}$$

where k_F is the Fermi wave vector of the electrons. If the width w of the constriction or the density of the 2DEG is increased, the dimensionless Fermi energy is also increased, and the eigenvalues λ_n of the Schrödinger equation for the transverse motion are lowered. If an eigenvalue λ_n was positive and close to zero, it becomes negative, thus increasing the number of open channels by one, and the conductance G by a conductance quantum times g, i.e., by $e^2/\pi\hbar$. It means that the conductance of the point contact at low temperatures is quantized in units of $ge^2/2\pi\hbar$. Its dependence on the electron density and the width of the constriction is stepped: it consists of plateaus and regions of cross-over between them.

The almost perfect quantization of the conductance G of the quantum point contact in Fig. 5.4 (the deviations of the conductance plateaus from whole numbers of quantum $ge^2/2\pi\hbar$ are small) may be attributed to its specific geometry owing to which the longitudinal and transverse coordinates are exactly separable. Glazman, Lesovik, Khmel'nitskii & Shekhter (1988) have shown that such quantization is characteristic for all quantum point contacts in which the constriction has a sufficiently smooth shape. Specifically, the variation of its width $w(x)$ with the longitudinal coordinate x must be smooth on the scale of the Fermi wavelength $\sim k_F^{-1}$. The condition of

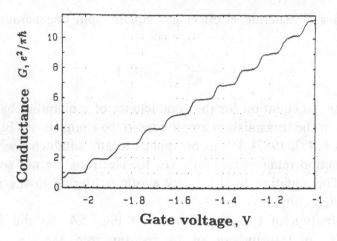

Fig. 5.5 Point .contact conductance in units of $e^2/\pi\hbar$ vs gate voltage. Conductance shows plateaus at multiples of $e^2/\pi\hbar$. From van Wees *et al.* (1988).

this 'adiabaticity' of the width's variation reads: $\pi^2\sqrt{2R/w} > 1$, where R and w are the radius of curvature and minimum width of the constriction. It is worth noticing that the confinement of the 2DEG by electrostatic fields (split-gate technique) is expected to be sufficiently smooth.

The effect of a stepped change of the conductance of the point contact was first found experimentally and correctly interpreted by van Wees, van Houten, Beenakker *et al.* (1988), and by Wharam, Thornton, Newbury *et al.* (1988). The dependence of G on the gate voltage which is used to change the electron density and the width of the constriction is illustrated in Fig. 5.5.

The stepped dependence of G on the electron density n_s and constriction width w is a purely quantum phenomenon (possibly better to say, wave-mechanical phenomenon). In a metal constriction, when the width w is very large as compared with the Fermi wave-length, G is simply proportional to the width and smoothly increases with the electron density.

In the semiclassical approximation, which is applicable to levels λ_n with large n, the quantization condition can be written as (Landau & Lifshitz, 1991 and Eq. (5.1.8)):

$$\int_{-v_m}^{v_m} dv\sqrt{\lambda_n + E\cos^2 v} = \pi\left(n + \frac{1}{2}\right). \qquad (5.1.25)$$

If in this equation λ_n is taken as zero, one obtains the number of negative eigenvalues λ_n, i.e., the number of open channels (if this number is large). For a point contact in a 2DEG and $E = E_F$, it reads:

$$N(\lambda \leq 0; E_F) = (2/\pi)\sqrt{E_F}\sin v_m = (2/\pi)k_F a\sin v_m = k_F w/\pi. \qquad (5.1.26)$$

If this number is substituted into Eq. (5.1.23) for the conductance, the equation for the conductance of a quasi-classical ballistic contact of width w between two 2DEG with electron density n_s is recovered ($g = 2$):

$$G = \frac{2e^2 n_s w}{\pi m v_F},$$

(5.1.27)

where v_F is the electron Fermi velocity.

The concept of channels of electron motion and the inference of the conductance quantization hold not only for ballistic contacts between two 2DEG but also for a ballistic contact between two 3DEG. In the case of a microcontact with axial symmetry between two bulk three-dimensional metals (Sharvin contact, Fig. 5.1) there are two transverse coordinates, the second being the azimuthal angle ϕ. The wave function depends on ϕ as $\exp(iM\phi)$, and $M = 0, \pm 1, \pm 2, \ldots$ is the second quantum number that numbers the conductance channels. A simple calculation yields an estimate for the number of open channels in the quasi-classical limit:

$$N(\lambda \leq 0; E_F) \simeq m E_F a^2 / \hbar^2 = (k_F a)^2 / 2.$$

(5.1.28)

If this number is substituted into Eq. (5.1.23), one obtains the equation for the inverse resistance of ballistic contact with area πa^2 between two bulk metals (Sharvin, 1965):

$$G = \frac{3e^2 n \pi a^2}{m v_F}.$$

(5.1.29)

5.2 Equilibrium noise and shot noise in quantum conductors

Spontaneous fluctuations of current arise even in purely ballistic systems through which electrons pass without any scattering at impurities or phonons. This conclusion follows from the fact that such systems have an active resistance given by the Landauer formula (Eq. (5.1.23)) and dissipate the energy of the electric field. According to the fluctuation–dissipation relation (Nyquist theorem, Sec. 2.2), the current fluctuates even in the state of thermodynamic equilibrium (certainly, if the circuit is open, corresponding voltage fluctuations arise).

We are studying below the simplest case of a two-electrode quantum conductor, i.e., a conductor within which the electron wave functions remain coherent (no inelastic scattering). The problem is to find the spectral density of noise at low frequencies f which are smaller than the following ones: $k_B T/h$, eU/h, and G/C, where U is the bias voltage, C is the capacitance of the system, h and e are the Planck's constant and the absolute value of the electron charge, respectively. The noise measurements in quantum point contacts are performed usually at low frequencies which meet these restrictions. In the absence of bias voltage U and mean current I the spectral density of the current noise is given by the Nyquist formula: $S_I(f) = 4k_B T G$. The main goal of this section is to find the change of the noise under nonzero mean U and I.

A simple derivation of the equation for the spectral density of noise in a ballistic system has been given by Landauer & Martin (1991) and Martin & Landauer (1992). They represented the random currents as a result of random transmissions of electron wave packets between the two electrodes. The following processes contribute to the noise at each electron energy E and in each channel m. (1) A pulse of current is measured when a wave packet incident from the left electrode is transmitted to an empty state in the right one. The rate of these events is proportional to $f_L(E)[1-f_R(E)]T_m(E)$, where $T_m(E)$ is the transmission probability (Sec. 5.1), the factor $1-f_R(E)$ accounts for the Pauli principle. (2) The reverse transmission of a wave packet, incident from the right electrode. It results in a current pulse of opposite sign. The rate of these processes is proportional to $f_R(E)[1 - f_L(E)]T_m(E)$. One has to take into account that the probabilities $T_m(E)$ for the $L \rightarrow R$ and $R \rightarrow L$ transmissions are equal.

If the voltage U and, consequently, the mean current I are nonzero, one has to subtract, in the expressions for the correlation function and the spectral density, the mean probability of transmission, $T_m(E)[f_L(E)-f_R(E)]$, squared (Eqs. (1.2.1) and (5.1.22)). Thus the contribution of electrons with energy E and with definite channel state m is proportional to:

$$T_m(E)[f_L(E)[1 - f_R(E)] + f_R(E)[1 - f_L(E)]] - T_m^2(E)[f_L(E) - f_R(E)]^2.$$

$$(5.2.1)$$

The coefficient can be found by using the fact that at $U = 0$ the Nyquist formula (2.2.4) must be recovered. According to Eq. (5.1.22) for G, the latter reads:

$$S_I(f = 0, U = 0) = 4k_B T G = 2\frac{ge^2}{2\pi\hbar} \int dE\, 2f(E)[1 - f(E)] \sum_m T_m(E).$$

$$(5.2.2)$$

At $U = 0$ the distribution functions $f_R(E) = f_L(E) \equiv f(E)$ are equal. Then the expression in Eq. (5.2.1) is simply $2f(E)[1 - f(E)]T_m(E)$. Comparing it with Eq. (5.2.2), one obtains the general equation for the spectral density of noise at zero frequency, which can be applied also at $U \neq 0$:

$$S_I(f = 0) = 2\frac{ge^2}{2\pi\hbar} \int dE \sum_m \{T_m(E)[f_L(E)[1 - f_R(E)]$$

$$+ f_R(E)[1 - f_L(E)]] - T_m^2(E)[f_L(E) - f_R(E)]^2\}. \tag{5.2.3}$$

We shall discuss this equation after rederiving it by a more rigorous method.

According to the general quantum definition (Eq. (1.2.11)), the correlation function of current fluctuations is equal to the mean value of a half of the anticommutator of the Heisenberg operators of current fluctuations. The operator of current fluctuation in a given electrode α equals simply $\delta\hat{I}_\alpha(t) = \hat{I}_\alpha(t) - \langle\hat{I}_\alpha\rangle$. At low frequencies the current fluctuations are quasistationary, i.e., are constant along the entire ballistic system including both electrodes. The calculation of current fluctuations can be performed for any of the two electrodes.

The current operator can be represented as an expansion in operators related to the quantum states of electrons in the ballistic system (compare with Eq. (5.1.13)):

$$\delta\hat{I}_\alpha(t) = \hat{I}_\alpha(t) - \langle\hat{I}\rangle$$

$$= \sum_{\beta m, \gamma n} \int dE\, dE'\, I_{\beta mE, \gamma nE'}(\alpha)[\hat{a}^+_{\beta mE}(t)\hat{a}_{\gamma nE'}(t) - \langle\hat{a}_{\beta mE}(t)\hat{a}_{\gamma nE'}(t)\rangle]. \tag{5.2.4}$$

The notations are the same as in the preceding section.

The correlation function of the currents equals (see Eq. (1.2.11)):

$$\frac{1}{2}\langle\{\delta\hat{I}_{\alpha_1}(t_1), \delta\hat{I}_{\alpha_2}(t_2)\}\rangle$$

$$= \sum_{\beta_1 m_1 \gamma_1 n_1} \int dE_1\, dE_1' \sum_{\beta_2 m_2 \gamma_2 n_2} \int dE_2\, dE_2'\, I_{\beta_1 m_1 E_1, \gamma_1 n_1 E_1'}(\alpha_1)$$

$$\times I_{\beta_2 m_2 E_2, \gamma_2 n_2 E_2'}(\alpha_2)\Big[\frac{1}{2}\Big\langle\Big\{\hat{a}^+_{\beta_1 m_1 E_1}(t_1)\hat{a}_{\gamma_1 n_1 E_1'}(t_1),$$

$$\times \hat{a}^+_{\beta_2 m_2 E_2}(t_2)\hat{a}_{\gamma_2 n_2 E_2'}(t_2)\Big\}\Big\rangle - \langle\hat{a}^+_{\beta_1 m_1 E_1}\hat{a}_{\gamma_1 n_1 E_1'}\rangle\langle\hat{a}^+_{\beta_2 m_2 E_2}\hat{a}_{\gamma_2 n_2 E_2'}\rangle\Big]. \tag{5.2.5}$$

As the effects of the interaction between the electrons are assumed to be negligible, the average of the product of four second quantization operators is a sum of only two terms, each of which is the mean product of only two such operators:

$$\langle\hat{a}^+_{\beta_1 m_1 E_1}(t_1)\hat{a}_{\gamma_1 n_1 E_1'}(t_1)\hat{a}^+_{\beta_2 m_2 E_2}(t_2)\hat{a}_{\gamma_2 n_2 E_2'}(t_2)\rangle$$

$$= \langle\hat{a}^+_{\beta_1 m_1 E_1}\hat{a}_{\gamma_1 n_1 E_1'}\rangle\langle\hat{a}^+_{\beta_2 m_2 E_2}\hat{a}_{\gamma_2 n_2 E_2'}\rangle$$

$$+ \langle\hat{a}^+_{\beta_1 m_1 E_1}(t_1)\hat{a}_{\gamma_2 n_2 E_2'}(t_2)\rangle\langle\hat{a}_{\gamma_1 n_1 E_1'}(t_1)\hat{a}^+_{\beta_2 m_2 E_2}(t_2)\rangle. \tag{5.2.6}$$

The first term in the r.h.s. is cancelled by an identical term in the r.h.s. of Eq. (5.2.5). Only the second term contributes to the correlation function.

Take into account also that, according to Eq. (1.2.9), the Heisenberg operators $\hat{a}_E^+(t)$ and $\hat{a}_E(t)$ depend on time as e^{iEt} and e^{-iEt}, respectively. The expectation (mean) values of the product of two operators equal (compare with Eq. (5.1.15)):

$$\langle \hat{a}_{\beta_1 m_1 E_1}^+ \hat{a}_{\gamma_2 n_2 E_2'} \rangle = \delta_{\beta_1 \gamma_2} \delta_{m_1 n_2} \delta(E_1 - E_2') f_{\beta_1}(E_1),$$

$$\langle \hat{a}_{\gamma_1 n_1 E_1'} \hat{a}_{\beta_2 m_2 E_2}^+ \rangle = \delta_{\beta_2 \gamma_1} \delta_{m_2 n_1} \delta(E_2 - E_1')[1 - f_{\beta_2}(E_2)]. \tag{5.2.7}$$

Let us substitute all these equations into Eq. (5.2.5) for the correlation function. According to the Wiener–Khintchine theorem (1.3.8), the spectral density is twice the Fourier transform of the correlation function:

$$S_{I\alpha_1\alpha_2}(f) = 4\pi\hbar \sum_{\beta m, \gamma n} \int dE_1 dE_2 \delta(E_2 - E_1 - hf) I_{\beta m E_1, \gamma n E_2}(\alpha_1)$$

$$\times I_{\beta m E_1, \gamma n E_2}^*(\alpha_2) \frac{1}{2}\{f_\beta(E_1)[1 - f_\gamma(E_2)] + f_\gamma(E_2)[1 - f_\beta(E_1)]\}. \tag{5.2.8}$$

This spectral density satisfies the Eqs. (1.3.11) and (1.3.12):

$$S_{I\alpha_1\alpha_2}(-f) = S_{I\alpha_2\alpha_1}(f). \tag{5.2.9}$$

Since we are studying the noise at low frequencies, we can take $f = 0$:

$$S_{I\alpha_1\alpha_2}(0) = 4\pi\hbar \sum_{\beta m, \gamma n} \int dE \, I_{\beta m E, \gamma n E}(\alpha_1) I_{\beta m E, \gamma n E}^*(\alpha_2)$$

$$\times \frac{1}{2}\{f_\beta(E)[1 - f_\gamma(E)] + f_\gamma(E)[1 - f_\beta(E)]\}. \tag{5.2.10}$$

The matrix elements of the current have to be calculated using the wave functions in the electrodes, i.e., the asymptotics of the wave functions. For the model presented in Fig. 5.4, these are given by Eq. (5.1.10). In the case of a two-electrode system we can take $\alpha_1 = \alpha_2$, and the sum over β and γ includes only 4 terms. The indexes R and L denote the right and left electrode, respectively. The matrix elements equal:

$$I_{RmE, RnE}(R) = \frac{e}{2\pi\hbar}[(\mathbf{r}^+(E)\mathbf{r}(E))_{Rm,Rn} - \delta_{mn}],$$

$$I_{LmE, LnE}(R) = \frac{e}{2\pi\hbar}(\mathbf{t}^+\mathbf{t})_{Lm,Ln}, \tag{5.2.11}$$

$$I_{RmE, LnE}(R) = \frac{e}{2\pi\hbar}(\mathbf{r}^+\mathbf{t})_{Rm,Ln}, \quad I_{LmE, RnE}(R) = \frac{e}{2\pi\hbar}(\mathbf{t}^+\mathbf{r})_{Lm,Rn}.$$

Upon substitution of these expressions into Eq. (5.2.10) and multiplication by the Kramers degeneracy g one obtains an equation for $S_I(0)$ which holds, as does the Landauer equation for the conductance (Eq. (5.1.23)), for any conductor in which the wave function is coherent, i.e., no inelastic scattering occurs (Lesovik, 1989; Büttiker, 1990):

$$
\begin{aligned}
S_I(0) \\
= 2\frac{ge^2}{2\pi\hbar} \int dE \Big\{ \mathrm{Tr}(\mathbf{t}^+\mathbf{t}\mathbf{t}^+\mathbf{t})_E \left[f_R(E)(1 - f_R(E)) + f_L(E)(1 - f_L(E)) \right] \\
+ (\mathrm{Tr}(\mathbf{t}^+\mathbf{t})_E - \mathrm{Tr}(\mathbf{t}^+\mathbf{t}\mathbf{t}^+\mathbf{t})_E) \left[f_R(E)(1 - f_L(E)) + f_L(E)(1 - f_R(E)) \right] \Big\}.
\end{aligned}
$$

$$(5.2.12)$$

This equation coincides with Eq. (5.2.3) obtained by a different method.

In the absence of voltage, when $f_R(E) = f_L(E)$, the terms containing $\mathrm{Tr}(\mathbf{t}^+\mathbf{t}\mathbf{t}^+\mathbf{t})$ drop out. The product $f(1 - f) = -k_B T(\partial f/\partial E)$, and therefore $S_I(0) = 4k_B T G$ in accordance with the Nyquist theorem (compare with Eq. (5.1.22) for G).

At nonzero voltage U the mean current I is also nonzero. The electron current may be considered as being injected from the negative electrode. It is expected intuitively that the spectral density of noise has to increase in excess of the equilibrium noise by the so called shot noise (Eq. (1.5.9)), the spectral density of which at low frequencies is $2eI$. This expectation is based on the intuitive idea that the electrons enter into the ballistic constriction randomly. However, it was shown (Lesovik, 1989) that, in general, this expectation is wrong. Consider the ballistic point contact between two degenerate semiconductors or metals at a bias voltage $U \gg k_B T/e$, that is, under just that condition at which the shot noise can be expected. The temperature T may in this case be put to zero. The products $f_R(1 - f_R)$ and $f_L(1 - f_L)$ are zero because $f_{R,L} = 0$ above the Fermi level and $1 - f_{R,L} = 0$ below it. The second expression in rectangular brackets in the r.h.s. of Eq. (5.2.12) can be transformed:

$$
f_R(E)(1 - f_L(E)) + f_L(E)(1 - f_R(E)) = \coth\left(\frac{eU}{k_B T}\right)[f_L(E) - f_R(E)]. \quad (5.2.13)
$$

Using Eq. (5.1.21) and the condition $|eU| \gg k_B T$, one obtains the equation for the noise linear in U (Lesovik, 1989; Büttiker, 1990):

$$
\begin{aligned}
S_I(0) - S_I(0)|_{U=0} &= 2\frac{ge^3}{2\pi\hbar} U \left[\mathrm{Tr}(\mathbf{t}^+\mathbf{t}) - \mathrm{Tr}(\mathbf{t}^+\mathbf{t}\mathbf{t}^+\mathbf{t}) \right]_{E=E_F} \\
&= 2\frac{ge^3}{2\pi\hbar} U \sum_m T_m(E_F)\Big(1 - T_m(E_F)\Big).
\end{aligned}
$$

$$(5.2.14)$$

This part of the current noise may be called 'quantum shot noise' by analogy with the shot noise in classical systems, the spectral density of which is proportional to the mean current (Sec. 1.5).

As was mentioned above, the eigenvalues T_m of the matrix $\mathbf{t}^+\mathbf{t}$ are the probabilities of electron transmission from one electrode to the other one in various conductance channels. Neither the open channels with $T_m = 1$ nor the closed ones ($T_m = 0$) contribute to that part of the noise spectral density which is linear in the voltage U. The only nonzero contributions to the r.h.s. of Eq. (5.2.14) come from channels with intermediate values of T_m. In particular, it means that in quantum ballistic point contacts no noise linear in U can be found at the plateaus of conductance G as a function of, say, the Fermi energy E_F: in these ranges of E_F the conductance channels are either open or closed. This noise is nonzero only in the regions of cross-over from one plateau to the next one, where the probability T_m of one of the channels changes from 0 to 1, or vice versa (Lesovik, 1989). The electrons in the channels with $T_m = 1$ pass through the point contact without any scattering at the constriction. Consequently, no randomness or noise arise.

The noise linear in voltage reaches the full value of the shot noise $2eI$ only in such ballistic conductors in which the probabilities of transmission in all channels are small, that is, $T_m \ll 1$. In fact, comparing Eqs. (5.2.14) and (5.1.22) one can see that in this case

$$S_I(0) - S_I(0)|_{U=0} = 2\frac{ge^3}{2\pi\hbar}U\sum_m T_m(E_F) = 2eGU = 2eI. \qquad (5.2.15)$$

The probabilities of transmission are all small, for example, when the electrons have to pass through a tunnel barrier or to surmount a high ($\gg k_B T$) barrier. It means that in tunnel junctions the noise linear in voltage is, at low temperatures, a full shot noise.

The proportionality of the augmentation of the current noise at non-zero applied voltage to the product $T_m(1 - T_m)$ indicates that the suppression of the shot noise is due to correlations imposed by the Fermi statistics of electrons, i.e., by the Pauli principle. This conclusion can be made also from the statistics of this noise in quantum conductors.

The statistics of the classical shot noise produced by uncorrelated electron pulses is Poissonian (Sec. 1.5). The probability distribution function of the number of electrons n_t passed through the anode circuit of a thermionic tube or through a biased tunnel junction in a time t is given by Eq. (1.1.17):

$$P(n_t) = \frac{\bar{n}_t^{n_t}}{n_t!}e^{-\bar{n}_t}, \qquad (5.2.16)$$

where $\bar{n}_t$ is the mean number of electrons passed in the time t. Levitov &

Lesovik (1993) have shown that, for a biased quantum conductor at low temperatures, the probability distribution $P(n_t)$ is quite different from Eq. (5.2.16):

$$P(n_t) = C_{\bar{N}_t}^{n_t} p^{n_t}(1-p)^{\bar{N}_t-n_t}. \qquad (5.2.17)$$

Here $p = T_m$ and $1-p$ are the probabilities of transmission of an electron and of its reflection, respectively, $\bar{N}_t = geUt/2\pi\hbar$ is the mean number of electrons which would pass through the quantum conductor if T_m were 1, and can be interpreted as the mean number of attempts made by electrons to pass through the conductor, and C_N^n are the known binomial coefficients.

The distribution given by Eq. (5.2.17) is a binomial one. It is a consequence of the fact that for each electron there are only two possible outcomes: to pass through the conductor (probability p) or not to pass (probability $1-p$). As was shown by Levitov and Lesovik (1993), the fluctuations of the number of attempts to pass through the conductor can be neglected if the inequality $k_B T \ll [(1-T_m)/T_m]eU$ is satisfied. Of course, the quantum distribution is reduced to the Poisson one in the limits $T_m \ll 1$ and $(1-T_m) \ll 1$ for transmitted and reflected electrons, respectively.

The statistics of the current noise given by Eq. (5.2.17) is determined by the correlations which stem from the Fermi statistics of electrons, i.e., ultimately, from the Pauli principle. The term 'shot noise' is, strictly speaking, inadequate if applied to current noise in quantum conductors. Therefore we used the term 'quantum shot noise' in order to emphasize the difference between the two statistics, Eq. (5.2.16) and Eq. (5.2.17).

Experiments confirm the suppression of the shot noise in ballistic contacts. Akimenko, Verkin & Yanson (1984) were, to the author's knowledge, the first to observe this phenomenon in metallic point contacts at $T = 1.7$ K (see p. 260 and Fig. 9 of their cited paper). In quantum ballistic point contacts it was observed by Li, Tsui, Heremans *et al.* (1990). They separated the noise with a white spectrum from $1/f$ resistance noise. The first part increased with the current, but even at the maximum current was far below the full shot noise (Fig. 5.6). Suppression of shot noise was observed also by Liefrink, Stok, Dijkhuis *et al.* (1993), and Liefrink, Dijkhuis, de Jong *et al.* (1994) in quantum point contacts with elastic scatterers (see the discussion in Sec. 5.4).

Reznikov, Heiblum, Shtrikman & Mahalu (1995) have measured the excess current noise in QPC at frequencies 8–18 GHz. They found that the part of the noise which is linear in the dc current, oscillates with the number of channels and its peaks correspond to the steps of the conductance. In the pinch-off regime, when the negative gate voltage creates a barrier in the

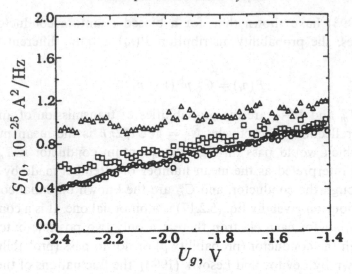

Fig. 5.6 Spectral density of current noise S_{I0} in a quantum point contact (the part proportional to I^2/f, where I is the mean current, subtracted) vs gate voltage. Circles: $I = 0$; squares: $I = 0.3$ μA; triangles: $I = 0.6$ μA. Solid line: the Johnson–Nyquist noise. Dash-dot line: theoretical full shot noise $2eI$ at $I = 0.6$ μA. From Li *et al.* (1990).

middle of the QPC, the excess noise is close to $2eI$ at small currents I and saturates at high currents, perhaps because, at such high currents, a portion of the injected electrons passes through the QPC over the barrier.

Equations (5.2.12) and (5.2.14) for the spectral density of noise in quantum conductors have been derived by Lesovik (1989) for the case of two electrodes and generalized to multiport systems by Büttiker (1990) (see also Beenakker & van Houten, 1991; Büttiker, 1992; Landauer & Martin, 1991, Martin & Landauer, 1992). Lesovik (1989) made the fundamental conclusion on the absence of shot noise when the probabilities T_m are equal to 1 or 0. However, it is very interesting and instructive that the conclusion of absence of shot noise in ballistic contacts between two metals had been made several years earlier by Kulik & Omel'yanchuk (1984) (see also Khlus, 1987). They examined a ballistic contact as depicted in Fig. 5.1 *b* and calculated the spectral density of noise using the quasi-classical method developed by Kogan & Shulman (1969) (Sec. 3.3). They found that a nonzero mean current does not affect the noise, which remains at the equilibrium level. It means that the absence of shot noise is characteristic not only of quantum ballistic contacts with dimensions of the order of the electron Fermi wavelength, it takes place also in metallic ballistic point contacts in which the wavelength of electrons at Fermi energies, $\sim 1/k_F$, being of atomic dimension, is many times smaller than the point contact dimension a, and no conductance plateaus or steps can be observed.

The absence of shot noise in metallic point contacts can be also easily understood from the standpoint of the quantum theory presented above in this section. According to Eq. (5.1.28), the number of open channels in a metallic contact with an area πa^2 is very large :$\sim (k_F a)^2 / 2 \gg 1$. The number of such channels whose probabilities T_m are intermediate between 0 and 1 is only a small part of this number, hence their contribution to the noise linear in U is negligible as compared with $2eI$: the electrons incident onto the constriction either pass freely through it or are reflected back, the number of those that are scattered at the contact's edges is small.

5.3 Modulation noise in quantum point contacts

The noise examined in the preceding section, whose spectral density is given by Eq. (5.2.12), is produced by random motion of electrons in the point contact. The properties of the point contacts, including the arrangement of defects, their charges, and the fields created by these defects, have been assumed to be constant. However, the motion of defects, in particular, their hopping motion, and the changes of the fields produced by electron traps after trapping or emission of an electron, affect the amplitudes and probabilities of charge carriers' transmission and reflection in the quantum point contact and modulate its conductance G. It results in a noise, the spectral density of which (S_I and/or S_U) is proportional to the mean current squared I^2 or to U^2. Such noise is usually called modulation noise (Chs. 6 and 8). The spectral density of conductance fluctuations due to random hopping of an impurity center between two states is given by Eq. (1.7.10):

$$S_G(f) = [\Delta G_1 - \Delta G_2]^2 \frac{4 w_1 w_2 \tau}{1 + \omega^2 \tau^2}. \tag{5.3.1}$$

Here w_1 and $w_2 = 1 - w_1$ are the probabilities of finding the center in the first and the second state, respectively, ΔG_1 and ΔG_2 are the corresponding changes of the conductance due to the presence of the center, τ^{-1} is the sum of transition probabilities per unit time (back and forth) between the two states.

It is interesting to elucidate how the modulation noise depends on the width w of the constriction and on the Fermi energy E_F of the electron gas in the point contact, that is, on the gate voltage V_G which is used to change E_F and w.

It was found in the preceding section that the shot noise is negligible in the regions of plateaus of G as a function of V_G, and is noticeable only in the narrow regions of V_G in which the conductance changes from one plateau to the next one. Does the modulation noise behave in the same manner?

Li, Tsui, Heremans *et al.* (1990) found in a quantum point contact that the spectral density S_I of $1/f$ current fluctuations, which is proportional to I^2 and can be considered as modulation noise, is nonmonotonic in the gate voltage V_G and has peaks between the plateaus of the conductance $G(V_G)$.

The modulation resistance noise was studied by Dekker, Scholten, Liefrink *et al.* (1991) and Cobden, Patel, Pepper *et al.* (1991). Dekker *et al.* studied two samples (Fig. 5.7). In one of them the spectrum of the noise linear in U^2 was close to $1/f$, in the other one at low temperatures $T < 15$ K it was Lorentzian (at higher temperatures it was also of $1/f$ type). Following the fluctuations in the time domain, the authors were able to reveal that the noise in the second sample was the so called random telegraph noise which is believed to be caused by random transitions of an electron trap between two of its states (Sec. 8.8). However, irrespective of the spectrum $S_G(f)$ of the conductance fluctuations, the spectral density S_G was found to be a periodic function of the gate voltage V_G: the maxima of the modulation noise are in line with the narrow regions of transition from one plateau of $G(V_G)$ to the next one, and, consequently, the minima coincided with the plateaus of $G(V_G)$. This effect is one more manifestation of the quantum motion of electrons in the point contact.

It is interesting to understand why identical changes of the fields created by defects produce such a different conductance response δG and, consequently, different noise $S_G(f)$ depending on the proximity of the contact's parameters to one of those narrow ranges in which the conductance changes from one quantized value to the next one. Dekker *et al.* (1991) suggested that the change of the defect's field changes the Fermi energy in the effective (narrowest) part of the contact, that is, acts in the same way as the gate voltage V_G. Therefore S_G is expected to be proportional to $(\partial G/\partial E_F)^2 \propto (\partial G/\partial V_g)^2$. The last quantity is, in fact, close to zero at the plateaus of $G(V_G)$ and is maximal in the transition range from one plateau to the next one. However, the assumption that the response to the change of the defect's field can be reduced to the change of the Fermi energy is likely to be valid only if the defect's field encompasses the point contact's constriction. It can not be applied when the defect's field is strongly localized as in the case of a point defect. Some general qualitative arguments that explain the absence of noise at plateaus of the conductance have been presented by Timp, Behringer & Cunningham (1990).

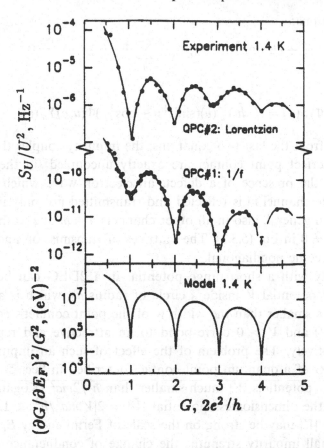

Fig. 5.7 Relative excess-noise spectral density vs conductance G in a quantum point contact (QPC). G is in units of $e^2/\pi\hbar$. QPC # 1: $1/f$ noise at $f = 100$ Hz. QPC # 2: Lorentzian low-frequency plateau value. Arrows indicate upper bounds. Solid lines are guides to the eye. Lower panel: $(\partial G/\partial E_F)/G^2$ vs G. From Dekker *et al.* (1991).

The problem of modulation noise is obviously reduced by Eq. (5.3.1) to the problem of the effect of an impurity, located in the constriction of a ballistic conductor, on its conductance G. For simplicity we consider the model depicted in Fig. 5.4. In the presence of defects the dimensionless potential energy in the Schrödinger Eq. (5.1.6), $\Phi(u,v) \neq 0$. In this case the electron wave function $\Psi(u,v)$ can be represented as a series in the functions $Y_n(v)$ of the unperturbed transverse motion in the absence of the defect:

$$\Psi(u,v) = \sum_n X_n(u)Y_n(v). \qquad (5.3.2)$$

It follows from Eq. (5.1.6) that the functions $X_n(u)$ satisfy a system of coupled equations:

$$X_n'' + (E \sinh^2 u - \lambda_n)X_n - \sum_{n'} \Phi_{nn'}(u)X_{n'}(u) = 0, \tag{5.3.3}$$

where

$$\Phi_{nn'}(u) = \int dv\, Y_n^*(v)(\sinh^2 u + \cos^2 v)\Phi(u,v)Y_{n'}(v). \tag{5.3.4}$$

As follows from the last two equations, the impurity couples the channels which in a perfect point contact are exactly uncoupled (in the model of Fig. 5.4). In the presence of a defect, an electron wave which is emitted from a definite channel m is reflected and transmitted not only in the same channel but, in general, also in all other channels $n \neq m$ due to the coupling terms $\Phi_{nn'}(u) \neq 0$ in Eq. (5.3.3). The matrices of transmission and reflection amplitudes become nondiagonal.

An impurity with a short-range potential in a 2DEG can be modelled by a constant potential V inside a circle of radius a_0, which is assumed to be many times smaller than the width w of the point contact's constriction. $V = -V_0 < 0$ and $V > 0$ correspond to an attractive and repulsive impurity, respectively. The problem of the effect of such an impurity on the conductance G of a quantum point contact described in Sec. 5.1 (Fig. 5.4) was solved for potentials $|V|$ much smaller than $\hbar^2/2ma_0^2$ (Kogan, 1994). In other words, the dimensionless potential $|\tilde{V}| = 2|V|ma_0^2/\hbar^2 \ll 1$. However, the potentials $|V|$ may be strong on the scale of Fermi energy E_F.

At very small impurity strengths the change of conductance due to the impurity is linear in V:

$$\frac{\Delta G}{e^2/\pi\hbar} = -4\pi\tilde{V}\sum_{m=1}^{n_m} |Y_m(v_i)|^2 \sqrt{T_m(1-T_m)}[(X_m^+)^2 - (X_m^-)^2]. \tag{5.3.5}$$

Here T_m is the probability of transmission in the channel m of a perfect contact, $n_m \gg 1$ is a number of the order of the ratio of the constriction width and impurity radius. The functions $X_m^{\pm}(u)$ are the even and odd solutions of the first Eq. (5.1.8) with λ equal to the eigenvalue λ_m of the second equation for $Y_m(v)$. They are taken at the location of the impurity. If it is in the center of the constriction, $X_m^- = 0$ and X_m^+ is exponentially small for closed channels and of order unity for open channels.

As follows from Eq. (5.3.5), at small $|V|$ an attractive impurity increases the conductance, a repulsive impurity decreases it. Due to the factor $\sqrt{T_m(1-T_m)}$ the open ($T_m = 1$) and closed ($T_m = 0$) channels do not contribute to the change of conductance in the linear approximation. The conduction is changed only in the regions of Fermi energy or constriction width where transition from one quantized value of G to the next one takes place. In these regions the transmission probability T_m of one of the channels changes from zero to unity (or vice versa).

If the transmission in only one (say, m-th) channel is of an intermediate value between 0 and 1, the spectral density of conductance noise follows from Eqs. (5.3.1) and (5.3.5):

$$S_G(f) = \left(\frac{e^2}{\pi\hbar}\right)^2 (4\pi\Delta\tilde{V})^2 [(X_m^+(u_i))^2 - (X_m^-(u_i))^2]^2 T_m^{(0)}(1 - T_m^{(0)})$$
$$\times \frac{4w_1 w_2 \tau}{1 + \omega^2\tau^2}.$$

(5.3.6)

This equation for the modulation noise contains the same product $T_m^{(0)}(1 - T_m^{(0)})$ as does the equation for the shot noise in quantum ballistic contacts (Eq. (5.2.14)). Like the shot noise, the modulation noise produced by modulating weak impurities is almost entirely confined to the regions of Fermi energy around the steps between quantized values of the conductance G.

As the strength of the impurity potential is increased, the change of the conductance becomes of the order of the height of a conductance step $e^2/\pi\hbar$. It also becomes negative for both repulsive and attractive impurities (Fig. 5.8). The maxima of $|\Delta G(E_F)|$ remain in the regions of conductance steps. However, at not too small $|V|$ the conductance change $|\Delta G|$ becomes significant also in the regions of conductance plateaus. The deviation of the plateaus from the quantized values progressively grows with the number of the plateau. Hence the modulation noise can be observed also in the regions of plateaus, especially at plateaus with high number.

5.4 Transition from a ballistic conductor to a macroscopic one

How does the current noise in a conductor vary as the conductor's length L is increased from small values, which are much smaller than the relaxation lengths of momentum and energy (ballistic regime), up to values greater than these relaxation lengths (macroscopic regime)? In order to answer this question we consider the most interesting case of a conductor at low temperatures when the electron gas is degenerate and the free-path length for elastic scattering by impurities l_{el} is many times smaller than the inelastic scattering length l_{in}.

As was shown above (Sec. 5.2), in an uniform ballistic conductor, that is, at $L \ll l_{el}, l_{in}$ and in the absence of barriers, the noise is close to the equilibrium (Nyquist) one even at large bias voltages $U \gg k_B T/e$. The deviation from

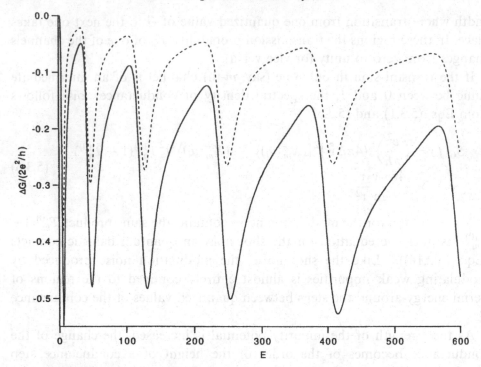

Fig. 5.8 The effect of a short-range impurity on the conductance of a quantum point contact shown in Figure 5.4 vs Fermi energy E. The impurity is located at the center of the constriction ($u = v = 0$). The dimensionless impurity potential: $\tilde{V} = -0.43$ (solid curve); $\tilde{V} = 0.43$ (dashed curve). From Kogan (1994).

the equilibrium noise linear in U (or I), i.e., the shot noise, appears only in definite narrow regions of Fermi energy where the transmission probability of one of the channels has an intermediate value between 0 and 1. When the length L of the sample becomes greater than l_{el}, the transmission of those channels which in the absence of scattering were 1 (open channels) is less than 1 due to the reflection of electrons by the scatterers. Owing to the interaction with impurities, the electron wave functions become extremely intricate. However, until $L \ll l_{in}$ this wave function remains coherent inside the sample, and Eq. (5.2.12) remains valid. Since the transmission probabilities of the channels are now not 0 or 1, one should expect that at $U \gg k_B T/e$ the shot noise would appear. The magnitude of the shot noise was independently calculated by Nagaev (1992) and Beenakker & Büttiker (1992) using different methods (see also Shimizu & Ueda, 1992). Their main result is that, in the case of wide conductors (width much greater than the Fermi wavelength) and in the absence of inelastic scattering, the spectral density of the excess noise is one third of the full shot noise $2eI$. The method used by Nagaev is presented in Sec. 3.3 and therefore we follow his paper.

If electrons do not emit or absorb phonons within the length L of the sample and electron–electron collisions may be neglected, the total energy E of each electron, i.e., the kinetic energy plus the potential energy in the electric field $\mathbf{F}$, is conserved. In the vicinity of the cathode ($z = -L/2$) the maximum total energy of the occupied states is $E_F + eU$, where E_F is the Fermi energy. It means that at $T = 0$ the states with $E < E_F + eU$ are occupied by electrons, and those with $E > E_F + eU$ are empty. Hence, the electron distribution function is $f_0(E - eU)$, where f_0 is the Fermi distribution function. At the opposite electrode ($z = L/2$) the maximum energy of the occupied states is E_F (the potential energy is zero), that is, the distribution function is $f_0(E)$. Between these extremes the distribution function, which satisfies the electron diffusion equation $D d^2 f/dz^2 = 0$, varies linearly with z:

$$f(E, z) = \left(\frac{1}{2} - \frac{z}{L}\right) f_0(E - eU) + \left(\frac{1}{2} + \frac{z}{L}\right) f_0(E). \tag{5.4.1}$$

This distribution function is essentially a nonequilibrium one.

Consider now the fluctuations of the distribution function, $\delta f(\mathbf{p}\mathbf{r}st)$, where s is the spin quantum number. This fluctuation as a function of momentum direction, $\mathbf{n} = \mathbf{p}/|\mathbf{p}|$, can be expanded in a series in spherical functions. When the scattering is strong the direction of the electron momentum is rapidly randomized. Then one can neglect the terms in this series with numbers $l > 1$. The part of the fluctuation which depends on $\mathbf{n}$ can be represented in the form $\mathbf{n} \cdot \delta \mathbf{f}(E\mathbf{r}t)$, where $\delta \mathbf{f}$ is independent of $\mathbf{n}$. In the same way the $\mathbf{n}$-dependent part of the random Langevin flux $\delta J(\mathbf{p}\mathbf{r}t)$ can be represented by $\mathbf{n} \cdot \delta \mathbf{J}(E\mathbf{r}t)$.

The current fluctuation averaged over the length of the sample is

$$\delta I(t) = e \sum_s \int dx\,dy \int \frac{dz}{L} \int 2\frac{d^3 p}{(2\pi)^3} v_z(\mathbf{n}) \delta f_z(E\mathbf{r}st)$$
$$= \frac{1}{3}\frac{e(N_F/2)v_F}{L} \sum_s \int d\mathbf{r}\,dE\,\delta f_z(E\mathbf{r}st). \tag{5.4.2}$$

Here N_F and v_F are the electron density of states, including spin states, and the velocity at the Fermi surface, respectively.

According to the Boltzmann–Langevin equation (3.3.1), in the case of purely elastic scattering and low frequencies $\omega \ll \tau_{el}^{-1}$, where τ_{el} is the elastic scattering time, the fluctuation δf_z is proportional to the corresponding random Langevin flux (an additional term linear in $\nabla \delta f(E\mathbf{r}t)$ disappears after averaging over the sample's volume):

$$\delta f_z(E\mathbf{r}st) = \tau_{el}\delta J_z(E\mathbf{r}st). \tag{5.4.3}$$

The correlation function of the Langevin sources (fluxes) can be easily found from Eq. (3.3.5). The last two terms in the r.h.s. of Eq. (3.3.5) do not contribute to this function and the first two terms are identical. Therefore

$$\langle \delta J_z(E_1\mathbf{r}_1s_1t_1)\delta J_z(E_2\mathbf{r}_2s_2t_2)\rangle$$
$$= \delta_{s_1s_2}\delta(E_1 - E_2)\delta(\mathbf{r}_1 - \mathbf{r}_2)\delta(t_1 - t_2)\frac{6}{\tau(N_F/2)}f(E_1\mathbf{r}_1)[1 - f(E_1\mathbf{r}_1)]. \tag{5.4.4}$$

The correlation function of the current fluctuations is found using Eqs. (5.4.2)–(5.4.4). The resulting expression is proportional to $\delta(t_1 - t_2)$. Using the Wiener–Khintchine theorem (1.3.8) one obtains the equation for the spectral density of the current noise:

$$S_I(0) = \frac{4}{LR}\int dz\, dE f(E,z)[1 - f(E,z)]. \tag{5.4.5}$$

Here

$$R^{-1} = \sigma\frac{A}{L} = \frac{e^2 N_F v_F^2 \tau_{el} A}{3L} \tag{5.4.6}$$

is the inverse resistance of the sample, σ is its conductivity, and A is the sample's cross-sectional area.

Upon integration in Eq. (5.4.5) with $f(E,z)$ given by Eq. (5.4.1) one obtains:

$$S_I(0) = \frac{4}{R}\left[\frac{2}{3}k_B T + \frac{1}{6}eU \coth\left(\frac{eU}{2k_B T}\right)\right]. \tag{5.4.7}$$

At small voltages, $eU \ll k_B T$, this spectral density reduces to the Nyquist noise $4k_B T/R$. In the opposite limit, $eU \gg k_B T$, the spectral density is proportional to the current $I = U/R$ and therefore may be considered as the shot noise:

$$S_I(0) = \frac{1}{3}2eI. \tag{5.4.8}$$

It is three times smaller than the full shot noise $2eI$ which can be found in tunnel junctions with the same current.

Nagaev (1992) explained this reduction of noise as follows. The non-equilibrium noise is due entirely to the deviation of the electron distribution function from the equilibrium one. However, in the vicinity of each electrode the distribution function (Eq. (5.4.1)) is close to an equilibrium one. It strongly deviates from the equilibrium function only in the middle part of the sample. The reduction of the shot noise may be also related to the values of the transmission probabilities in the 'dirty' but otherwise uniform sample: they are on average not as small as in tunnel junctions (Beenakker & Büttiker, 1992).

As the sample's length becomes greater, the inelastic processes of emission and absorption of phonons become important. They tend to reduce the deviation of the electron distribution function from the equilibrium one and, consequently, to reduce the nonequilibrium noise at, say, a given current. The inelastic processes have to be included in the kinetic equation for the symmetric distribution function $f_0(E\mathbf{r})$. At low phonon temperatures T, one may take into account only the spontaneous emission of phonons by the electrons and neglect their absorption. Then the noise depends on the parameter $\gamma = \alpha_{\mathrm{ph}} k_B \Theta_D L^2 / \hbar D$, where α_{ph} is the dimensionless electron–phonon coupling parameter, $k_B \Theta_D$ is the maximum energy of acoustic phonons, $D = v_F^2 \tau_{\mathrm{el}}/3$ is the electron diffusion coefficient. At $L \ll (\hbar D / \alpha_{\mathrm{ph}} k_B \Theta_D)^{1/2}$ the parameter γ is small, hence at $k_B T \ll eU < k_B \Theta_D$ the shot noise given by Eq. (5.4.8) dominates. It dominates even at $L > (\hbar D / \alpha_{\mathrm{ph}} k_B \Theta_D)^{1/2}$, if the voltage across the sample, being greater than $k_B T / e$ is, nevertheless, small enough, specifically,

$$k_B T \ll eU \ll \gamma^{-1/3} k_B \Theta_D. \qquad (5.4.9)$$

Both conditions (5.4.9) may be met at lengths $L \ll \sqrt{D\tau_e}$, where $\tau_e \simeq \hbar \Theta_D^2 / \alpha_{\mathrm{ph}} k_B T^3$ is the energy relaxation time.

At larger $L \gg (\hbar D / \alpha_{\mathrm{ph}} k_B \Theta_D)^{-1/2}$ and $eU \gg \gamma^{-1/3} k_B \Theta_D$ the sample may be considered as a macroscopic one. At very low phonon temperatures $(T \to 0)$ the Joule power σF^2 is counterbalanced by the emission of phonons. The edge of the electron distribution in energy is smeared at the Fermi level, i.e., the electrons are 'heated'. This heating corresponds to an effective electron temperature T_{eff} which depends on the electric field in the sample, $F = U/L$:

$$k_B T_{\mathrm{eff}}(F) = \left[\frac{0.08 e^2 F^2 (k_B \Theta_D)^2 \hbar D}{\alpha_{\mathrm{ph}}} \right]^{1/5}. \qquad (5.4.10)$$

The spectral density of current noise equals (Nagaev, 1992):

$$S_I = \frac{2.4}{R} \left[\frac{e^2 U^2 (k_B \Theta_D)^3}{\alpha_{\mathrm{ph}}} \right]^{1/5} = 4 k_B T_{\mathrm{eff}} R^{-1}. \qquad (5.4.11)$$

Like the shot noise, it is a nonequilibrium noise because it appears due to the nonequilibrium distribution in electron energies. However, it is much smaller than that shot noise which would occur at the same bias voltage U.

Their ratio is small, being a product of two small quantities:

$$\frac{4k_B T_{\text{eff}} R^{-1}}{2eI} \simeq \left(\frac{k_B T}{eU}\right)^{3/5} \left(\frac{D\tau_e}{L^2}\right)^{1/5} \ll 1. \qquad (5.4.12)$$

The spectral density of the relative current fluctuations in a long sample is

$$\frac{S_I}{I^2} = \frac{4k_B T_{\text{eff}}(F) R^{-1}}{(R^{-1}U)^2} = \frac{1}{V} \frac{4k_B T_{\text{eff}}(F)}{\sigma F^2}. \qquad (5.4.13)$$

It is inversely proportional to the sample's volume V, as it should be in the case of a macroscopic body, the dimensions of which are much larger than the relaxation (correlation) length.

In contrast to the electron–phonon scattering, the electron–electron scattering in long, $L \gg l_{el}$, samples increases the shot noise in excess of its value in the absence of any inelastic scattering (Nagaev, 1995). The effect depends on the parameter

$$\eta_{ee} = \frac{(eU)^2 L^2}{\hbar k_F E_F D r_{\text{scr}}}, \qquad (5.4.14)$$

where r_{scr} is the screening radius, $D = v_F^2 \tau_{el}/3$ is the electron diffusion coefficient. As η_{ee} grows, the shot noise increases from $1/3$ to $\sqrt{3}/4$ of the full shot noise.

The shot noise in 'dirty' quantum point contacts was measured by Liefrink, Stok, Dijkhuis *et al.* (1993), and Liefrink, Dijkhuis, de Jong *et al.* (1994). They found that the shot noise was suppressed, but the suppression factor varied with the gate voltage V_G from 0.45 at high V_G down to 0.2 at low V_G, i.e., the theoretical value $1/3$ (see above) was in the middle of this interval (decreasing V_G resulted in decreasing the electron density, point contact width, and effective mean free path). Fig. 5.9 is an illustration of the variation of shot noise.

5.5 Noise in tunnel junctions

In the simple model of a tunnel junction which does not take into account the interaction of the tunneling electrons with various excitations inside the tunnel barrier, e.g. the emission of phonons, the tunnel junction may be viewed as a ballistic conductor, all transmission probabilities of which are very small ($\ll 1$). As was shown in Sec. 5.2, in this case the spectral density of the current noise at small bias voltages $U \ll k_B T/e$ is given by the Nyquist theorem, and at $U \gg k_B T/e$ is equal to full shot noise $2eI$. However, the tunnel junction is so important as a physical system that it is

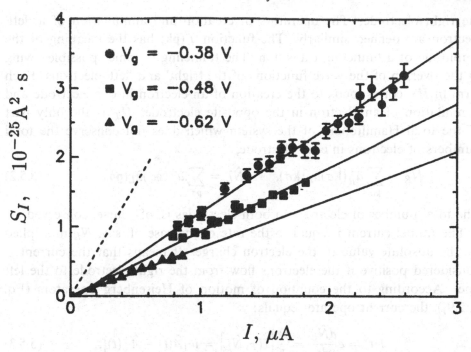

Fig. 5.9 Spectral density of current fluctuations vs current in a diffusive point contact, the length of which is less than the length of inelastic scattering. V_g is the gate voltage. The dashed line corresponds to full shot noise $2eI$. From Liefrink, Dijkhuis *et al.* (1994).

worth presenting a different, in some respects more general, derivation of the same results. This derivation is used later in Ch. 9.

The theory of the tunnel current is usually based on the notion of 'right' and 'left' electrons. These are the electrons in the right and left electrode, respectively, which in the absence of tunneling are confined to their electrodes. The tunneling event is a transformation of a 'right' electron into a 'left' one (or vice versa). In other words, in the case of two degenerate electron gases, a tunneling event is a simultaneous creation of an electron in the left electrode and a hole under the Fermi surface in the right electrode (or vice versa). The effect of tunneling is described by a tunneling Hamiltonian $\hat{H}_T$ which is added to the sum of the Hamiltonians of the 'right' and 'left' electrons, $\hat{H}$ (Bardeen, 1961; Cohen, Falikov & Philips, 1962):

$$\hat{H}_T = \sum_{\mathbf{p}\mathbf{k}\sigma}\{T(\mathbf{p}\mathbf{k})\hat{a}_L^+(\mathbf{p}\sigma)\hat{a}_R(\mathbf{k}\sigma) + T^*(\mathbf{p}\mathbf{k})\hat{a}_R^+(\mathbf{k}\sigma)\hat{a}_L(\mathbf{p}\sigma)\}. \tag{5.5.1}$$

Here $\hat{a}_R(\mathbf{k}\sigma)$ and $\hat{a}_R^+(\mathbf{k}\sigma)$ are, respectively, the operators of annihilation and creation of an electron with momentum $\hbar\mathbf{k}$ and spin quantum number σ in the right electrode. The operators of creation and annihilation of a 'left' electron are defined similarly. The function $T(\mathbf{pk})$ has the meaning of the amplitude of a tunneling transition. The tunneling becomes possible owing to the overlap of the wave functions of the 'right' and 'left' electrons. Each term in $\hat{H}_T$ corresponds to the creation of an electron in one electrode and annihilation of an electron in the opposite electrode. $\hat{H}_T$ is the only part of the total Hamiltonian of the system which does not conserve the total numbers of electrons in each electrode,

$$\hat{N}_R = \sum_{\mathbf{k}\sigma} \hat{a}_R^+(\mathbf{k}\sigma)\hat{a}_R(\mathbf{k}\sigma), \qquad \hat{N}_L = \sum_{\mathbf{p}\sigma} \hat{a}_L^+(\mathbf{p}\sigma)\hat{a}_L(\mathbf{p}\sigma). \qquad (5.5.2)$$

The total number of electrons in both electrodes is, of course, conserved.

The tunnel current is equal to the rate of increase of, say, N_L multiplied by the absolute value of the electron charge. It means that the current is considered positive if the electrons flow from the right electrode to the left one. According to the equation of motion of Heisenberg operators (Eq. (1.2.9)), the current operator equals:

$$\hat{I}(t) = e\frac{d\hat{N}_L}{dt} = \frac{ie}{\hbar}\left[\hat{H}_T, \hat{N}_L\right] = ie[\hat{A}(t) - \hat{A}^+(t)], \qquad (5.5.3)$$

where for brevity the operator $\hat{A}(t)$ and its Hermitian conjugate $\hat{A}^+(t)$ have been introduced:

$$\hat{A}(t) = \frac{1}{\hbar}\sum_{\mathbf{pk}\sigma} T^*(\mathbf{pk})\hat{a}_R^+(\mathbf{k}\sigma t)\hat{a}_L(\mathbf{p}\sigma t). \qquad (5.5.4)$$

The Heisenberg operators in the r.h.s. of Eq. (5.5.3) satisfy the equations of motion with the total Hamiltonian, $\hat{H} + \hat{H}_T$ (Eq. (1.2.9)). If one neglects $\hat{H}_T$ in these equations, the operator $\hat{I}$ given by Eq. (5.5.3) (it can be denoted by $\hat{I}^{(1)}$) upon averaging over states with definite N_R and N_L yields zero. In order to find the tunnel current, which is of second order in $\hat{H}_T$, one has to find the first-order, in $\hat{H}_T$, correction to the r.h.s. of Eq. (5.5.3). According to the perturbation theory for Heisenberg operators, the expression for this term, i.e., for the tunnel current operator, reads:

$$\hat{I}^{(2)} = \frac{i}{\hbar}\int_{-\infty}^{t} dt'\left[\hat{H}_T(t'), \hat{I}(t)\right] = e\int_{-\infty}^{t} dt'\{[\hat{A}(t), \hat{A}^+(t')] + [\hat{A}(t'), \hat{A}^+(t)]\}. \qquad (5.5.5)$$

Terms with $\hat{a}_L\hat{a}_L$, $\hat{a}_L^+\hat{a}_L^+$, $\hat{a}_R\hat{a}_R$, $\hat{a}_R^+\hat{a}_R^+$ have been omitted in Eq. (5.5.5) because in normal metals their average values are zero (in superconductors they are not zero, see Sec. 9.1.1). The mean value of $\hat{I}^{(2)}$ obtained by averaging over states with definite numbers of electrons in the electrodes is not zero, unlike the mean value of $\hat{I}^{(1)}$.

The tunnel current operator has to be averaged over the state of the electrodes, each of which is assumed to be in equilibrium and to have its own electrochemical potential, μ_R and μ_L. The difference of these electrochemical potentials is equal to $\mu_R - \mu_L = eU$, where U is the bias voltage across the junction. In systems with variable numbers of particles, the probabilities of states w_m are determined by the difference $E'_m = E_m - \mu N_m$, where E_m is the energy of the state, N_m is the number of particles in this state of the system. Therefore it is convenient to use the Heisenberg operators which satisfy the equations of motion, not with the Hamiltonian $\hat{H}$, but with the operator $\hat{H} - \mu \hat{N}$. In the tunnel junction at nonzero bias this effective Hamiltonian has to be

$$\hat{H}' = \hat{H} - \mu_L \hat{N}_L - \mu_R \hat{N}_R. \tag{5.5.6}$$

This formalism yields the same results as the standard one. The only difference is that the energies of the quasi-particles, electrons and Fermi holes, are counted in each electrode from the local Fermi level, μ_R or μ_L. Taking into account that the operator $\hat{A}(t)$ increases the number of 'right' electrons and decreases the number of 'left' electrons by one, and the action of $\hat{A}^+(t)$ is just the opposite, one can easily prove that the mean values of the products of these operators, satisfying the equations with $\hat{H}$ and with $\hat{H}'$, are related by the equation:

$$\langle [\hat{A}(t_1), \hat{A}^+(t_2)] \rangle_H = \exp\left[\frac{ieU}{\hbar}(t_1 - t_2)\right] \langle [\hat{A}(t_1), \hat{A}^+(t_2)] \rangle_{H'}. \tag{5.5.7}$$

Upon averaging the current operator (Eq. (5.5.5)) and using Eq. (5.5.7) one obtains the following equation for the tunneling current:

$$I = \langle \hat{I}^{(2)} \rangle = e \int_{-\infty}^{+\infty} dt \, \exp\left[i\frac{eU}{\hbar}t\right] \langle [\hat{A}(t), \hat{A}^+(0)] \rangle_{H'}. \tag{5.5.8}$$

A more explicit expression is obtained by substitution of Eq. (5.5.4) into the last equation. The averaged products of four operators may be decomposed into the product of two averaged pairs of operators, each related to the right or left electrode. The tunnel current equals:

$$I = \frac{2\pi e}{\hbar} \sum_{\mathbf{k}\mathbf{p}\sigma} |T(\mathbf{k}\mathbf{p})|^2 [f_R(E(\mathbf{k})) - f_L(E(\mathbf{p}))] \delta(E(\mathbf{p}) - E(\mathbf{k}) - eU). \tag{5.5.9}$$

Here $E(\mathbf{k})$ and $E(\mathbf{p})$ are the energies of the 'right' and 'left' electrons counted from the local Fermi levels, $f_R(E)$ and $f_L(E)$ are electron Fermi distribution functions for the right and left electrodes, respectively.

If the electron gases in both electrodes are highly degenerate and $k_B T \ll eU \ll \mu_L, \mu_R$, only electrons in the vicinity of the Fermi level contribute

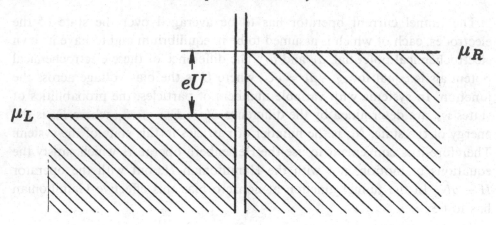

Fig. 5.10 Schematic diagram of a tunnel junction. U is the voltage across the junction, μ_L and μ_R are the electrochemical potentials of the left and right electrodes. The energy regions occupied by electrons in each electrode are shown by hatching. The electrons in this junction are flowing from right to left.

to the tunneling current owing to the factor $f_R - f_L$ and delta-function in Eq. (5.5.9). Specifically, at $U > 0$ 'right' electrons with energies $E(\mathbf{k}) < 0$ in the range between $E(\mathbf{k}) = 0$ and $E(\mathbf{k}) = -eU$ (under the Fermi surface) are tunneling into the empty states of 'left' electrons with energies $E(\mathbf{p}) > 0$ up to eU (Fig. 5.10). If the tunneling amplitudes are almost constant within the narrow range of electron energies around the Fermi energy (a very plausible assumption), the difference between the Fermi functions in the electrodes can be represented in the form:

$$[f(E(\mathbf{k} - eU/2) - f(E(\mathbf{p}) + eU/2)]\delta(E(\mathbf{p}) - E(\mathbf{k}) - eU)$$
$$\approx eU\delta[E(\mathbf{k})]\delta[E(\mathbf{p})]. \tag{5.5.10}$$

Under these conditions, the tunneling current is proportional to the voltage, i.e., the current–voltage characteristic is Ohmic:

$$I = R^{-1}U, \tag{5.5.11}$$

where R is the resistance of the tunnel junction:

$$R^{-1} = \frac{2\pi e^2}{\hbar} \sum_{\mathbf{pk}\sigma} |T(\mathbf{pk})|^2 \delta(E_R)\delta(E_L). \tag{5.5.12}$$

The r.h.s. of Eq. (5.5.12) is proportional to the tunneling amplitude squared, taken at the Fermi energy and averaged over the orientations of the 'right' and 'left' electrons' momenta.

The correlation function of the tunnel currents' fluctuations in the lowest order in the tunneling amplitudes can be found by substituting the current operator (Eq. (5.5.3)) into the general Eq. (1.2.10). In this approximation the operators in Eq. (5.5.3) can be considered as satisfying the equations of motion with Hamiltonian $\hat{H}$, i.e., $\hat{H}_T$ neglected. The correlation function can be represented in terms of the operators $\hat{A}$ and $\hat{A}^+$ (the products $\hat{A}\hat{A}$ and $\hat{A}^+\hat{A}^+$ can be omitted for normal conductors, compare with Sec. 9.1.2):

$$\psi_I(t_1 - t_2) = \frac{e^2}{2} \langle \{\hat{A}(t_1), \hat{A}^+(t_2)\} + \{\hat{A}(t_2), \hat{A}^+(t_1)\} \rangle. \tag{5.5.13}$$

According to the Wiener–Khintchine theorem (Eq. (1.3.8)), the spectral density of current fluctuations equals (see also Eq. (5.5.7)):

$$S_I(f) = 2e^2 \int_{-\infty}^{+\infty} d(t_1 - t_2) \cos[\omega(t_1 - t_2)] \langle \{\hat{A}(t_1), \hat{A}^+(t_2)\} \rangle_H$$

$$= 2e^2 \int_{-\infty}^{+\infty} d(t_1 - t_2) \cos[\omega(t_1 - t_2)] \exp\left[\frac{ieU}{\hbar}(t_1 - t_2)\right] \langle \{\hat{A}(t_1), \hat{A}^+(t_2)\} \rangle_{H'} \tag{5.5.14}$$

The mean value of the anticommutator in Eq. (5.5.14) can be expressed in terms of the mean value of the commutator of the same operators (as in Sec. 2.1):

$$\int_{-\infty}^{+\infty} d(t_1 - t_2) \exp[i\omega(t_1 - t_2)] \langle \{A(t_1), A^+(t_2)\} \rangle$$

$$= \coth\left(\frac{\hbar\omega}{2k_B T}\right) \int_{-\infty}^{+\infty} d(t_1 - t_2) \exp[i\omega(t_1 - t_2)] \langle [A(t_1), A^+(t_2)] \rangle. \tag{5.5.15}$$

The last Fourier integral can be expressed in terms of the conductance R^{-1} (Eqs. (5.5.8) and (5.5.12)). Thus the spectral density of current noise can be represented in the form (Dahm, Denenstein, Langenberg et al., 1969):

$$S_I(f) = \frac{1}{R} \left\{ (eU + \hbar\omega) \coth\left(\frac{eU + \hbar\omega}{2k_B T}\right) + (eU - \hbar\omega) \coth\left(\frac{eU - \hbar\omega}{2k_B T}\right) \right\}$$

$$= e \left\{ I(U + hf/e) \coth\left(\frac{eU + hf}{2k_B T}\right) + I(U - hf/e) \coth\left(\frac{eU - hf}{2k_B T}\right) \right\}. \tag{5.5.16}$$

Here $I(U)$ is the dependence of the current on the voltage.

This general and simple relation resembles the fluctuation–dissipation relation (Secs. 2.1 and 2.2). It holds even though at nonzero voltage the tunnel junction is not an equilibrium system. This is a consequence of the fact that even at $U \neq 0$, when the electrochemical potentials of the electrodes are different, each electrode remains in an equilibrium state. It follows once more from Eq. (5.5.16) that at $U \ll k_B T/e$ the noise is an equilibrium one, and in the opposite case the noise at $f = 0$ is a full shot noise $2eI$.

At very small area the capacitance C of the tunnel junction is so small that the energy of its charging by a single electron, $e^2/2C$, may be greater than the thermal energy $k_B T$. This regime is called the Coulomb blockade. The mean characteristics of such small junctions are different from the ones of large junctions. In particular, the current is not proportional to the voltage (see reviews on the Coulomb blockade in Grabert & Devoret, 1992). The noise in these nanostructures has been studied by Hershfield, Davies, Hyldgaard, Stanton & Wilkins (1993) and Korotkov (1994).

The noise examined above is produced by random transitions of electrons across the barrier, the properties of which have been assumed constant. An additional current noise, usually observed at low frequencies, is produced by the fluctuations of the barrier's resistance, $\delta R(t)$. They are produced by the motion of defects at the boundaries or inside the barrier, by random trapping of electrons in the barrier (fluctuations of the fields created by the traps' charges affect the transmission of electrons). This noise is called 'modulation noise' and is usually proportional to the mean current squared, I^2. Mechanisms of this type of noise are discussed in Sec. 5.3 and Chs. 6 and 8.

6

Resistance noise in metals

In many systems the measured spectral density of noise above the equilibrium one increases with the current I as I^2 (or with voltage U as U^2). This part of the noise is interpreted as the modulation noise which is caused by fluctuations (random 'modulation') of the sample's resistance, $\delta R(t)$. According to the Kirchhoff law, the spectral densities of current noise at fixed voltage and voltage noise at fixed current are given by simple equations:

$$\left(\frac{S_I(f)}{I^2}\right)_{U=\text{const}} = \left(\frac{S_U(f)}{U^2}\right)_{I=\text{const}} = \frac{S_R(f)}{R_d^2}, \tag{6.1}$$

where R_d is the differential (small-signal) resistance.

A great number of mechanisms of resistance fluctuations are known: generation–recombination fluctuations of the number of charge carriers in semiconductors (Ch. 4), the fluctuations of temperature in a conductor with temperature-dependent resistance, etc. The resistance noise, usually with a $1/f$ type spectrum, was found in metals in a great number of experiments. An interesting problem is: what are the mechanisms of resistance fluctuations in metals? The generation–recombination fluctuations seem to be impossible in metals. Obviously, if the effects of temperature fluctuations are neglected and if the structure of the sample, including the scattering centers, is fixed, the resistance is also fixed and does not fluctuate. However, in a real sample some of the defects are mobile. The hopping motion of the scattering centers changes their arrangement and affects the resistance. Below, in this chapter, the mechanisms which are widely used for the interpretation of experiments are discussed.

6.1 Incoherent scattering of electrons by mobile defects

To clarify how the random hopping motion of randomly arranged scattering
centers, i.e., impurities and defects, can produce the low-frequency fluctuations of resistance, consider a simple model of defects, the point-symmetry
of which is lower than that of the crystal (Kogan & Nagaev, 1982, 1984).
Such defects can be in several, say s, states with equal free energies but
different orientation. The best known examples are interstitial impurities
located at octahedral positions in a bcc lattice (for instance, carbon or nitrogen in α-iron). They have tetragonal symmetry with the fourfold axis
along one of the three cubic axes of the crystal (Fig. 6.1). The number of
states in this case is $s = 3$. When an impurity atom hops to an adjacent
equivalent interstitial site, the orientation of the defect changes, its fourfold
axis becomes parallel to another cubic axis. In thermodynamic equilibrium
in the absence of any strain, all s possible orientations are equally probable
and the mean numbers of defects in each state with definite orientation α
are equal: $\langle N_\alpha \rangle = N/s$, where N is the total number of these defects in the
sample. However, the hops of impurity atoms are random, and the number
of those with a definite orientation, N_α, fluctuates around its mean value:
$N_\alpha(t) = \langle N_\alpha \rangle + \delta N_\alpha(t)$.

The defects contribute to the scattering of electrons and to the resistivity
ρ of the sample. Since the symmetry of the defect is lower than the point
symmetry of the host crystal, its contribution to the resistivity ρ consists
of two parts. One part is independent of the defect's orientation, and in a
cubic crystal it is isotropic. However, there is also an anisotropic part that is
different for different orientations α. It can be interpreted in simple terms if
one views each scattering center as an ellipsoid (not a sphere) with different
cross-sections along different directions. The effective cross-section is at each
instant that one which is normal to the current, hence it depends on the
orientation of the scattering center.

The fluctuations of the concentrations of defects $\delta n_\alpha(t) = \delta N_\alpha(t)/V$,
where V is the volume of the sample, lead to fluctuations of the resistivity
tensor:

$$\delta \rho_{ij}(\mathbf{r}, t) = \sum_\alpha \frac{\partial \rho_{ij}}{\partial n_\alpha} \delta n_\alpha(\mathbf{r}, t). \tag{6.1.1}$$

The concentration of defects is supposed here to be averaged over volumes
with dimensions greater than the free-path length l.

As the sum $\sum_\alpha \delta n_\alpha = 0$, one can subtract from the derivative in Eq.
(6.1.1) the part which is independent of the orientation α and substitute this

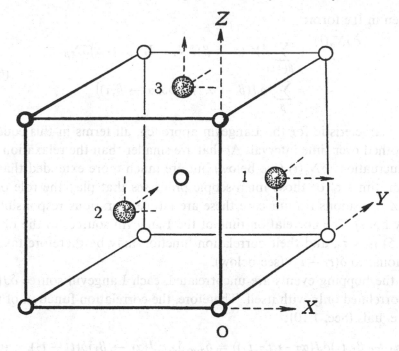

Fig. 6.1 Three different positions of an impurity in a bcc crystal. The impurity is located at the center of a cubic face. Its octahedral symmetry is lower than the cubic symmetry of the host crystal.

derivative by its anisotropic part (the isotropic part drops out):

$$\left(\frac{\partial \rho_{ij}}{\partial n_\alpha}\right)_a = \frac{\partial \rho_{ij}}{\partial n_\alpha} - \frac{1}{s}\sum_\beta \frac{\partial \rho_{ij}}{\partial n_\beta}. \tag{6.1.2}$$

The anisotropic part satisfies the equation:

$$\sum_\alpha \left(\frac{\partial \rho_{ij}}{\partial n_\alpha}\right)_a = 0. \tag{6.1.3}$$

It follows immediately from Eqs. (6.1.1) and (6.1.3) that, in the model under consideration,

$$\operatorname{Tr}\delta\rho_{ij} \equiv \sum_i \delta\rho_{ii} = 0, \tag{6.1.4}$$

that is, the resistivity fluctuations do not contain any isotropic part and may be considered as purely anisotropic (traceless).

Let $W(\alpha \to \beta)N_\alpha$ be the rate of transitions of defects from the state α into the state β. The Langevin kinetic equation for the fluctuations $\delta N_\alpha(t)$ can be written in the form:

$$\frac{\partial \delta N_\alpha(t)}{\partial t} + \sum_{\beta(\neq\alpha)} \left[W(\alpha \to \beta)\delta N_\alpha - W(\beta \to \alpha)\delta N_\beta \right]$$

$$= \sum_\beta \left[\delta J(\beta \to \alpha, t) - \delta J(\alpha \to \beta, t) \right]. \tag{6.1.5}$$

As is characteristic for the Langevin approach, all terms in this equation are smoothed over time intervals Δt that are smaller than the relaxation time of the fluctuations $\delta N_\alpha(t)$ (see below) but are much more extended than the relaxation times τ_c of those microscopic processes that play the role of the source of transitions (in our case these are lattice vibrations responsible for impurity hops). The correlation time of the Langevin sources in the r.h.s. of Eq. (6.1.5) is $\sim \tau_c$, and their correlation function may be therefore taken as proportional to $\delta(t_1 - t_2)$ (see below).

Since the hopping events are uncorrelated, each Langevin source $\delta J(\alpha \to \beta, t)$ is correlated only with itself. Therefore, the correlation function of these sources equals (Sec. 1.10):

$$\langle \delta J(\alpha_1 \to \beta_1, t_1)\delta J(\alpha_2 \to \beta_2, t_2) \rangle = \delta_{\alpha_1\alpha_2}\delta_{\beta_1\beta_2} J(\alpha_1 \to \beta_1)\delta(t_1 - t_2), \tag{6.1.6}$$

It is nonzero only if the initial states and, separately, the final states of the hops coincide. As for shot noise (Sec. 1.5), the correlation function is proportional to the mean rate of transitions $J(\alpha_1 \to \beta_1)$.

Since the free energies of all the states $\alpha = 1, \ldots, s$ are equal, all transition probabilities $W(\alpha \to \beta)$ ($\beta \neq \alpha$) are also equal. Taking into account the conservation of defects, one obtains that the sum

$$\sum_{\beta(\neq\alpha)} \delta N_\beta = -\delta N_\alpha. \tag{6.1.7}$$

Therefore the Langevin kinetic equation (6.1.5) can be rewritten:

$$\frac{\partial \delta N_\alpha(t)}{\partial t} + \frac{1}{\tau}\delta N_\alpha(t) = \sum_\beta \left[\delta J(\beta \to \alpha, t) - \delta J(\alpha \to \beta, t) \right]. \tag{6.1.8}$$

Here $\tau^{-1} = sW(\alpha \to \beta)$ is the inverse relaxation time of the fluctuations.

Using Eqs. (6.1.6) and (6.1.8) one obtains the spectral density of the fluctuations $\delta N_\alpha(t)$:

$$S_{\alpha\beta}(f) = \left(\delta_{\alpha\beta} - \frac{1}{s} \right)\frac{N}{s}\frac{4\tau}{1 + \omega^2\tau^2}. \tag{6.1.9}$$

If the effects of diffusion of impurities on the specimen's resistance are small (they may be observed at much lower frequencies f), the fluctuations of the concentrations $\delta n_\alpha(\mathbf{r}, t)$ at different points $\mathbf{r}_1$ and $\mathbf{r}_2$ are uncorrelated, and their correlation function is proportional to $\delta(\mathbf{r}_1 - \mathbf{r}_2)$. From Eq. (6.1.9) one can find that the spectral density, i.e., the Fourier transform of the correlation function doubled, equals:

$$2\langle \delta n_\alpha(\mathbf{r}_1 t_1)\delta n_\beta(\mathbf{r}_2 t_2)\rangle_\omega = \delta(\mathbf{r}_1 - \mathbf{r}_2)\left(\delta_{\alpha\beta} - \frac{1}{s}\right)\frac{n_d}{s}\frac{4\tau}{1 + \omega^2\tau^2}, \qquad (6.1.10)$$

where $n_d = N/V$.

The spectral density of resistivity fluctuations is easily found from Eqs. (6.1.1), (6.1.2), and (6.1.10):

$$2\langle \delta\rho_{ij}(\mathbf{r}_1 t_1)\delta\rho_{kl}(\mathbf{r}_2 t_2)\rangle_\omega = \delta(\mathbf{r}_1 - \mathbf{r}_2)F_{ijkl}(f), \qquad (6.1.11)$$

where

$$F_{ijkl}(f) = \frac{n_d}{s}\frac{4\tau}{1 + \omega^2\tau^2}\sum_{\alpha,\beta=1}^{s}\left(\delta_{\alpha\beta} - \frac{1}{s}\right)\left(\frac{\partial\rho_{ij}}{\partial n_\alpha}\right)_a\left(\frac{\partial\rho_{kl}}{\partial n_\beta}\right)_a. \qquad (6.1.12)$$

Owing to Onsager symmetry relations, the tensor F_{ijkl} in the absence of magnetic field is symmetric against the interchange of indexes $i \leftrightarrow j, k \leftrightarrow l$. It is invariant also under the interchange of these two pairs of indexes. From the theory of tensor symmetry it is well known that in cubic crystals the nonzero components of such a 4-th rank tensor can be expressed in terms of three components: F_{1111}, F_{1122}, and F_{1212}. The condition of zero trace (Eq. (6.1.4)) results in a relation: $F_{1122} = -\frac{1}{2}F_{1111}$, reducing the number of independent components to two.

A simple estimate shows that

$$\frac{1}{\rho}\left|\left(\frac{\partial\rho_{ij}}{\partial n_\alpha}\right)_a\right| \sim l\sigma_s. \qquad (6.1.13)$$

Here l is the mean free path of electrons, $\sigma_s \sim a_0^2$ is the scattering cross-section of the defects, and a_0 is the lattice parameter.

As follows from Eqs. (6.1.11)–(6.1.13), the spectral density of the fractional fluctuations of the sample's resistance is of the order of

$$\frac{S_R(f)}{R^2} \simeq \frac{n_d(l\sigma_s)^2}{V}\frac{4\tau}{1 + \omega^2\tau^2}. \qquad (6.1.14)$$

Complexes of point defects (vacancies, interstitials) are of frequent occurrence in solids. Electron scattering by such complexes is anisotropic even when the scattering by each of the point defects, taken separately, is isotropic. Relative motion of the complex's constituents changes the total scattering cross-section and, consequently, produces a fluctuation of the resistance (Pelz & Clarke, 1987). The anisotropy of scattering by the complex is determined by the interference of electron waves scattered by each point defect. Therefore Pelz and Clarke coined the term 'local interference model' for the entire model of resistance fluctuations produced by mobile defects with symmetry lower than the point symmetry of the crystal.

Besides low-symmetry point defects and defect complexes, dislocations can also be a source of resistance fluctuations (Vinokur & Obukhov, 1989). Many real solids contain not only one type of mobile defects, but a wide variety of such defects with various relaxation times τ. The resulting resistance noise spectrum may therefore strongly differ from the simple Lorentzian spectrum (Eq. (6.1.14)). In Ch. 8 this theory of modulation noise is applied to $1/f$ noise in metals.

The outlined theory assumes that electrons are scattered by different impurity centers or different impurity complexes independently, and there is no interference between scattering events. Therefore this theory, in general, can not be applied to low temperatures when, in metals, the free-path length for elastic scattering by impurities is many times shorter than the free-path length for inelastic scattering, which destroys the coherence of the electron wave function. In this case a hop of an impurity does not merely change its scattering cross-section, but changes the wave function interference pattern in a region that encompasses many scattering centers. These phenomena are discussed in the next section.

6.2 Effect of mobile scattering centers on the electron interference pattern

The last decade has witnessed fundamental revision of the basic concepts of electrical conductivity of metals at low temperatures. According to the previous concepts developed by Lorentz, Drude and Sommerfeld, each defect in a metal scatters the electrons independently of all other defects, and all such scattering centers contribute additively to the resistance. Since the scattering centers are arranged at random, it was assumed that the resistance does not depend on their particular positions at a fixed concentration of the centers, and does not change if one of the centers is displaced by, say, a lattice constant. The quantum mechanics of the system was taken into account only in two respects: (1) the quantum Fermi–Dirac statistics for electrons was used, (2) the scattering of electrons by a defect was considered as a quantum-mechanical process. These concepts leaned upon the fact that in typical metals the wavelength $2\pi/k_F$ of an electron at the Fermi surface is many times smaller than the free-path length of the electrons, l, that is, $k_F l \gg 1$. It was assumed also that the scattering by the defects and the scattering by the lattice vibrations are mainly independent and their contributions to the resistance are additive (Mattisen's rule).

The understanding of the fundamental difference between the effect of electron scattering by the defects, on one hand, and the effect of scattering by phonons, on the other hand, on the electron kinetic phenomena is one of the important recent successes of the theory of conductivity in metals. Randomly arranged defects create a static field acting on electrons. The motion of electrons in this field is coherent, i.e., is represented by definite stationary wave functions which are, obviously, extremely intricate functions of electron coordinates, due to complex interference patterns created by multiple scattering of electrons by the scattering centers. On the contrary, the absorption or emission of a phonon interrupts the stationary motion of the electron and breaks the evolution of the wave function's phase. Due to inelastic scattering, the wave function's coherence is lost. It means that in impure metals at low temperatures, when the time between two successive inelastic scattering events, τ_{in}, is much greater than the time between successive elastic scattering events, τ_{el}, the electrons are for long times in definite stationary quantum states, the wave function of each of which is determined by the specific arrangement of scattering centers in the specimen. The shape of the wave functions, i.e., of the interference patterns, can be drastically changed by a hop of only one such center by a lattice parameter, that is, by a distance of the order of the electron's wavelength (Altshuler & Spivak, 1985; Feng, Lee & Stone, 1986). This phenomenon underlies the mechanism of resistance fluctuations presented in this section.

The idea of the effect of a displacement of only one impurity center on the electron interference pattern in metals and on the resistance of macroscopic samples appeared as a logical development of the concepts of so called universal conductance fluctuations (Altshuler, 1985; Lee & Stone, 1985). The conductances (inverse resistances) G of two identical samples, with equal numbers of scattering centers, which differ only by the particular arrangement of these centers, may differ by a quantity of the order of the conductance quantum e^2/h (Sec. 5.1), because the electron interference patterns in the samples are different. In the experiments, of course, different samples are not compared: the interference pattern in one and the same sample is varied by varying the magnetic field, B, or the Fermi energy, E_F. The conductance G varies in a very intricate, random-like, manner by amounts $\sim e^2/h$. These deviations of G from an average value are called universal conductance fluctuations. Unlike the noise-producing fluctuations considered in this book, the universal conductance fluctuations are static and quite reproducible, that is, the functions $G(E_F)$ or $G(B)$ measured in two experimental runs coincide if in the time between these two runs no defect in the sample happened to hop.

It is well known that a clear light interference pattern is obtained only if the coherence length of the light is long enough. Similarly, the effects of interference of electrons moving from one electrode to the opposite one are more pronounced, the greater the coherence length of the electron quantum states, L_ϕ. At high enough concentrations of impurities and low temperatures, when $\tau_{el} \ll \tau_{in}$, the motion of electrons between two successive inelastic scattering events is diffusive with the diffusion coefficient $D = v_F^2 \tau_{el}/d$, where d is the Euclidean dimension of the space in which the diffusion of electrons takes place ($d = 1, 2$ or 3), and v_F is the electron Fermi velocity. The length of displacement by diffusion in a time τ_{in} is $\sim \sqrt{D\tau_{in}}$. It determines the linear dimensions of the coherence region.

However, even at very long τ_{in} the interference effects are negligible if the electrons are not 'monochromatic', i.e., the spread of their energies δE in the vicinity of the Fermi energy E_F is too great. The phase difference between waves, the frequencies of which differ by $\delta E/\hbar$, reaches $\simeq \pi$ in a time $\simeq \hbar/\delta E$. Because the spread of electron energies is purely thermal, $\delta E = k_B T$. The linear dimension of the region of diffusive spreading of electrons in a time $\hbar/k_B T$ is $L_T = \sqrt{\hbar D/k_B T}$. It means that under conditions when $\tau_{el} \ll \tau_{in}$ the coherence length equals:

$$L_\phi = \min\{\sqrt{D\tau_{in}}, \quad \sqrt{\hbar D/k_B T}\}. \tag{6.2.1}$$

Theoretical considerations and measurements show that at low temperatures in metals L_ϕ is a macroscopic length which may be as much as several μm. On one hand, the number of atoms and even the number of scattering centers in a sample with linear dimensions $\sim L_\phi$ is huge, i.e., macroscopic. On the other hand, the conductance of this sample depends on the specific microscopic arrangement of these scattering centers. Such systems have been called mesoscopic (the Greek word mesos means medium), that is, being between macroscopic and microscopic.

The change of the conductance after a displacement of a scattering center depends on the arrangement of all scattering centers in the sample and on the position of the displaced center. Obviously, it is a random quantity. A nonrandom, definite, quantity is the change of the conductance squared and averaged over all possible arrangements of all scattering centers, both fixed and mobile:

$$(\delta G_1)^2 = \frac{1}{2}\langle\{G(\mathbf{R}_1 + \delta\mathbf{R}_1, \ldots, \mathbf{R}_N) - G(\mathbf{R}_1, \ldots, \mathbf{R}_N)\}^2\rangle. \tag{6.2.2}$$

The angle brackets denote averaging over possible arrangements of all the N scattering centers $\mathbf{R}_1, \ldots, \mathbf{R}_N$ and all possible $\delta\mathbf{R}_1$.

Let L_z be the length of the conductor along the direction of current. The other two, transverse, dimensions are denoted by L_x and L_y. The metal is assumed to be 'dirty', that is, with a high concentration n_i of scattering centers. The free-path length for the elastic scattering, l, is assumed to be many times smaller than any dimension of the sample, but is much greater than the de-Broglie wavelength of the electrons at the Fermi surface:

$$k_F^{-1} \ll l \ll L_i \qquad i = x, y, z. \tag{6.2.3}$$

The maximum effect of the electron interference occurs, obviously, when all the sample's dimensions are smaller than the coherence length L_ϕ (Eq. (6.2.1)), that is, when the motion of the electrons in the entire sample is coherent. Under these conditions the following equation was obtained for $(\delta G_1)^2$ (Altshuler & Spivak, 1985; Feng, Lee & Stone, 1986):

$$(\delta G_1)^2 = 48\left(\frac{e^2}{h}\right)^2 \frac{1}{L_z^4} \sum_{\mathbf{q}} \left[\frac{1}{q^4} - \frac{1}{(q^2 + \alpha_1/l^2)^2}\right]. \tag{6.2.4}$$

Here $\mathbf{q}$ are the wave vectors of the modes (waves) of electron diffusion. Due to boundary conditions (reflection at the surface) the components of these wave vectors are

$$q_i = \pi m_i/L_i, \quad i = x, y, z; \quad m_{x,y} = 0, 1, 2, \ldots; \quad m_z = 1, 2, \ldots. \tag{6.2.5}$$

The quantity α_1 is a characteristic of the displacement of one of the scattering centers, the potential of which is assumed to be short-range:

$$\alpha_1 = \frac{2}{N} C(k_F \delta R), \quad C(x) = 1 - \left[\frac{\sin(x/2)}{x/2}\right]^2. \tag{6.2.6}$$

The derivation of Eq. (6.2.4) is not presented here: the description of the diagrammatic methods used in this derivation is beyond the scope of this book.

At small values of α_1, specifically, when

$$\alpha_1 \ll (\pi l/L_z)^2, \tag{6.2.7}$$

Eq. (6.2.4) becomes simpler. Since the total number of scattering centers in the sample $N = n_i L_x L_y L_z$, and $n_i l = \sigma_s^{-1} \sim k_F^2$, where σ_s is the scattering cross-section of the impurities, the equation for $(\delta G_1)^2$ takes the form (Altshuler & Spivak, 1985; Feng, Lee & Stone, 1986):

$$(\delta G_1)^2 = \left(\frac{e^2}{h}\right)^2 \frac{l}{(k_F l)^2 L_z^4} C(k_F \delta R) W(L_x, L_y, L_z). \tag{6.2.8}$$

Here

$$W(L_x, L_y, L_z) = \frac{256\pi}{V} \sum_{\mathbf{q}} q^{-6}. \tag{6.2.9}$$

For a wire ($L_x = L_y \ll L_z \ll L_\phi$) the main contribution comes from the vectors $\mathbf{q} = (0, 0, \pi m_z/L_z)$. Therefore $W \simeq L_z^5/L_x L_y$ and (Feng, 1991)

$$(\delta G_1)^2 \simeq 0.88 \left(\frac{e^2}{h}\right)^2 C(k_F \delta R) \frac{l L_z}{(k_F l)^2 L_x L_y}, \qquad \frac{\delta G_1^2}{G^2} \propto \left(\frac{L_z}{L_x L_y}\right)^3. \tag{6.2.10}$$

It is assumed in this equation that the last factor in the r.h.s. is small ($\ll 1$), i.e., the effect of the scattering center's displacement on conductance is smaller than the conductance quantum e^2/h. In the opposite case of a large effect it, nevertheless, can not be greater than the typical magnitude of universal conductance fluctuations $\simeq e^2/h$ (see below).

For a thin film, i.e., $l \ll L_x \ll L_y = L_z \ll L_\phi$, Eq. (6.2.4) yields (Feng, 1991):

$$(\delta G_1)^2 = 0.96 \left(\frac{e^2}{h}\right)^2 C(k_F \delta R) \frac{l}{(k_F l)^2 L_x}, \qquad \frac{\delta G_1^2}{G^2} \propto \frac{1}{L_x^3} \frac{L_z^2}{L_y^2}. \tag{6.2.11}$$

It is interesting that in this case the absolute change of the conductance and its relative change are independent of the area of the film (if $L_y = L_z$).

For a three-dimensional conductor (cube), $l \ll L_x = L_y = L_z \equiv L \ll L_\phi$, a similar calculation gives (Feng, 1991):

$$(\delta G_1)^2 = 0.96 \left(\frac{e^2}{h}\right)^2 C(k_F \delta R) \frac{l}{(k_F l)^2 L}, \qquad \frac{\delta G_1^2}{G^2} \propto \frac{1}{L^3}. \tag{6.2.12}$$

Equations (6.2.10)–(6.2.12), which specify the dependence of $(\delta G_1/G)^2$ on the sample's dimensions for three different geometries, hold only until these dimensions are smaller than the phase-breaking length L_ϕ, which in this system plays the role of the correlation length.

The change of the conductance induced by a hop of a scattering center and the resulting change of the electron interference pattern at large α_1 (Eq. (6.2.6)) can not exceed the magnitude of universal conduction fluctuations $\sim e^2/h$, i.e., can not exceed the effect of a complete change of the arrangement of all scattering centers in the sample. By the same reasons, the effect of a displacement of two scattering centers separated by a distance smaller than L_ϕ is, in general, not additive, i.e., it is not equal to the sum of the effects of displacements of two individual centers.

The change of conductivity due to a displacement of a scattering center and the corresponding change of the electron interference pattern are confined to a region, the greatest dimension of which does not exceed L_ϕ. If even only one dimension of the sample is greater than L_ϕ, the sample may be imagined as divided into incoherent (uncorrelated) blocks: in the quasi-one-dimensional case (wire) $(L_x = L_y \ll L_\phi, L_z > L_\phi)$ into segments L_ϕ long, in the case of a thin film $(L_x \ll L_\phi, \quad L_y, L_z \gg L_\phi)$ into squares with each side $\simeq L_\phi$, and in the case when all dimensions are greater than L_ϕ into cubes with each side $\simeq L_\phi$. The conductance of only that block which contains the displaced scattering center is changed. Equating the total Joule power dissipated in the sample to the sum of Joule powers dissipated in each block, one obtains an expression for the change of the total conductance of the sample:

$$\delta G_1(L_z) = \sum_i (U_i^2/U^2)\delta G_1^{(i)}(L_\phi). \tag{6.2.13}$$

Here U and U_i are the total voltage across the length L_z of the sample and the voltage across the i-th block, respectively, $\delta G_1^{(i)}(L_\phi)$ is the conductance change of the i-th block (linear dimension L_ϕ) due to displacement of a scattering center inside this block. On the derivation of equations for the conductance of inhomogeneous conductors, like Eq. (6.2.13), see also Secs. 7.1 and 7.2.

Since only the conductance of one block is changed (by $\delta G_1(L_\phi)$), and $U_i = (L_\phi/L_z)U$, the change of the conductance equals:

$$\delta G_1(L_z) = (L_\phi/L_z)^2 \delta G_1(L_\phi). \tag{6.2.14}$$

This equation may be obtained more rigorously by substituting in Eq. (6.2.4) $q^2 + L_\phi^{-2}$ for q^2.

The most probable candidates for those scattering centers that are mobile at low temperatures and produce low-frequency fluctuations of the conductance are the so called 'two-level tunneling systems' (Sec. 8.3.3): in each such center an atom or a group of atoms is tunneling back and forth between two states with different positions in the lattice. According to Eq. (1.7.10), the spectral density of fluctuations of the conductance of one block equals:

$$S_G^{(i)}(f) = [\delta G_1(L_\phi)]^2 \frac{4w_1 w_2 \tau}{1 + \omega^2 \tau^2}. \tag{6.2.15}$$

Here $\delta G_1(L_\phi)$ is the change of the conductance of the i-th block due to a transition of a scattering center from one state to another one, τ is the relaxation time of the population of any of the two states, w_1 and $w_2 = 1 - w_1$ are the probabilities of finding the scattering center in the first and the second state, respectively. The product $4w_1 w_2$ equals:

$$4w_1 w_2 = \left[\cosh\left(\frac{\Delta E}{2k_B T}\right)\right]^{-2}, \tag{6.2.16}$$

where ΔE is the difference of the impurity's two energy levels.

Since fluctuations in different blocks are uncorrelated, the spectral density of the total sample's conductance fluctuations is a sum over all blocks (Eq. (6.2.13)):

$$S_G(f) = \sum_i \frac{U_i^4}{U^4} S_G^{(i)}(f).$$
(6.2.17)

At small concentrations n_{mi} of mobile centers, each block is expected to contain no more than one such center. Then the number of nonzero terms in the r.h.s. of Eq. (6.2.17) is $n_{mi}V$, where V is the volume of the sample. Thus

$$S_G(f) = n_{mi} V \left(\frac{L_\phi}{L_z}\right)^4 [\delta G_1(L_\phi)]^2 \left[\cosh\left(\frac{\Delta E}{2k_B T}\right)\right]^{-2} \frac{\tau}{1 + \omega^2 \tau^2}.$$
(6.2.18)

This equation applies to samples at least one dimension of which is much greater than L_ϕ. These samples can not be called 'mesoscopic'. However, the fluctuations of conductance due to the changes of interference patterns in small blocks $\sim L_\phi$ may be essential even in large samples which in other respects do not exhibit mesoscopic behaviour.

If all dimensions of a metal sample are greater than L_ϕ, the indefiniteness of G on the scale of e^2/h is comparatively small and may be neglected. Then a simple equation holds:

$$G = \frac{L_x L_y}{L_z} \sigma = \frac{L_x L_y}{L_z} \frac{ne^2 l}{\hbar k_F} = \frac{L_x L_y}{L_z} \frac{e^2 k_F^2 l}{3\pi^2 \hbar},$$
(6.2.19)

where $n = k_F^3/3\pi^2$ is the electron density. The spectral density of the fractional conductance fluctuations (modulation noise) equals:

$$\frac{S_G(f)}{G^2} = \frac{n_{mi}}{V} \left[\frac{3\pi^2 L_\phi^2}{k_F^2 l} \frac{\delta G_1(L_\phi)}{e^2/\hbar}\right]^2 \left[\cosh\left(\frac{\Delta E}{2k_B T}\right)\right]^{-2} \frac{\tau}{1 + \omega^2 \tau^2}.$$
(6.2.20)

This equation is qualitatively valid even if in each coherent block there are several hopping scattering centers. It is important only that the change of the conductance of a block has to be smaller than e^2/h, i.e., it has to be far from 'saturation'.

The spectral density of relative fluctuations is inversely proportional to all that sample's dimensions which are greater than L_ϕ, in accordance with what should be expected. In fact, these dimensions are not included in $\delta G_1(L_\phi)$, and S_G/G^2 depends on these dimensions only through the volume V. Such dependence of the spectral density of relative fluctuations is specific for macroscopic bodies with dimensions greater than the correlation length (Sec. 1.2).

The most striking feature of the conduction fluctuations mechanism discussed in this section is its sensitivity: an atomic displacement of only one scattering center changes the resistance of a region $\sim L_\phi$ around this center. At low temperatures ~ 1 K the coherence length L_ϕ is a macroscopic length, much greater than the free-path length. This feature is often called the fluctuations' enhancement.

A distinctive feature of this mechanism of conductance fluctuations is also its significant and unusual temperature dependence: as the temperature decreases the spectral density of noise *increases*. This effect is due to the increase of L_ϕ as the temperature T is lowered. In the case of a thin film $(L_x < L_\phi, L_y = L_z \gg L_\phi)$ the change of conductance δG_1 is independent of L_z (Eq. (6.2.11)). Therefore, $\delta G_1(L_\phi)$ is independent of L_ϕ, and $S_G/G^2 \propto L_\phi^4$.

Similarly, using Eq. (6.2.12) one finds that in the case of a cubic sample, each side of which $L \gg L_\phi$, the spectral density of relative fluctuations is $\propto L_\phi^3$.

Obviously, the length L_ϕ and $\delta G_1(L_\phi)$ are not the only temperature-dependent quantities in the r.h.s. of Eq. (6.2.20). The concentration of mobile defects, n_{mi}, and the product of the last factors in Eq. (6.2.20) may vary with temperature in the opposite direction. Nevertheless, the unusual temperature dependence of $1/f$ noise was observed in some experiments and is one of the main arguments in favor of just this mechanism of conductance noise in metals at sufficiently low temperatures (Sec. 8.4.2).

It is instructive to understand the effect of electron waves' interference also in terms of the Feynman interpretation of quantum mechanics (see Feynman & Hibbs, 1965). The application of this interpretation to reproducible universal conductance fluctuations and to noise is presented by Kane, Serota & Lee (1988) and Stone (1989). The basic concept of Feynman interpretation is the amplitude $A(\mathbf{rr}')$ of the particle's propagation from one point (say, $\mathbf{r}$) to another point (say, $\mathbf{r}'$). For example, in an electron diffraction experiment the point $\mathbf{r}$ is the source of electrons and $\mathbf{r}'$ is the location of the detector of electrons or a definite point on the photographic film. According to the Feynman interpretation, the total amplitude $A(\mathbf{rr}')$ is a sum of contributions $A_i(\mathbf{rr}')$ from all possible classical trajectories connecting $\mathbf{r}$ and $\mathbf{r}'$. Each $A_i \propto \exp[(i/\hbar)S_i(\mathbf{rr}')]$, where $S_i(\mathbf{rr}')$ is the classical action for the path i. The probability of the particle passing from $\mathbf{r}$ to $\mathbf{r}'$ equals:

$$P(\mathbf{rr}') \propto |A(\mathbf{rr}')|^2 = |\sum_i A_i(\mathbf{rr}')|^2 = \sum_{ij} A_i(\mathbf{rr}')A_j^*(\mathbf{rr}'). \qquad (6.2.21)$$

Here A^* is the complex conjugate of A.

In the problem under consideration, the two points, $\mathbf{r}$ and $\mathbf{r}'$, can be chosen, for instance, at two opposite leads. Of course, it is assumed that the distance between these leads is smaller than the coherence length L_ϕ and the motion between the two points is coherent. As the electron propagates along a trajectory, the phase of the amplitude grows as $k_F L_i$, where L_i is the length of the i-th trajectory (path). Due to the interaction with the randomly arranged scattering centers (the interaction energy enters into the electron's classical mechanical action) the phases of the amplitudes A_i vary at random. The main contributions to the r.h.s. of Eq. (6.2.21) give the diagonal terms in the last sum, i.e., terms with $i = j$, in which the phase of each A_i is exactly cancelled by the phase of A_i^*. Only these terms survive after averaging of Eq. (6.2.21) over all possible arrangements of the scattering centers. The terms with $i \neq j$ retain in their phases the concrete, random, arrangement of scattering centers in a given specimen. Their total contribution to $P(\mathbf{rr}')$ and, consequently, to the conductance of a specific specimen, not averaged, is, in general, nonzero. Just these terms correspond to the universal conductance fluctuations.

The magnitude of the conductance fluctuations (their variance) and of the noise produced by mobile scattering centers can be expressed in terms of the correlation function:

$$\langle P(\mathbf{r}_1\mathbf{r}_1')P(\mathbf{r}_2\mathbf{r}_2')\rangle - \langle P(\mathbf{r}_1\mathbf{r}_1')\rangle\langle P(\mathbf{r}_2\mathbf{r}_2')\rangle$$
$$\propto \left\{ \left\langle \sum_{ijkl} A_i(\mathbf{r}_1\mathbf{r}_1')A_j^*(\mathbf{r}_1\mathbf{r}_1')A_k(\mathbf{r}_2\mathbf{r}_2')A_l^*(\mathbf{r}_2\mathbf{r}_2')\right\rangle\right.$$
$$\left. - \langle A_i(\mathbf{r}_1\mathbf{r}_1')A_j^*(\mathbf{r}_1\mathbf{r}_1')\rangle\langle A_k(\mathbf{r}_2\mathbf{r}_2')A_l^*(\mathbf{r}_2\mathbf{r}_2')\rangle\right\}. \qquad (6.2.22)$$

Here the angle brackets denote averaging over all possible arrangements of the scattering centers.

Only those combinations of 4 trajectories contribute to the correlation function (Eq. (6.2.22)) in which the phases of the amplitudes are cancelled out. Two types of such trajectories are presented in Fig. 6.2 *a* and 6.2 *b* (Stone, 1989). In both types the trajectories i and j, as well as the trajectories k and l, coincide only near their common starting and final points. In the intermediate region the trajectory j coincides with k, and l with i. This coincidence guarantees the cancellation of the amplitudes' phases and a nonzero contribution to the averaged r.h.s. of Eq. (6.2.22).

The trajectories in Fig. 6.2 *a* form a loop. Within this loop the directions of both trajectories j and k as well as the directions of the trajectories l and i coincide. On the contrary, in the case of the combination in Fig. 6.2 *b*, within the loop the direction of the trajectory j is opposite to that of trajectory k, and the same takes place for the trajectories i and l. In the absence of

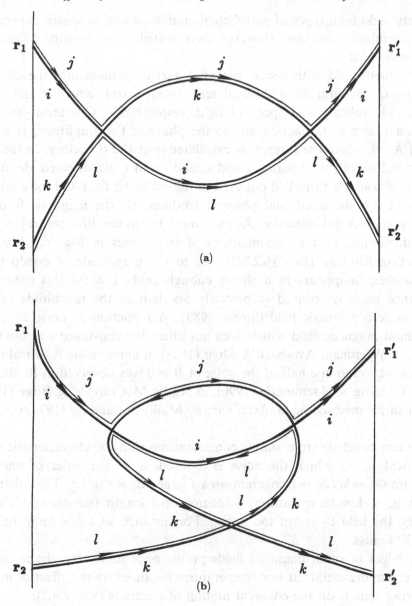

Fig. 6.2 Electron trajectories that contribute to the correlation function of the conductance fluctuations. Four trajectories are labelled by i, j, k, l. $\mathbf{r}_1, \mathbf{r}_2, \mathbf{r}_1', \mathbf{r}_2'$ are initial and final points of the trajectories. Trajectories i and k are intersecting. (a) In the loop the directions of i and l, as well as j and k, coincide. (b) In the loop the direction of i is opposite to that of l, the direction of j is opposite to that of k (see Stone, 1989).

magnetic field both types of paths' combinations, a and b, contribute equally to the correlation function. However, their contributions become different in a magnetic field.

A magnetic field with vector potential $\mathbf{A(r)}$ introduces into the classical Lagrangian function an additional term $(e/c)\mathbf{v} \cdot \mathbf{A(r)}$, where $\mathbf{v}$ and c are the particle velocity and speed of light, respectively. This term yields an additional term to the action and to the phase of the amplitude. It equals $(e/\hbar) \int \mathbf{A} \cdot d\mathbf{l}$, where the integral is calculated over the trajectory. In the case of Fig. 6.2 a both trajectories, j and k, and l and i, are directed identically and the phases are cancelled out even in the magnetic field. On the contrary, in Fig. 6.2 b the additional phases introduced by the magnetic field are additive and are not cancelled. As they are different for different trajectories, the contribution of the combinations of trajectories in Fig. 6.2 b to the correlation function (Eq. (6.2.22)), i.e., to the magnitude of conductance fluctuations, disappears in a strong enough field. Due to this effect, the resistance noise is reduced to precisely one half of the magnitude of this noise in zero magnetic field (Stone, 1989). A reduction of noise by 15 % in a small magnetic field which does not affect the resistance was observed by Alers, Weissman, Averback & Shyu (1989) in amorphous Ni–Zr alloy. A clear reduction to one half of the noise at $B = 0$ was observed in Bi films by Birge, Golding & Haemmerle (1990), in Ag by McConville & Birge (1993), and in single mesoscopic GaAs:Si wire by Mailly & Sanquer (1992) (see Sec. 8.4.2).

One can conclude from simple considerations that the characteristic magnetic field B_c, in which the noise is reduced, is of the order of one flux quantum $\Phi_0 = hc/2e$ per coherent area L_ϕ^2, i.e., $B_c = \Phi_0/L_\phi^2$. The coherence length is, at low temperatures, a macroscopic length (see above). Consequently, the field B_c is not too strong. For instance, at $L_\phi \simeq 1\,\mu m$ the field $B_c \simeq 20$ Gauss.

The effect of small magnetic fields proves once more that the resistance noise in impure metals at low temperatures is caused by the effect of mobile scattering centers on the coherent motion of electrons (Sec. 8.4.2).

6.3 Fluctuations of the number of diffusing scattering centers

One of the possible mechanisms of resistance noise is fluctuations of the number of impurities that scatter the electrons in the specimen and hence affect the resistance. Scofield & Webb (1985) found this phenomenon in thin Nb films containing hydrogen ions H^+, which are known to scatter electrons in Nb. They measured the resistance fluctuations in segments of a

Nb film. The number of hydrogen impurities N_i in each segment fluctuates due to diffusion of H^+ ions, i.e., due to the exchange of ions between the segment and the rest of the film through the segment's ends. To understand the results of the experiments, we have to find the spectral density of the impurities' number fluctuations δN_i in a segment of an infinite strip.

The hopping of the impurities, which results in their diffusion, is a typical Markov process. In fact, the waiting time between two consecutive hops of an impurity atom is usually greater by many orders than the time of crystal relaxation after each such hop. It means that each impurity atom quickly forgets its previous 'history'. Let $\delta n_i(\mathbf{r}t)$ be the fluctuation of the density of impurities at a point $\mathbf{r}$. In the Langevin approach (Sec. 1.10) the kinetics of this fluctuation is governed by the continuity equation:

$$\frac{\partial \delta n_i(\mathbf{r}t)}{\partial t} + \text{div } \delta \mathbf{j}(\mathbf{r}, t) = 0. \tag{6.3.1}$$

Here $\delta \mathbf{j}(\mathbf{r}, t)$ is the fluctuation of the impurities' flow density. The zero r.h.s. of Eq. (6.3.1) is equivalent to an assumption that processes of the mobile centers' trapping to and detrapping from various defects may be neglected.

The flow density consists of two parts:

$$\delta j_\alpha(\mathbf{r}, t) = -D_{\alpha\beta}\nabla_\beta \delta n_i(\mathbf{r}t) + \delta j_\alpha^{\text{int}}(\mathbf{r}t). \tag{6.3.2}$$

The first term in the r.h.s. is the diffusion flow proportional to the gradient of the density fluctuation, $D_{\alpha\beta}$ is the tensor of diffusion coefficients. The second term is the Langevin source of fluctuations (Sec. 1.10). The problem is, as usual, reduced to the calculation of the correlation function of these sources. Their correlation time is the relaxation time of those atomic vibrations that 'push' the impurity atom. It is smaller by many orders than the times between hops.

Consider for simplicity a simple cubic or tetragonal lattice and let $\mathbf{a}_\alpha$ be the mutually perpendicular vectors that connect a site with its three neighbours ($\alpha = x, y, z$). The net flow along a direction α at the lattice site $\mathbf{r}$ is proportional to the difference between the random rates of hopping along $\mathbf{a}_\alpha$ and in the opposite direction:

$$\delta j_\alpha^{\text{int}}(\mathbf{r}, t) = s_\alpha^{-1}[\delta J(\mathbf{r} \to \mathbf{r} + \mathbf{a}_\alpha, t) - \delta J(\mathbf{r} \to \mathbf{r} - \mathbf{a}_\alpha, t)]. \tag{6.3.3}$$

Here s_α is the area of the elementary lattice cell face perpendicular to $\mathbf{a}_\alpha$.

The random Langevin rates, say, $\delta J(\mathbf{r} \to \mathbf{r} + \mathbf{a}_\alpha, t)$ and $\delta J(\mathbf{r}' \to \mathbf{r}' + \mathbf{a}_\beta, t')$ are correlated only if they correspond to the same type of hop, i.e., only if $\mathbf{r} = \mathbf{r}'$ and $\alpha = \beta$, and to one instant of time, $t' = t$. Different hops are uncorrelated, of course, only if the atomic concentration of the impurities is small and their interaction is negligible. The correlation function of the random flows equals:

$$\langle \delta J(\mathbf{r}_1 \to \mathbf{r}_1 + \mathbf{a}_\alpha, t_1) \delta J(\mathbf{r}_2 \to \mathbf{r}_2 + \mathbf{a}_\beta, t_2) \rangle$$
$$= \delta_{\alpha\beta} \delta_{\mathbf{r}_1 \mathbf{r}_2} \delta(t_1 - t_2) J(\mathbf{r}_1 \to \mathbf{r}_1 + \mathbf{a}_\alpha),$$

(6.3.4)

where $J(\mathbf{r}_1 \to \mathbf{r}_1 + \mathbf{a}_\alpha)$ is the mean flow. Hence,

$$\langle \delta j_\alpha^{\text{int}}(\mathbf{r}_1 t_1) \delta j_\beta^{\text{int}}(\mathbf{r}_2 t_2) \rangle$$
$$= s_\alpha^{-2} \delta_{\alpha\beta} \delta_{\mathbf{r}_1 \mathbf{r}_2} \delta(t_1 - t_2) [J(\mathbf{r}_1 \to \mathbf{r}_1 + \mathbf{a}_\alpha) + J(\mathbf{r}_1 \to \mathbf{r}_1 - \mathbf{a}_\alpha)].$$

(6.3.5)

In equilibrium the opposite flows are equal. Each of them can be expressed in terms of the diffusion coefficients of the impurities, $D_{\alpha\beta}$. One can easily find that

$$J(\mathbf{r} \to \mathbf{r} + \mathbf{a}_\alpha) = |\mathbf{a}_\alpha|^{-2} D_{\alpha\alpha} f(\mathbf{r}),$$

(6.3.6)

where $f(\mathbf{r}) = n_i v_c$ is the mean occupancy number of impurities at the site $\mathbf{r}$ of the lattice, n_i is their density, $v_c = s_\alpha |\mathbf{a}_\alpha|$ is the lattice cell volume. Upon switching to continuous coordinates, $v_c^{-1} \delta_{\mathbf{r}_1 \mathbf{r}_2} = \delta(\mathbf{r}_1 - \mathbf{r}_2)$, one obtains:

$$\langle \delta j_\alpha^{\text{int}}(\mathbf{r}_1 t_1) \delta j_\beta^{\text{int}}(\mathbf{r}_2 t_2) \rangle = \delta_{\alpha\beta} \delta(\mathbf{r}_1 - \mathbf{r}_2) \delta(t_1 - t_2) 2 D_{\alpha\alpha} n_i.$$

(6.3.7)

This equation is an analogue of Eqs. (3.5.4) and (6.4.4).

Since the side boundaries of the film are impermeable for the mobile impurities, one can average the density fluctuation over the cross-section of the film with dimensions L_y and L_z. This averaged fluctuation, $\delta \bar{n}(xt)$, satisfies the one-dimensional diffusion equation which is obtained by substitution of Eq. (6.3.2) into Eq. (6.3.1). The Fourier component of the fluctuation is easily found:

$$\delta \bar{n}(k\omega) = \frac{-ik\delta \bar{j}_x^{\text{int}}(k\omega)}{-i\omega + Dk^2}.$$

(6.3.8)

The fluctuation of the total number of impurities in the segment of length L_x equals:

$$\delta N_i(t) = V \frac{1}{L_x} \int_{-L_x/2}^{L_x/2} dx \delta \bar{n}_i(xt).$$

(6.3.9)

The spectral density of these fluctuations readily follows from Eqs. (6.3.7) and (6.3.8) after some simple but lengthy calculations:

$$S_{N_i}(f) = \frac{\pi N_i}{f_c(f/f_c)^{3/2}} \left[1 - \left(\cos \sqrt{f/f_c} + \sin \sqrt{f/f_c} \right) \exp(-\sqrt{f/f_c}) \right]. \quad (6.3.10)$$

Here the characteristic frequency $f_c = D/\pi L^2$ is the inverse time of diffusion of impurities over the length L. One can check that the integration of

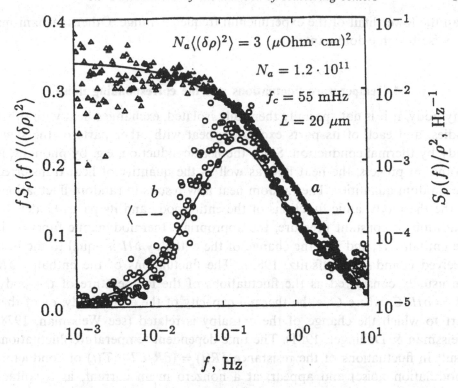

Fig. 6.3 Frequency dependence of the spectral density of fractional resistivity fluctuations in thin Nb films with dissolved hydrogen ions acting as electron scatterers. Curve a: $N_a S_\rho(f)/\rho^2$ (right, logarithmic scale); curve b: $f S_\rho(f)/\rho^2$ (left scale). $N_a = 2n_a L A$, where n_a is the atomic density, L the length of the film segment, A the film cross section. $\langle\delta\rho^2\rangle$ is the variance of resistivity fluctuations found by integration of $S_\rho(f)$. From Scofield & Webb (1985).

$S_N(f)$ over all frequencies yields the total mean number of impurities N_i in accordance with the statistical result for the variance: $\langle(\delta N_i)^2\rangle = N_i$.

At high frequencies, $f \gg f_c$, the spectral density $S_N(f)$ falls off as $f^{-3/2}$. This is a very general and distinctive property of noise caused by diffusion processes (Lax & Mengert, 1960). Indeed, the spectral density of the resistance noise in the experiments by Scofield & Webb (1985) followed this characteristic law (Fig. 6.3).

The noise produced by the number fluctuations of dissolved hydrogen impurities was found also in Pd (Zimmerman & Webb, 1988) and amorphous alloys Ni−Zr (Alers, Weissman, Averback & Shyu, 1989). In Pd at high temperatures ($T > \approx 155$ K) and high frequencies $f > f_c$ the spectral density falls off as $f^{-3/2}$. At low temperatures the diffusion coefficient of impurities D becomes very small, and the characteristic frequency f_c becomes lower than the lower limit of the experimental frequency range. Other mechanisms of noise become dominant.

6.4 Temperature fluctuations and the corresponding noise

Any body, if it is not specially thermally isolated, exchanges heat with other bodies, and each of its parts exchanges heat with other parts of the same body by thermal conduction. Since the heat conduction, e.g. by phonons, is a random process, the heat flow as well as the quantity of heat transferred are random quantities. The random heat flow results in random fluctuations of the thermodynamic functions of the entire body and its parts. Under the condition of constant pressure, the appropriate thermodynamic function is the enthalpy H, and just the change of the enthalpy δH is equal to the heat received (Landau & Lifshitz, 1985). The fluctuations of the enthalpy δH are usually considered as the fluctuations of the temperature of the body, $\delta T = \delta H/C$, where C is the thermal capacity of the entire body, or of that part to which the change of the enthalpy is related (see Weissman, 1978; Weissman & Dollinger, 1981). The time-dependent temperature fluctuations result in fluctuations of the resistance $\delta R(t) = (\partial R/\partial T)\delta T(t)$ of conductors (modulation noise) and appear, at a nonzero mean current, as a voltage noise.

According to the theory of thermodynamic fluctuations (Landau & Lifshitz, 1985), the variance of the temperature fluctuations in a body or in part of it equals:

$$\overline{(\delta T)^2} = \frac{k_B T^2}{C}. \tag{6.4.1}$$

Here T is the mean absolute temperature of the body, C is the heat capacity of the body or the part over which the fluctuations are averaged, k_B is the Boltzmann constant.

The temperature fluctuations satisfy the continuity equation:

$$c\frac{\partial \delta T}{\partial t} + \text{div}\,\delta\mathbf{q} = 0, \tag{6.4.2}$$

where c is the specific heat, $\delta\mathbf{q}(\mathbf{r}, t)$ is the random heat flow. It consists of two parts:

$$\delta\mathbf{q} = -\kappa\vec{\nabla}\delta T + \delta\mathbf{q}^{\text{int}}. \tag{6.4.3}$$

The first term in the r.h.s. is the flow caused by the random temperature gradient, κ is the thermal conductivity. The last term in Eq. (6.4.1) is the Langevin (intrinsic) flow (Sec. 1.10). The application of the Langevin approach is based on the fact that the temperature relaxation time is greater by many orders than the correlation time of microscopic processes of thermal conduction (the free-path-time of phonons, the time of energy exchange between electrons, and so on). According to the theory of hydrodynamic fluctuations (Lifshitz & Pitaevskii, 1980), the correlation function of the components of the Langevin thermal flows in an isotropic medium equals:

$$\langle \delta q_\alpha^{\text{int}}(\mathbf{r}_1 t_1) \delta q_\beta^{\text{int}}(\mathbf{r}_2 t_2)\rangle = 2k_B T^2 \kappa \delta_{\alpha\beta}\delta(\mathbf{r}_1 - \mathbf{r}_2)\delta(t_1 - t_2). \qquad (6.4.4)$$

As usual, the correlation function of the Langevin sources is proportional to delta-functions, because the sources are correlated only on microscopic lengths and times. The correlation function depends only on the local thermal conductivity. Of course, Eq. (6.4.4) is valid only within the hydrodynamic approximation. Specifically, it is assumed that the characteristic lengths of spatial variations of the mean temperature, fluctuations of the temperature, and of the parameters of heat conduction are many times greater than the free-path length of the heat carriers (e.g., phonons, electrons), and the frequencies of both the variation of the mean temperature and the fluctuations measured are many times smaller than the inverse relaxation time of the heat carriers.

The equation for the temperature fluctuations follows from Eqs. (6.4.2) and (6.4.3):

$$c\frac{\partial \delta T}{\partial t} - \kappa \Delta \delta T = -\operatorname{div} \delta \mathbf{q}^{\text{int}}. \qquad (6.4.5)$$

The source of fluctuations in this equation is $-\operatorname{div} \delta \mathbf{q}^{\text{int}}$. It is the only Langevin source in systems in which the temperature fluctuations are caused only by randomness of heat conduction.

One can express $\delta T(\mathbf{r}t)$ in terms of $\delta \mathbf{q}^{\text{int}}(\mathbf{r}'t')$ at instants $t' < t$ by solving Eq. (6.4.5), taking definite boundary conditions into account. Since the correlation function of the Langevin flows is known, the desired correlation function of the temperature fluctuations, $\psi_T(\mathbf{r}_1 t_1, \mathbf{r}_2 t_2)$, at different points $\mathbf{r}_1$ and $\mathbf{r}_2$ and instants t_1 and t_2, is then easily found (for a review of fluctuations due to diffusion processes see van Vliet & Fassett (1965); see also Lax & Mengert (1960); Voss & Clarke (1976)). Obviously, the correlation function depends on the boundary conditions for the temperature fluctuations. How-ever, some conclusions can be made from qualitative physical considerations. If the temperature fluctuations are averaged over a specimen of length L, and the fluctuations arise from random heat flow through the ends of the specimen, the correlation time of the fluctuations is of the order of the time of heat diffusion over the length L, i.e., $\tau_T \sim L^2/D$, where $D = \kappa/c$ is the thermal diffusivity coefficient. The behaviour of the Fourier transform of the correlation function $\psi(\omega, |\mathbf{r}_1 - \mathbf{r}_2|)$ as a function of the distance $|\mathbf{r}_1 - \mathbf{r}_2|$ depends, of course, on the frequency ω of the fluctuations.

Since $2\pi/\omega$ is the period of fluctuations, the temperature fluctuations become uncorrelated if the distance $|\mathbf{r}_1 - \mathbf{r}_2|$ is much greater than the length of diffusion in this period. It means that the distance $\lambda(\omega) \simeq \sqrt{2\pi D/\omega}$ is the correlation length of temperature fluctuations with frequency ω.

The measured temperature fluctuations are always averaged over a finite volume V of the specimen:

$$\delta T(t) = \frac{1}{V} \int_V d\mathbf{r} \delta T(\mathbf{r}t). \tag{6.4.6}$$

Consider a specimen of volume $V = L_1 \times L_2 \times L_3$. If it is not isolated thermally, i.e., is a part of an infinite medium, the spectral density can be easily found. Eq. (6.4.5) is solved by Fourier transformation. Using the correlation function of Langevin sources (Eq. (6.4.4)), the Wiener–Khintchine relation (1.3.8) and Eq. (6.4.6), one obtains:

$$S_T(f) = \frac{4k_B T^2 D}{c} \int \frac{d\mathbf{k}}{(2\pi)^3} \frac{k^2}{\omega^2 + D^2 k^4} \prod_i \frac{\sin^2(k_i L_i/2)}{(k_i L_i/2)^2}. \tag{6.4.7}$$

The case of $L_1 \gg L_2 \gg L_3$ was studied by Voss & Clarke (1976). There are three characteristic frequencies (inverse correlation times) $\omega_i = 2D/L_i^2$. At high frequencies $\omega \gg \omega_2$, the fluctuations in different segments $\sim \lambda(\omega)$ of the lengths L_1 and L_2 are uncorrelated. The values of $|k_1|$ and $|k_2|$ that contribute to the spectral density are smaller than those of $|k_3|$, i.e., $k^2 \approx k_3^2$. The integrals over components of $\mathbf{k}$ are easily calculated (compare with Eq. (6.3.10)):

$$S_T(f) \simeq \frac{2k_B T^2 \sqrt{\omega_3}}{Vc\omega^{3/2}} \{1 - \exp(-\sqrt{\omega/\omega_3}) [\cos\sqrt{\omega/\omega_3} + \sin\sqrt{\omega/\omega_3}]\}. \tag{6.4.8}$$

At high frequencies $\omega \gg \omega_3$, this spectral density function decays as $\omega^{-3/2}$. At smaller frequencies $\omega \ll \omega_3$ (but still $\omega \gg \omega_2$), the spectral density decays as $\omega^{-1/2}$. At $\omega \ll \omega_1$ the random heat flow is three-dimensional. It follows from Eq. (6.4.7) that $S_T(f) \to$ const. at $\omega \to 0$.

An analogue of the fluctuation–dissipation relationship (FDR, Sec. 2.1) is known for temperature fluctuations (Voss & Clarke, 1976; Weissman, 1978). Let us assume that at an initial instant of time heat ΔH is imparted to the body in the form of a short pulse. The initial increment of temperature $\Delta T(0) = \Delta H/C$. Owing to the heat conduction, this change of temperature decays in time. In the linear limit the ratio $\Delta T(t)/\Delta H$ is independent of

ΔH. The correlation function $\psi_T(t)$ of the temperature fluctuations in the absence of external sources of heat and the relaxation function $\Delta T(t)/\Delta H$ are related:

$$\psi_T(t) \equiv \langle \delta T(t)\delta T(0) \rangle = k_B T^2 \frac{\Delta T(t)}{\Delta H}. \qquad (6.4.9)$$

This relationship means that the calculation and/or measurement of the correlation function or spectral density of temperature fluctuations can be reduced to, respectively, calculation and/or measurement of the temperature response $\Delta T(t)$ to a heat pulse produced, for instance, by a current pulse.

The power introduced into the sample may be periodical in time, i.e., $\Delta P(t) = \Delta P(\omega)e^{-i\omega t}$. Using Eq. (6.4.9) and the Wiener–Khintchine relation (1.3.8), one can express the spectral density of temperature fluctuations in terms of the response of the temperature to this power, i.e., $\Delta T(\omega)e^{-i\omega t}$:

$$S_T(f) = 4k_B T^2 \mathrm{Re}\left(\frac{\Delta T(\omega)}{\Delta P(\omega)}\right). \qquad (6.4.10)$$

To keep track of the sample's temperature in the experiment, it is convenient to measure the sample's resistance R and to use the equation $\Delta R(t) = (\partial R/\partial T)\Delta T$. The correlation function of resistance fluctuations caused by the temperature fluctuations, according to Eq. (6.4.9), equals (Weissman, 1978):

$$\psi_R(t) = k_B T^2 \frac{\partial R}{\partial T}\frac{\Delta T(t)}{\Delta H}. \qquad (6.4.11)$$

Some calculations and measurements of the spectral density of temperature fluctuations have been performed to verify the hypothesis, according to which the source of fluctuations with an $1/f$-type spectrum are the fluctuations of temperature (Sec. 8.8). The conclusion was, in general, negative: the $1/f$ noise is not generated by the temperature fluctuations.

The equilibrium temperature fluctuations only have been examined above. When the temperatures at the ends of the conductor differ considerably, and hence a mean gradient of the temperature and a thermal flow between these ends appears, a more general approach to the calculation of temperature fluctuations is required. The main idea is that the correlation function of the Langevin sources is given by Eq. (6.4.4), where T is substituted by the local temperature $T(\mathbf{r})$. For a review of nonequilibrium hydrodynamic fluctuations see Schmitz (1988).

7
Noise in strongly disordered conductors

An enormous variety of conductors can be called strongly disordered. First of all, these are conductors prepared of macroscopically randomly inhomogeneous materials, i.e., various composites: dielectrics (polymers, plastics) into which in some proportion fine particles of metal or graphite are introduced, say, before hardening. Such material becomes conductive due to contacts between the particles of the conducting substance forming continuous chains between the electrodes attached to the whole sample. One may attribute to such conductors also cermets, polycrystalline solids in which each crystallite is strongly anisotropic (the material as a whole may be isotropic owing to random orientation of the crystallites), metallic films subjected to sandblasting (sand particles randomly remove some parts of the film), etc. In the simplest case the conducting particles are homogeneous within their boundaries, their dimensions are many times larger than the free-path length of the charge carriers, and the contacts between them are perfect. Real disordered materials, of course, are usually much more complex, and the simplest case described above can be considered merely as a model.

Semiconductors with hopping conductivity may also be attributed to strongly disordered conductors. The charge carriers are hopping between impurity centers by tunneling. Owing to the random distribution of impurity centers and strong, exponential, dependence of the tunneling probability on the distance between the impurities, the current paths are extremely intricate, as in randomly inhomogeneous media.

The static conductivity of disordered mixtures, one of the components of which is an insulator, tends to zero as the fraction p of the conductive component approaches from above a critical *nonzero* value p_c. This phenomenon is a kind of metal–insulator transition. Many of its features are similar to those of a second-order phase transition in the vicinity of its critical point. It is the most striking phenomenon in this field.

The current noise produced by resistance fluctuations was measured in a great number of strongly disordered systems as a function of the composition of the conductor, that is, of the fraction of the conductive component in the dielectric host, and also of the electric and magnetic fields and temperature. The dependence of the noise on the concentration of the conductive component is especially pronounced in the vicinity of the metal–insulator transition. The spectrum of the resistance fluctuations $S_R(f)$ in strongly disordered conductors is usually of $1/f$ type (Ch. 8).

In Sec. 7.1 the basic ideas of the theory of conduction in strongly disordered conductors are outlined. Sec. 7.2 is devoted to the theory of resistance noise in such conductors. In Sec. 7.3 a brief review of experiments in this field is given.

7.1 Basic ideas of the percolation theory

The physics of the mean electric conductivity and of conductivity fluctuations in real strongly disordered conductors is, in general, very complicated and specific for each type of such conductors. However, there are common features which can be understood by examining the simplest models. We restrict our consideration to two of them only:

1. The model of a random lattice (network) of resistors which exhibits the percolation phenomenon. The mathematical problems and the very term 'percolation' have been introduced by Broadbent & Hammersley (1957) in connection with the problems of percolation of gases or liquids through porous filters. Comprehensive explications of the theory of percolation and its applications can be found in several books and reviews (Shklovskii & Efros, 1984; Isichenko, 1992; Stauffer & Aharoni, 1992, and references therein).

2. The model of a continuous medium whose local conductivity (tensor) $\sigma_{\alpha\beta}(\mathbf{r})$ (or local resistivity $\rho_{\alpha\beta}(\mathbf{r})$) is a random function of the radius-vector $\mathbf{r}$.

The simplest model of a strongly disordered conductor is a two-dimensional square lattice of conductors (Fig. 7.1). In each site of the lattice four conducting bonds come together connecting this site with four neighbouring ones. Consider a large lattice (the number of periods along each dimension $L \gg 1$) to the opposite edges of which electrodes are soldered and a voltage is applied across these electrodes. Let us break *at random* the bonds connecting the lattice sites. Each broken bond is assumed to have zero conductance. Let p be the fraction of unbroken bonds. As p decreases from 1, the resistance of the lattice increases. It has been rigorously proved in the theory of

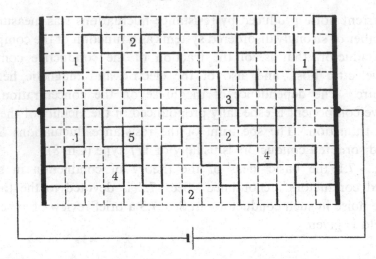

Fig. 7.1 The random resistor network (bond problem). Thick and thin lines correspond to unbroken and broken bonds (conductors), respectively. 1: 'bottlenecks';
2: isolated clusters; 3: a segment of the IC; 4: 'dead ends'; 5: 'shunts'.

percolation that the resistance $R(p)$ becomes infinitely large (current between
the electrodes disappears) at a definite critical value of p which is denoted
by p_c. It depends on the type of the lattice and its Euclidean dimension d.
It follows from symmetry considerations that in the case of a square lattice
($d = 2$) with randomly broken bonds $p_c = 1/2$.

In another lattice model the intersections (contacts) of the bonds, i.e., the
lattice sites, are broken at random. The bonds that initially converged in the
broken site become 'dead ends' and are eliminated from current conduction.
The models with broken bonds and broken sites are called the bond and the
site percolation problem, respectively.

In order to understand the mechanism of the dependence $R(p)$ let us,
starting from $p = 0$ (all bonds broken), successively increase p by connecting
the lattice sites at random. At small $p \ll 1$, obviously, almost all bonds are
lonely (without adjacent neighbours), and there is no conductance through
the entire lattice between the electrodes. At greater values of p the number
of complexes (clusters) of two, three and more connected unbroken bonds
increases. It is reasonable to consider an unbroken bond as a channel which
is able to pass a liquid from one site to its neighbour. If a site is wetted, all
sites connected with it by unbroken bonds, that is all sites belonging to the
same cluster, are also wetted. As p is increased, the number of large clusters
is also increased. However, at $p < p_c$ all clusters, even in an infinite lattice
($L \to \infty$), are finite, i.e., consist of a finite number of sites. Only at $p = p_c$
does an infinite cluster (IC) appear which threads the entire lattice. If a site

Table 7.1. *Percolation thresholds for some lattices*

Lattice	Bond problem	Site problem
Square	$1/2^a$	$0.59273(6)^b$
Triangular	0.347296^a	$1/2^a$
Honeycomb	0.652701^a	0.697^c
Simple cubic	0.2488^d	0.3117^e
Body-centered cubic	0.18025^d	0.2464^e
Face-centered cubic	0.1198^e	0.1998^e
Diamond	0.3886^e	0.4299^e

[a] Sykes & Essam (1963, 1964)
[b] Kertesz (1986)
[c] Sykes, Gaunt & Glen (1976)
[d] Adler, Meir, Aharony & Harris (1990)
[e] Gaunt & Sykes (1983)

Table 7.2. *Critical indexes for lattice percolation systems*

	$d = 2$	$d = 3$
β	$5/36^a$	0.405^b
ν	$4/3^a$	0.905^b
t	1.32 ± 0.05^c	1.95 ± 0.1^c
κ	1.339^d	2.332^d

[a] Nienhuis (1982)
[b] Adler, Meir, Aharony & Harris (1990)
[c] Murat, Marianer & Bergman (1986)
[d] Rammal, Tannous, Breton & Tremblay (1985)

which happened to belong to the IC is wetted, an infinite number of sites in the lattice become wet. When this is applied to electrical conductivity, it means that a current appears between infinitely distant electrodes ($L \to \infty$) just at $p = p_c$, therefore p_c is called the percolation threshold. The values of p_c for some lattices, and for two lattice models, bond and site problems, are presented in Table 7.1. Only the exact or the latest calculated values are given. For a comprehensive summary of data see the review by Isichenko (1992).

In the vicinity of the threshold, but above it, the IC is a sparse network (Fig. 7.1) in the pores of which finite clusters are located. As $p \to p_c$, the fraction $P(p)$ of all bonds that belong to the IC tends to zero:

$$P(p) \propto (p - p_c)^{\beta}. \tag{7.1.1}$$

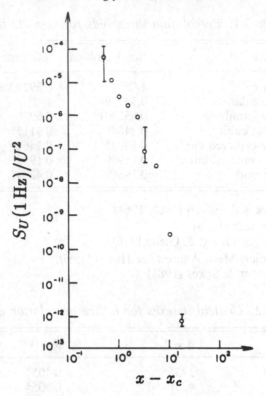

Fig. 7.2 Logarithmic plot of spectral density of relative voltage fluctuations in wax-carbon mixtures versus $x - x_c$, where x is the volume concentration of carbon, $x_c = 0.108$ is the critical value of x. From Chen & Chou (1985).

The power, or critical index, β was calculated for various types of lattices. The results are presented in Table 7.2. As in Table 7.1, only the exact or the latest calculated values are given. For a detailed summary of the data see Isichenko (1992).

A very important characteristic of any percolation system is its correlation length L_c. It is defined as follows. Choose two lattice sites: one at the point $\mathbf{r}_i$, the other at $\mathbf{r}_i + \mathbf{r}$, where $\mathbf{r}$ is a definite vector. Let the function $g(\mathbf{r}_i, \mathbf{r}_i + \mathbf{r})$ be 1 if the two sites belong to one cluster (wetting one site results in wetting the second), and is zero if the sites do not belong to one cluster. Let us sum this function over all sites $\mathbf{r}_i$ of the lattice, at a given $\mathbf{r}$, and divide by the number of sites, that is, average it at a constant $\mathbf{r}$. The result is the correlation function $G(r)$ which depends only on $r = |\mathbf{r}|$ because the lattice is, on average, isotropic. $G(r)$ is a decreasing function of r. The characteristic value at which the power-law decrease turns into an exponential one is called the correlation length L_c. As $p \to p_c$ from above, the correlation length increases according to a power law:

$$L_c \propto (p - p_c)^{-\nu}. \tag{7.1.2}$$

The power ν is called the index of the correlation length. The properties of the disordered lattice scale with its linear dimension L in the same way as the properties of a macroscopic body only if $L \gg L_c$. In the opposite case, $L < L_c$, the dependence of the properties on L is different.

A formal expression for the resistance of the entire lattice is derived by equating the entire Joule power $I^2 R$ to the sum of Joule powers dissipated in the resistances r_m which form the lattice network:

$$R = \sum_m r_m i_m^2 = \sum_m r_m^{-1} U_m^2 / I^2. \tag{7.1.3}$$

Here i_m is the current passing through the resistance r_m divided by the total current I through the lattice, U_m is the voltage across the resistance r_m. Equation (7.1.3) is only a formal relation until the currents i_m or voltages U_m are not calculated. The second Eq. (7.1.3) is used for computer modeling of the percolation lattice and, thus, for calculation of the resistance R.

The conductivity of the percolation lattice is determined only by the IC. Therefore it tends to zero as $p \to p_c$ because the IC becomes more and more sparse (Eq. (7.1.1)). The conductance tends to zero according to a power law:

$$G = R^{-1} \propto (p - p_c)^t. \tag{7.1.4}$$

The power t is called the index of conductivity. If the IC were a regular network, as a superlattice, and almost all sites and bonds belonging to the IC participated in current conduction, the conductivity R^{-1} would fall, as $p \to p_c$, according to the same law as the number of sites belonging to the IC, that is, the index t would be equal to β. However, only a part of the IC, the so called 'backbone', is involved in the conduction of current. The so called 'dead ends' (side branches of the IC 'backbone' which carry no current), and 'shunts' which do not change noticeably the conductivity of the parts of the IC they are shunting (Fig. 7.1), in fact, do not participate in the conduction. Therefore t is significantly greater than β (see Table 7.2).

For definiteness, the lengths L of the lattice are usually taken identical in all d directions, where d is the Euclidean dimension of the lattice. If the length $L \gg L_c$, the resistance of the entire lattice $R \propto L^{2-d}$ as the resistance of any macroscopic uniform conductor with all linear dimensions equal. However, if $L \ll L_c$ (but, certainly, $L \gg 1$) the resistance in the vicinity of the percolation threshold depends on L according to a different power law, $R \propto L^{-\beta_L}$. The index β_L differs from $d - 2$.

In the model of random continuous media the characteristic length of the resistivity $\rho_{\alpha\beta}(\mathbf{r})$ variations in space is assumed to be much greater than the free-path length of charge carriers, the screening length of electric fields and other microscopic lengths. The Joule power I^2R dissipated by the current I in the conductor with resistance R can be represented equivalently in terms of the local current density $\mathbf{j}(\mathbf{r})$ and local electric field $\mathbf{E}(\mathbf{r})$ (compare with a similar procedure in lattice systems, Eq. (7.1.3)):

$$I^2R = \int dV \mathbf{j}(\mathbf{r})\mathbf{E}(\mathbf{r}) = \int dV\, j_\alpha(\mathbf{r})\rho_{\alpha\beta}(\mathbf{r})j_\beta(\mathbf{r}) = \int dV\, E_\alpha(\mathbf{r})\sigma_{\alpha\beta}(\mathbf{r})E_\beta(\mathbf{r}).$$

(7.1.5)

Two identical expressions for the resistance of this randomly inhomogeneous sample follow from Eq. (7.1.5):

$$R = \int dV e_\alpha(\mathbf{r})\sigma_{\alpha\beta}(\mathbf{r})e_\beta(\mathbf{r}) = \int dV i_\alpha(\mathbf{r})\rho_{\alpha\beta}(\mathbf{r})i_\beta(\mathbf{r}).$$

(7.1.6)

Here $\mathbf{e}(\mathbf{r}) = \mathbf{E}(\mathbf{r})/I$, $\mathbf{i}(\mathbf{r}) = \mathbf{j}(\mathbf{r})/I$. Equation (7.1.6) is general but formal (as Eq. (7.1.3)) because the quantities $\mathbf{e}(\mathbf{r})$ and $\mathbf{i}(\mathbf{r})$ are not expressed in terms of local conductivities or local resistivities, and R is not expressed in terms of the statistical characteristics of the local conductivities or resistivities. This is accomplished, for instance, by computer simulations.

7.2 Resistance fluctuations in percolation systems

The theory of resistance fluctuations in lattice percolation systems was developed by Rammal (1985), and by Rammal, Tannous, Breton & Tremblay (1985) (see also the review by Bergman & Stroud, 1992). The fluctuations $\delta R(t)$ of the entire lattice resistance are assumed to be produced by the fluctuations $\delta r_m(t)$ of the bonds' resistances r_m.

A formal equation for δR can be derived by linearization of Eq. (7.1.3). One has to take into account that the Joule power is extremal in the variations δi_m of the currents at given resistances r_m. Hence the term linear in δi_m is zero. Thus

$$\delta R(t) = \sum_m i_m^2 \delta r_m(t).$$

(7.2.1)

In those disordered systems, the models of which are the lattice percolation systems, the microscopic correlation lengths (e.g., the free-path length of the charge carriers) are usually smaller than the linear dimensions of the particles of which the heterogeneous conductor is composed. That is why the fluctuations in different conductors of the lattice network can be considered as uncorrelated. It is assumed also that the resistances of all unbroken bonds are identical and equal to r. The spectral density of fluctuations of these resistances is twice the Fourier transform of the correlation function:

$$2\langle\delta r_m(t_1)\delta r_n(t_2)\rangle_\omega = \mathfrak{z}(f)r^2\delta_{mn}. \tag{7.2.2}$$

Here $\mathfrak{z}(f)$ is the spectral density of fractional fluctuations of the resistance r. The spectral density of the fractional fluctuations of the entire lattice resistance follows from Eqs. (7.2.1) and (7.2.2):

$$\tilde{S}_R(f) \equiv \frac{S_R(f)}{R^2} = \mathfrak{z}(f)\frac{\sum_m i_m^4}{\left(\sum_m i_m^2\right)^2}. \tag{7.2.3}$$

The function $\mathfrak{z}(f)$ is determined by the mechanism of fluctuations in the resistances forming the lattice (the dependence of this function on f is usually of $1/f$ type, see Ch. 8). It does not depend on p and is beyond the scope of the theory presented in this chapter. The dependence on p is determined by the last factor in the r.h.s. of Eq. (7.2.3), which can be calculated by computer simulation of the random lattices.

As p approaches the percolation threshold p_c, the spectral density of fractional resistance fluctuations diverges. If the dimensions L of the conductor are large as compared with the correlation length L_c,

$$\tilde{S}_R \propto (p - p_c)^{-\kappa}. \tag{7.2.4}$$

The exponent κ is a critical index which has been found by computer simulations (Table 7.2).

Let us discuss this phenomenon qualitatively. If the conductor is strongly disordered, the electric field and the current density are strongly inhomogeneous, especially in the vicinity of the metal–insulator transition. The most significant contribution to the resistance and to the noise comes from 'bottlenecks' or 'hot spots', in which the local electric field and current density are much greater than the average $\langle E\rangle$ and $\langle j\rangle$, respectively. Occupying a relatively small volume, they significantly determine the total resistance R of the conductor. As is seen from Eq. (7.1.6), the contribution of such a region to R is proportional to the local electric field squared. The contribution of this region to the current noise is, comparatively, even more important: it is proportional to the fourth power of the local field. Computer simulations revealed that in the vicinity of p_c the main contribution to the noise comes from those single bonds which connect large parts of the infinite cluster (IC) and through which a large current i_m passes (Rammal, Tannous, Breton & Tremblay, 1985).

This effect is easily understood. Far from the metal–insulator transition, the number of current paths in the sample is large. As the fluctuations of resistance along different paths are uncorrelated, if they have no common segments, the contributions to the fluctuation of the total resistance tend to cancel one another. As the system approaches the percolation threshold, the number of effective current paths decreases and the effect of their cancellation becomes smaller. This effect is similar to the inverse dependence of the spectral density of relative fluctuations in macroscopic bodies on their volume (Sec. 1.2). In other words, in a strongly disordered conductor the fluctuations of R are determined not by the entire volume V of the conductor but by an essentially smaller volume, and therefore the noise is large.

Let us compare the real dependence of $\tilde{S}_R$ on $p - p_c$ given by Eq. (7.2.4) with the dependence which would be if the IC were a regular superlattice. Suppose that the bonds belonging to the IC form regular chains in the cubic lattice. If that part of all bonds which belong to the IC is $P(p)$, the number of chains of unbroken bonds that connect the electrodes would be $P(p)L^2/3$, L unbroken bonds in each. The current in each such chain divided by the total current equals:

$$i = \frac{3}{P(p)L^2}. \tag{7.2.5}$$

According to Eqs. (7.2.3) and (7.1.1), in this case:

$$\tilde{S}_R = \tilde{s}\frac{3}{P(p)L^2} \propto (p - p_c)^{-\beta}. \tag{7.2.6}$$

However, the critical indexes $\kappa \gg \beta$ (see Table 7.2). It means that as $p \to p_c$ the real $\tilde{S}_R$ increases much faster than in the assumed model of regular chains: a much smaller part of the bonds (much smaller than $P(p)$) determines the noise of the entire disordered lattice.

The correlation length L_c depends on $p - p_c$: it diverges as $p \to p_c$ (Eq. (7.1.2)). If the linear dimension of the conductor $L \gg L_c$, the spectral density $\tilde{S}_R$ (Eq. (7.2.4)) has to be inversely proportional to the volume, as in any macroscopic conductor with dimensions greater than the correlation length, i.e., $\tilde{S}_R \propto L^{-d}$, where d is the Euclidean dimension of the lattice. However, at $L \ll L_c$ the dependence on L is different (Rammal *et al.*, 1985):

$$\tilde{S}^R \propto L^{-b}. \tag{7.2.7}$$

The exponent (index) b is related to the index v of the correlation length (Eq. (7.1.2)) and index κ, by the equation $b = d - \kappa/v$. It means that the relative resistance noise decreases with the lattice dimension L less strongly than in the case of greater $L \gg L_c$. According to the data in Table 7.2, the index b equals ≈ 1.18 and ≈ 1.09 for $d = 2$ and $d = 3$, respectively.

The difference $p - p_c$ in the vicinity of the percolation threshold is very difficult to measure in experiments with real percolation systems: it requires an extremely accurate measurement of, say, the concentration of conductive component of a composite. It is more convenient to consider $\tilde{S}_R$ as a function of the resistance R. Comparing Eqs. (7.1.4) and (7.2.4) one obtains:

$$\tilde{S}_R \propto R^w, \qquad w = \kappa/t. \tag{7.2.8}$$

In the model of a continuous random conductor, the current noise is generated by the fluctuations of the local resistivity $\delta\rho_{\alpha\beta}(\mathbf{r}, t)$. The fluctuations of the entire conductor's resistance equal:

$$\delta R(t) = \int dV \, i_\alpha(\mathbf{r})\delta\rho_{\alpha\beta}(\mathbf{r}, t)i_\beta(\mathbf{r}) = \int dV e_\alpha(\mathbf{r})\delta\sigma_{\alpha\beta}(\mathbf{r}, t)e_\beta(\mathbf{r}). \tag{7.2.9}$$

The fluctuations of the local resistivity produce, of course, fluctuations $\delta\mathbf{j}(\mathbf{r}t)$ of the local current density. However, the stationary distribution of the current $\mathbf{j}(\mathbf{r})$ corresponds to the minimum of Joule power (at given I). Therefore, the integrals linear in $\delta\mathbf{j}$ in the full expression for the variation of R in Eq. (7.1.6) are zero and only Eq. (7.2.9) remains (compare with the derivation of Eq. (7.2.1)).

In the model considered it is assumed (see above) that the length of resistivity variations is many times larger than all lengths of the charge carriers' relaxation. Within the framework of this model the resistivity fluctuations $\delta\rho_{\alpha\beta}(\mathbf{r}, t)$ at different points $\mathbf{r}_1$ and $\mathbf{r}_2$ ($\mathbf{r}_1 \neq \mathbf{r}_2$) can be considered as uncorrelated, that is,

$$\langle\delta\rho_{\alpha\beta}(\mathbf{r}_1, t_1)\delta\rho_{\gamma\delta}(\mathbf{r}_2, t_2)\rangle = F_{\alpha\beta\gamma\delta}(\mathbf{r}_1, t_1 - t_2)\delta(\mathbf{r}_1 - \mathbf{r}_2). \tag{7.2.10}$$

The function $F_{\alpha\beta\gamma\delta}(\mathbf{r}, t_1 - t_2)$ is determined by the mechanism of resistivity fluctuations.

The spectral density of the conductor's resistance fluctuations follows from Eqs. (7.2.9) and (7.2.10) and the Wiener–Khintchine relation (1.3.8):

$$S_R(f) = 2\int dV i_\alpha(\mathbf{r})i_\beta(\mathbf{r})i_\gamma(\mathbf{r})i_\delta(\mathbf{r})F_{\alpha\beta\gamma\delta}(\mathbf{r}, \omega). \tag{7.2.11}$$

In the model of noise in highly disordered percolation systems presented above, the resistances fluctuate (Eq. (7.2.2)) but the fraction p of unbroken bonds (conductors) is assumed to be nonfluctuating in time. In some disordered systems, even the number of unbroken bonds can be fluctuating (Kiss & Svedlindh, 1993). An example is a high-T_c superconductor composed of superconductive grains coupled by Josephson junctions (Sec. 9.2), which can be in a superconductive or in a normal state. Fluctuations of the number of superconductive junctions are equivalent in the percolation model to fluctuations $\delta p(t)$. The dependence of the fractional spectral density $S_R(f)/R^2$

on the specimen's resistance R in this model depends on the specific statistical properties of the fluctuations $\delta p(t)$ and is different from the one in the common percolation model.

A special example of a strongly disordered conductor is a semiconductor with hopping mechanism of conductivity (for a review on hopping conduction see Shklovskii & Efros, 1984). Because the probability of tunneling from one impurity center to another one depends exponentially on the distance between the two centers, the hopping probability for many pairs of impurity centers, even neighbours, is extremely small. Hence, some paths of current flow are, in fact, cut just as in the disordered lattice of resistors. The current flows mainly along paths which form a network similar to the infinite cluster backbone in the lattice model (Sec. 7.1).

At least two mechanisms of hopping conductivity fluctuations can be suggested. In one of them the noise stems from fluctuations of the number of electrons in those impurity centers that belong to the infinite cluster network. These fluctuations are akin to the generation–recombination ones (Ch. 4) and are produced by the electron exchange between the impurity centers belonging to the infinite cluster and those beyond the IC. As the relaxation time of hopping depends exponentially on the hopping distance, it varies for different impurity pairs in a very broad range. The spectral density of resistance noise $S_R(f)$ is a power-law function, $\propto f^{-\gamma}$, in a wide frequency range (Kogan & Shklovskii, 1981). The calculated exponent $\gamma \simeq 0.6$, i.e., it considerably differs from unity. The longest relaxation times in this system correspond to electron exchange with such impurities that are almost isolated, i.e., the distance between them and their nearest neighbours is much greater than the mean distance between impurity centers, $N_D^{-1/3}$, where N_D is the concentration of impurities. As the fraction of such 'isolated' impurities is exponentially small, the frequency $f_{\min}$, at which the power-law behaviour of $S_R(f)$ turns to a constant, depends on N_D. The last can not be too small, otherwise the resistivity becomes immeasurably high. It means that this mechanism can not yield the extremely small $f_{\min}$ necessary to explain $1/f$ noise (Ch. 8).

Another mechanism of resistance fluctuations is the modulation of the energy levels of the impurities belonging to the current path and the potential relief between these impurities by random variations of the charges of nearby traps which, in general, differ from the impurities that determine the hopping conductivity. One can assume that these traps are distributed uniformly in random among the main, doping, impurities. This model is in some respects similar to the disordered lattice model, in which the spectral densities of resistance fluctuations in all resistors are equal but the total current noise is sensitive to the fluctuations of only those resistors which belong to the current path, especially, to 'bottlenecks'. The spectrum of relaxation times of the traps is not limited by the concentration of the main impurities. Hence, such traps are more likely to be the source of $1/f$ noise in hopping conductivity.

The main conclusion of this section is that strongly disordered conductors exhibit, as a rule, a much higher resistance noise than do the homogeneous or only weakly disordered ones with the same mean characteristics, i.e., with the same dimensions and resistance. This phenomenon is a result of a strongly nonuniform distribution of current density and electric field which are confined to some narrow paths, the effective volume of which is a small part of the total volume of the conductor. This effect is extremely pronounced in the vicinity of the percolation threshold, where the resistance steeply increases as a function of the conductor's composition.

7.3 Experiments

The carbon-membrane microphone was, perhaps, the first strongly disordered system in which the current noise was investigated (Christensen & Pearson, 1936). Some authors studied the noise in thin island-like metallic films (Williams & Burdett, 1969; Williams & Stone, 1972; Celasko, Masoero, Mazetti & Stepanescu, 1978; Celasko, Masoero & Stepanescu, 1972) and in metal–insulator composites (Mantese, Goldburg, Darling *et al.*, 1981). The distinctive feature of these conductors is a very high noise level.

The steep increase of the current noise as the portion of the conducting component decreased and approached the percolation threshold was observed first, to the author's knowledge, by Chen & Chou (1985). The conductor was a composite: particles of conducting carbon introduced into wax (insulator). The percolation threshold was found to be at volume concentration of carbon $p_c = 0.108 \pm 0.001$. (Fig. 7.2). The resistance followed a power law, Eq. (7.1.4), with a critical index $t = 2.3 \pm 0.4$. The dependence of the spectral density of relative resistance fluctuations on the mean resistance followed a power law, Eq. (7.2.8), with $w = 1.7 \pm 0.2$ (the spectrum was of $1/f$ type, see Ch. 8). Therefore, the measured value of the index κ in this system is 4 ± 0.6.

Pierre, Deltour, Van Bentum *et al.* (1990) studied conduction and noise in polymer–copper-particle composites in the limit of low concentration of metallic particles, i.e., near the percolation threshold (this paper contains also a review of the field). The spectral density of relative fluctuations S_U/U^2 is a power function of the resistance of the specimen, $\sim R^w$ (Eq. (7.2.8)).

However, in the low resistance range the exponent $w \approx 1.5$, while in the high resistance range $w \approx 1.0$. This behaviour is interpreted as follows. In the low-resistance range (higher metal densities) the dominating electrical contacts are of the metal–metal junction type in the ballistic (Sharvin) limit (Sec. 5.1). The value of $w \approx 1.5$ is just that one which corresponds to such microcontacts between particles. The value of $w \approx 1.0$ corresponds to the expected one for poor highly resistive contacts with a dielectric layer between the particles.

Sandblasting makes initially continuous metallic films on dielectric substrates randomly discontinous. Increasing the dose of sandblasting is qualitatively equivalent to decreasing the fraction p of metal in the film. The current noise in sandblasted Al, Cr, and In films was measured by Garfunkel & Weissman (1985). The values of w found were in the range between 3.4 and 6.

The treatment of metallic films by an ion beam also makes them strongly disordered. The measurements of the resistance R and current noise $\tilde{S}_R$ in Au films yield the value of $w = 2.1 \pm 0.1$, as well as in Ag films (Koch, Laibowitz, Alessandrini & Viggiano, 1985). In the case of Ag films grown by sputtering, $w \approx 0.9$.

A different method for variation of the disorder was used by Cohen, Ovadyahu & Rokni (1992) and Cohen & Ovadyahu (1994). The resistance of thin ($\simeq 13$ nm) In_2O_{3-x} films decreases upon exposure to ultraviolet radiation due to reversible removal of oxygen from the oxide lattice. The measured noise was, as usually, of $1/f$-type (Ch. 8). The dimensionless characteristic of $1/f$ noise magnitude, the parameter $\alpha = fN_aS_U/U^2$, where N_a is the number of atoms in the sample, was found to increase from $\sim 10^4$ up to $\sim 10^7$ when the samples' resistance increased by only a factor of $\simeq 2$. The samples are believed to be metallic in this range of resistance. The noise parameter α increased further with resistance less steeply up to $\sim 10^8$ (in common crystalline metals $\alpha \sim 10^{-3} - 10^{-2}$).

The dramatic increase of resistance noise by three orders in a narrow range of resistance is interpreted by Cohen, Ovadyahu & Rokni and by Cohen & Ovadyahu as a result of Anderson transition, i.e., a metal–insulator transition due to the increase of disorder of the potential in which the electrons are moving (Anderson, 1958; for a review see Lee & Ramakrishnan, 1985).

8
$1/f$ noise and random telegraph noise

For this, indeed, is the true source of our ignorance—
the fact that our knowledge can only be finite,
while our ignorance must necessarily be infinite.
Karl Popper, Lecture to the British Academy,
January 20, 1960.

8.1 Introduction

In 1925 J.B. Johnson, studying the current fluctuations of electronic emission
in a thermionic tube with a simple technique, found, apart from the shot
noise whose spectral density was independent of frequency and was in
agreement with the Schottky formula (1.5.10) (Schottky, 1918), also a noise
whose spectral density increased with decreasing frequency f (Johnson,
1925). Schottky (1926) suggested that this last noise arises from slow random
changes of the thermocathode's surface, and proposed for this kind of noise
the name 'flicker effect', or 'flicker noise'. The same type of current noise
spectrum was found also in carbon microphones (Christensen & Pearson,
1936), and later, in the 1940s and 1950s, in various semiconductors and
semiconductor devices. It has become evident that the flicker noise is a very
often encountered, if not universal, phenomenon in conductors.

Up to the present, measurements of the current noise spectra have been
performed on a vast number of semiconductors, semiconductor devices,
semimetals, metals in normal state, superconductors and superconductor
devices, tunnel junctions, strongly disordered conductors etc. One observes,
in practically all cases, an increase of the spectral density of current noise
with decreasing frequency f approximately proportional to $1/f$, down to the
very lowest frequencies at which the measurements of the spectral density
have been performed. Therefore this current noise is usually called $1/f$ noise
(or of $1/f$ type). The term proposed by Schottky (flicker noise) and the term
'excess noise' are now more rarely used.

$1/f$ noise poses many problems. Why does the spectral density increase
infinitely with decreasing frequency: does a minimum nonzero frequency of
this growth exist? The variance of the fluctuating quantity which is equal to
the integral of spectral density over all frequencies seems to be infinite if the

$1/f$ behaviour holds down to zero frequency: how to resolve this paradox? What are the sources of $1/f$ noise? Why does the $1/f$ noise (or noises) found in very different physical systems (insulators, semiconductors, normal metals, superconductors, various structures, etc.) have so many common features, why are their low-frequency spectra so similar? Does any general equation exist for the intensity of $1/f$ noise in terms, for instance, of the number of charge carriers in the sample? Is $1/f$ noise indeed universal or, in other words, is it an inherent property of all systems, and does some universal minimum level of this noise exist? Does $1/f$ noise arise in a conductor in which the charge carriers are scattered only by phonons and by *immobile* (fixed) scattering centers? The fluctuation processes are known, in general, to be connected with dissipation and relaxation processes in the same physical systems: to what other physical phenomena is $1/f$ noise related, in what other physical phenomena do the physical mechanisms of $1/f$ noise exhibit itself? Is $1/f$ noise a property of some chaotic dynamical systems with a finite number of degrees of freedom and governed by purely dynamical equations, or is $1/f$ noise found, in fact, only in kinetic systems each of which is composed of a huge number of atoms and has a great number of degrees of freedom? In the author's opinion, answers for most of these questions are already known. They will be presented in this chapter.

Many properties of $1/f$ noise are well known for many classes of conductors and electronic devices. To these properties belong, first of all, the frequency dependence of the spectral density and the time dependence of the correlation function (Sec. 8.2.1), the statistical properties of this noise and its stationarity (Sec. 8.2.2). Although usually the observed quadratic dependence of the spectral density on the applied voltage in uniform ohmic conductors indicates that the noise is caused by fluctuations of the resistance which are independent of the mean current (the current only 'reveals' these fluctuations), this fact was proved by special experiments (Sec. 8.2.3). An important property of the noise, the symmetry (anisotropy) of the conductance fluctuations in crystals, was also examined in interesting experiments (Sec. 8.2.4). In almost all experiments aimed to measure the correlation of $1/f$ fluctuations at different points it has not been detected. It means that the correlation length (radius) proved to be very small (Sec. 8.2.5).

There is another part of the problem of $1/f$ noise: what precisely is producing this noise and causing the observed current or magnetic noise? In the author's opinion, in some systems, for instance, in spin glasses and in metals, the mechanism of $1/f$ noise is already, to some extent, known (Secs. 8.4 and 8.6). However, for the absolute majority of systems exhibiting $1/f$ noise, this part of the problem has been solved to a much lesser degree, and the search of mechanisms of this noise is being intensively continued to the present day.

The great interest in studies of 1/f noise (the number of papers on 1/f noise is huge, and it is no less than one third of all papers on noise problems) stems from the fact that this problem is very general and to a considerable degree unsolved. The great practical importance of these studies also plays a significant role, since 1/f noise is a serious interference that limits the parameters of many electronic devices at low frequencies. It also affects the operation of such high-frequency devices as frequency standards (quartz resonators and atomic standards): their operation frequency undergoes random fluctuations with a 1/f-type spectrum. For this reason, for instance, the relative inaccuracy of the most accurate time measurements in physics cannot be reduced below a certain value as one increases the measured time interval, and this limit is imposed by 1/f noise in the electronic equipment of the devices (Sec. 8.9).

The list of reviews on 1/f noise includes: Malakhov (1959); Press (1978); van der Ziel (1979); Bell (1980); Dutta & Horn (1981); Hooge, Kleinpenning & Vandamme (1981); Keshner (1982); Kogan (1985); Weissman (1988); Weissman (1993).

8.2 Some general properties of 1/f noise

8.2.1 Frequency-dependence of the spectral density and time-dependence of the correlation function

The principal feature of 1/f noise is that its spectral density $S(f)$ increases with decreasing frequency f down to the lowest frequencies to which one can conduct measurements, and no levelling of $S(f)$ onto a plateau can be seen. Caloyannides (1974) has measured the excess noise spectrum of operational amplifiers down to the record low frequency of $\simeq 0.5\,\mu\text{Hz}$ (Fig. 8.1). Dutta & Horn (1981) cite an unpublished study in which the 1/f spectrum has been followed down to 10^{-7} Hz. Such measurements require times of averaging of the order of a month, and are technically very difficult.

At high frequencies 1/f noise 'sinks' into the generation–recombination or shot noise, whose spectral density at the same frequencies is independent of f. The frequency at which the 1/f noise becomes negligibly small depends on the intensity of this noise, and in various systems ranges from $\sim 10^2$ to $\sim 10^6$ Hz.

Often the experimental spectrum is approximated by a power function $f^{-\gamma}$. A typical spectrum of current noise in a VO_2 specimen is shown in Fig. 8.2.

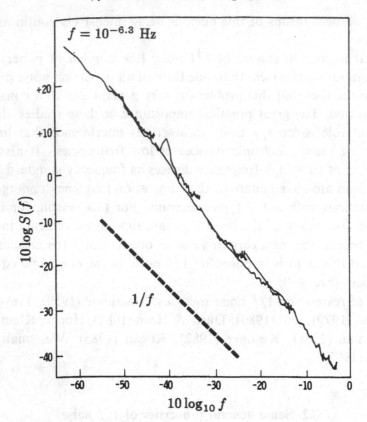

Fig. 8.1 Spectral density of noise in an operational amplifier vs frequency f. Dashed line: ideal $1/f$ spectrum. The minimum frequency $f_{min} = 10^{-6.3}$ Hz. From Caloyannides (1974).

With a high accuracy it can be fitted to a power law with $\gamma = 1.188 \pm 0.002$. Usually this exponent is very close to unity. This feature justifies the term $1/f$ noise. For example, in continuous metal films (of thickness $\simeq 0.1\,\mu$m) at room temperature, the values of γ averaged over many specimens are (Fleetwood & Giordano, 1983): 1.19 ± 0.07 (Ag), 1.17 ± 0.09 (Cu), 1.08 ± 0.08 (Au), 1.20 ± 0.08 (Au–Ag alloy), 1.14 ± 0.08 (In), 1.16 ± 0.11 (Sn), 1.15 ± 0.10 (Pb), 1.15 ± 0.07 (Pt). The exponent is very close to unity in carbon resistors and in many other systems. However, often γ differs appreciably from unity, reaching $\simeq 0.8$ in island films of Pt (Williams & Stone, 1972), in inversion layers on silicon (Table 8.1), etc., and $\simeq 1.45$ in Au film on sapphire at $\simeq 150$ K (Eberhard & Horn, 1978). Values of γ in various systems are presented in Table 8.1.

The spectra of low-frequency noise depend on temperature. Values of γ measured in Ag, Au, Ni, and Cu in the temperature range from 200 to

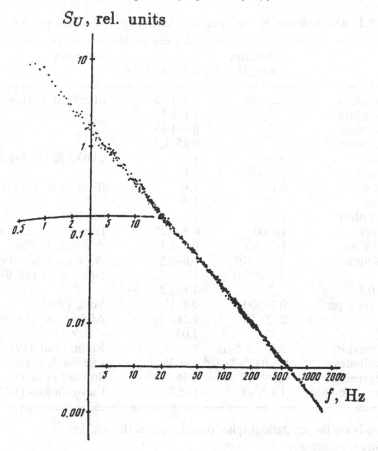

Fig. 8.2 Spectral density of low-frequency voltage noise in VO_2. The frequency exponent $\gamma = 1.188 \pm 0.002$. From Andreev *et al.* (1980).

600 K are shown in Fig. 8.3 (Eberhard & Horn, 1978). The inaccuracy in determining γ, which was found to be ± 0.06, is considerably smaller than the observed variations of γ. The difference between the values of γ (1.188 $\pm$ 0.002 and 1.05 $\pm$ 0.01) measured in different specimens of VO_2 (Andreev, Zakharchenya, Kapshin *et al.*, 1980) also exceeds the experimental inaccuracy.

In many noise spectra one finds also small 'ripples' and 'waviness' (Vande Voorde, Iddings, Love *et al.*, 1979). Such nonmonotonic behaviour of the derivative dS/df most often is within the limits of experimental accuracy, but it sometimes exceeds these limits. Fig. 8.4 shows the current noise spectra of several *n*-InSb specimens having similar electrophysical parameters, while Fig. 8.5 shows the noise spectra of the same specimens in a magnetic field (Vande Voorde & Love, 1981). Despite the similarity of the electrophysical

Table 8.1. *Dependence of the spectral density of 1/f noise on frequency*

Noise source	Frequency range, Hz	$\gamma = -d\ln S/d\ln f$	Reference
Ag on sapphire	0.25–50	0.9–1.15	Eberhard & Horn (1978)
Cu on sapphire	"	1.1–1.3	"
Au on sapphire	"	0.9–1.45	"
Ni on sapphire	"	0.85–1.2	"
Sn on glass	0.1–50	1	Clarke & Hsiang (1976)
Pb on glass	1–1000	1.1	"
Cr on glass	0.1–50	1.1	Black *et al.* (1981)
Bi on glass	"	1.0	"
Au on sapphire	1–100	1.0–1.1	Scofield *et al.* (1981)
Bi whiskers	1–100	0.9–1.2[a]	Leemann *et al.* (1980)
Cr island films	$1\text{-}2\cdot10^5$	0.9–1.3	Palenskis (1976)
Pt island films	$30\text{-}3\cdot10^4$	0.8–1.2	Williams *et al.* (1972)
Ge	$3\cdot10^{-2}\text{-}1$	1	Dilmi *et al.* (1979)
Ge, $B = 0.5$ T	"	1.1–1.2	"
Si inversion layers	0.2–500	0.8–1[b]	Voss, 1978b
VO_2	$2\text{-}2\cdot10^4$	1.188	Andreev *et al.* (1980)
VO_2	"	1.05	"
Carbon resistors	$2.5\cdot10^{-4}\text{-}10$	1	Rollin *et al.* (1953)
Ge photodiodes	$2\cdot10^{-3}\text{-}2\cdot10^4$	1–1.28	Lifshits & Pervova (1959)
Bipolar transistor	$10^{-4}\text{-}10^{-1}$	0.86	Stoisiek *et al.* (1980)
Operational amplifier	$10^{-6.3}\text{-}1$	1–1.3[c]	Caloyannides (1974)

[a] γ depends on the crystallographic orientation of the whisker
[b] Hopping conduction
[c] In the greater part of the frequency range $\gamma = 1.3$

parameters, the noise spectra of different specimens differ not only in the noise intensity, but even in form. For some specimens the noise spectra do not fit a power-law dependence on f. The magnetic field alters not only the magnitude but also the very form of the noise spectra.

One may conclude that the experimental data indicate rather that there is no universal spectrum of low-frequency current noise. The term 'spectra of $1/f$ type' is therefore more justified.

As soon as the spectral density $S_x(f)$ of a stationary noise $x(t)$ is known, one can use the Wiener–Khintchine theorem (Eq. (1.3.8)) to find the correlation function $\psi_x(t_1 - t_2)$ (Eqs. (1.2.1) and (1.2.2)) as a function of time difference $t_1 - t_2$:

$$\psi_x(t) = \langle \delta x(t)\delta x(0)\rangle = \int_0^\infty df \cos(\omega t)S_x(f). \qquad (8.2.1.1)$$

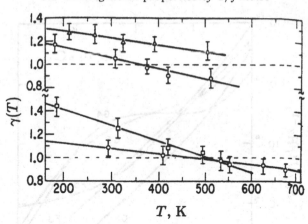

Fig. 8.3 Frequency exponent γ ($S_U(f) \propto f^{-\gamma}$) versus temperature for several metal films. Upper part: Cu (triangles), Ag (ovals). Lower part: Au (squares), Ni (circles). From Eberhard & Horn (1978).

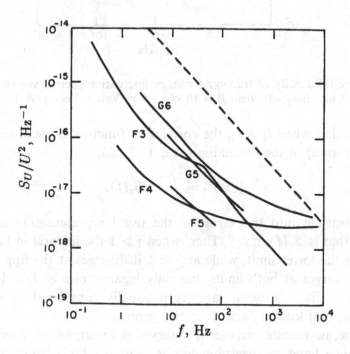

Fig. 8.4 Spectral density of fractional voltage fluctuations in several bar-shaped samples of n-InSb in the absence of a magnetic field. The dotted line represents $S_U(f)/U^2 = \alpha_H/Nf$, where N is the total number of charge carriers in the sample, $\alpha_H = 2 \cdot 10^{-3}$. From Vande Voorde & Love (1981).

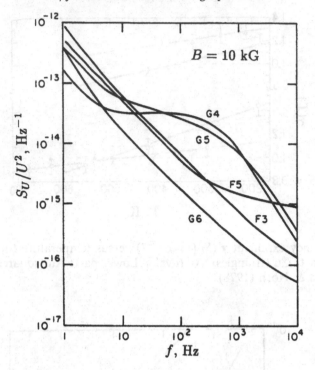

Fig. 8.5 Spectral density of fractional voltage fluctuations across the same samples as in Fig. 8.4 in a magnetic field $B = 10$ kG. From Vande Voorde & Love (1981).

In particular, when $t_1 = t_2$, the correlation function is equal to the mean square (variance) of the fluctuations (Eq. (1.3.16)):

$$\langle (\delta x)^2 \rangle = \int_0^\infty df\, S_x(f). \tag{8.2.1.2}$$

Let us substitute into this equation the usual representation of the $1/f$ spectrum, that is, $S_x(f) \propto f^{-\gamma}$. Then, when $\gamma > 1$ the integral in Eq. (8.2.1.2) diverges at the lower limit, while at $\gamma < 1$ it diverges at the upper limit. If $\gamma = 1$, it diverges at both limits, but only logarithmically, i.e., slowly. The divergence of the $1/f$ noise variance is usually considered as a paradox, which is the best known paradox in this problem.

Of course, no infinite variance is observed in experiments, if only because the frequency band of measurements is bounded both below and above. It is bounded below either by the frequency filter or simply by the finite duration of each realization of the random process. If this time equals t_m (time of measurement), then the measurable frequencies are $\omega > t_m^{-1}$. At high frequencies $1/f$ noise is smaller than the noise of any other origin and becomes immeasurable. The measurements of the characteristics of $1/f$ noise itself are always restricted to a range of frequencies in which it predominates, and the higher frequencies are simply cut off.

Let us denote by f_1 and f_2 the minimum and maximum frequencies passed by the measuring system, and let us study the properties of the noise that is actually measured, that is, with the allowance for filtration. We can assume that the spectral density is zero outside the range from f_1 to f_2, while in this range it is equal to $C_{1/f}/f$, $C_{1/f}$ being a coefficient that determines the intensity of the noise:

$$S_x(f) = \begin{cases} C_{1/f}f^{-1}, & \text{if } f_1 < f < f_2; \\ 0, & \text{if } f < f_1, f > f_2. \end{cases} \tag{8.2.1.3}$$

According to Eqs. (8.2.1.2) and (8.2.1.3), the variance of the noise is

$$\langle x^2 \rangle = C_{1/f} \ln \frac{f_2}{f_1}. \tag{8.2.1.4}$$

If the lower frequency $f_1 \sim t_m^{-1}$ (see above), the variance increases logarithmically with t_m. This is a distinctive feature of $1/f$ noise: the variance of any noise, the spectral density of which is constant at frequencies $f \ll f_c$, does not depend on t_m if $t_m \gg f_c^{-1}$. In terms of the $1/f$ noise model of a continuous spectrum of relaxation modes (Sec. 8.3.1) both the increase of the spectral density with decreasing f and the growth of the variance with increasing time t_m may be explained by the growth of the number of relaxation modes contributing to the measured noise characteristics.

Usually, and it is the most interesting case, $f_2 \gg f_1$. According to Eqs. (8.2.1.1) and (8.2.1.3), the correlation function $\psi_x(t)$ declines with t from the initial value, given by Eq. (8.2.1.4). At small $t \ll (2\pi f_2)^{-1}$ it falls off parabolically. At greater t, when $f_2^{-1} \ll 2\pi t \ll f_1^{-1}$, it decays according to a logarithmic law:

$$\frac{\psi_x(t)}{\langle x^2 \rangle} \approx 1 - \frac{1}{\ln(f_2/f_1)}[C + \ln(2\pi f_2 t)]. \tag{8.2.1.5}$$

Here $C = 0.577\ldots$ is Euler's constant. Finally, when $t \gg (2\pi f_1)^{-1}$, the correlation function has the form of damped oscillations that arise from the sharp cutoff of the spectral density at $f = f_1$. Plots of $\psi_x(t)$ for several values of f_2/f_1 are shown in Fig. 8.6.

Thus, in the most important range of time from $\sim f_2^{-1}$ to $\sim f_1^{-1}$, the correlation function decays logarithmically with time. It means that the correlation function of $1/f$ noise is strongly nonexponential. Some experimental substantiation of this behavior of the correlation function will be presented below (Secs. 8.2.2 and 8.6).

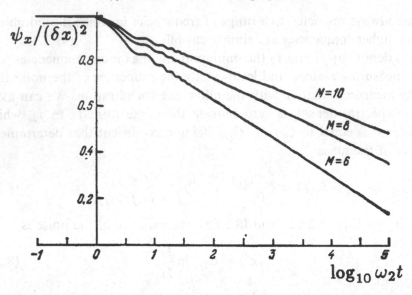

Fig. 8.6 Autocorrelation function ψ_x of a stationary random process $x(t)$ whose spectral density is proportional to $1/f$ in the frequency range from f_1 to f_2 and equals zero outside this range. $\overline{x^2} = \psi_{xx}(0)$ is the variance of the noise. The numbers at the curves are the values of $M = \log_{10}(f_2/f_1)$. Angular frequency $\omega_2 = 2\pi f_2$.

8.2.2 Some statistical properties of $1/f$ noise and its stationarity

The noise is Gaussian when the fluctuating quantity is composed of a large number of independent and identically distributed random quantities (Sec. 1.1). Therefore, to elucidate the nature of $1/f$ noise, it is important to know whether it is a Gaussian random process. Strictly speaking, in order to establish this one should measure arbitrarily high moments of the fluctuating physical quantity and compare them with the second moments. Since this is practically impossible, one should content oneself with testing the Gaussian properties of the first several moments. Of course, one must measure the first probability density function of the random fluctuation $\delta x(t)$ and find whether it has a Gaussian form (Eq. (1.1.13)):

$$w_1(\delta x) = \frac{1}{\sqrt{2\pi\langle(\delta x)^2\rangle}} \exp\left[-\frac{(\delta x)^2}{2\langle(\delta x)^2\rangle}\right]. \qquad (8.2.2.1)$$

Measurements of $w_1(\delta x)$ and other statistical properties of $1/f$ noise have been made by many authors (Brophy, 1968; Brophy, 1969; Greenstein & Brophy, 1969; Hooge & Hoppenbrouwers, 1969; Brophy, 1970; Purcell, 1972; Moore, 1974; Voss, 1978a; Restle, Weissman & Black, 1983; Restle, Hamilton, Weissman & Love, 1985). Very often $w_1(\delta x)$ is a Gaussian

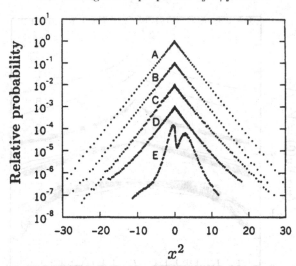

Fig. 8.7 Distribution functions of fluctuations. x is the fluctuation in units of its rms value. Successive distributions have been offset by one decade. A, B, C, D, E are five different sources of noise described in the text. From Voss (1978a).

function, especially when $1/f$ noise is observed in sufficiently pure form, that is, not mixed with the so called burst, or pop-corn, noise. However, deviations from Gaussian behaviour are also observed in many systems.

Voss (1978a) has performed measurements on five different sources of $1/f$ noise: (A) a MOSFET; (B) a carbon resistor; (C) a reverse-biased collector-base junction of a Si *n-p-n* transistor (current fluctuations); (D) a Si *n-p-n* transistor (fluctuations of the output voltage); (E) a reverse-biased *p-n* junction (current fluctuations). In contrast with the rest of the sources, the source E showed considerable burst noise. The noise spectra in the studied frequency range (0.03 Hz–5 kHz) were all of $1/f$ type. The corresponding probability density functions of the fluctuations $w_1(\delta x)$ are shown in Fig. 8.7. One can see that in the case of the sources A, B, and C, $w_1(\delta x)$ is a Gaussian function, while in the case of D and especially E, the deviations from a normal distribution (Eq. (8.2.2.1)) are significant. One should conclude that the mechanisms of $1/f$ noise may be different, and their statistical properties may strongly differ.

Voss also measured for each of these five sources the quantity which was denoted in Eq. (1.2.5) by $\langle \delta x(t) | \delta x_0, 0 \rangle$. It is the mean value of the fluctuation δx at instant $t > 0$ under the condition that its value at instant $t_0 = 0$ is given and equal to δx_0. Through Eq. (1.2.6) this quantity is related to the correlation function $\psi_x(t, 0)$ of the fluctuations. In the experiment, the computer selected those recorded realizations of the noise (series of

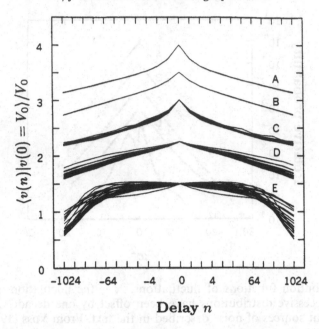

Delay n

Fig. 8.8 The function $\phi(t|V_0) = \langle v(n)|v(0) = V_0 \rangle/V_0$ (Eq. (8.2.2.2)) vs $t = n\Delta t$ for different values of the fluctuation V_0 at $t = 0$. In the experiments $\Delta t = 0.2$ msec. The five different noise sources are described in the text. From Voss (1978a).

measurements each of a definite length) for which the magnitude of the fluctuation in the middle of the series lay in a small interval around a given value δx_0. The sought quantity was obtained by averaging over all realizations satisfying the given condition.

The idea of Voss (1978a) consisted of the following. If the kinetics of the fluctuations is linear, then the ratio

$$\phi(t|\delta x_0) = \frac{\langle \delta x(t)|\delta x_0, 0 \rangle}{\delta x_0} \qquad (8.2.2.2)$$

should not depend on the magnitude of δx_0, that is, it should be an identical function of time t for all δx_0. The results of the experiments are shown in Fig. 8.8. For the sources A and B with noise spectra very close to $1/f$, the functions $\phi(t|\delta x_0)$ lie on a single curve for all δx_0. In the cases C and D, the curves differ appreciably for different δx_0, while for the source E, the dependence on the magnitude of the initial fluctuation is very significant. Voss concluded that the kinetics of different systems exhibiting $1/f$ noise may be different, i.e., it can be either linear or nonlinear, and hence no universal mechanism of $1/f$ noise exists.

This interpretation has to be analyzed more precisely. As was noted by Nelkin & Tremblay (1981), Voss's conclusion is not pertinent to the linearity or nonlinearity of the dynamics of the microscopic processes which produce $1/f$ noise. One can speak only of the linearity or nonlinearity of the phenomenological stochastic equations for the fluctuating quantity. Moreover, if the stationary noise is a Gaussian random process, then Eq. (1.2.7) holds (Nelkin & Tremblay, 1981):

$$\phi(t|\delta x_0) = \frac{\psi_x(t)}{\langle(\delta x)^2\rangle}, \qquad (8.2.2.3)$$

and obviously $\phi(t|\delta x_0)$ does not depend on the value of δx_0. Therefore, one can interpret the result by Voss (1978a) as follows: whenever the distribution function $w_1(\delta x)$ is Gaussian, the $1/f$ noise also satisfies another necessary condition of the random process to be Gaussian: the independence of $\phi(t|\delta x_0)$ of δx_0.

It is interesting to note that, in the case of sources A and B, the function $\phi(t)$, which is in this case proportional to the correlation function $\psi_x(t)$ (Eq. (8.2.2.3)), follows a logarithmic law, as is expected for a random process with an $1/f$ spectrum (Sec. 8.2.1).

The aim of the experiments performed by Stoisiek & Wolf (1976) was to verify the stationarity of $1/f$ noise. However, Nelkin & Tremblay (1981) noticed that these experiments are also checking the Gaussian property of the noise. Let $\delta x(t)$ be a stationary random process with a correlation function $\psi_x(t_1 - t_2)$. The quantity

$$y(t) = \frac{1}{T} \int_0^\infty dt' [\delta x(t - t')]^2 e^{-t'/T} \qquad (8.2.2.4)$$

is the square of the fluctuations averaged over a time interval such that the end of this interval coincides with the instant t of time, while its effective duration is T. The quantity $y(t)$ is random. The definition (8.2.2.4) implies that its mean value is the variance of $\delta x(t)$, i.e., $\bar{y} = \overline{(\delta x)^2}$. The correlation function of the fluctuations $y(t) - \bar{y}$ is of fourth order in the fluctuations $\delta x(t)$. However, if the process $x(t)$ is Gaussian, the correlation function of the fluctuations $y(t) - \bar{y}$ can be expressed in terms of ψ_x, which is of second order in $\delta x(t)$:

$$\overline{(y(t+\tau) - \bar{y})(y(t) - \bar{y})} = \int_0^\infty du\, e^{-u} [\psi_x^2(uT + \tau) + \psi_x^2(uT - \tau)]. \qquad (8.2.2.5)$$

The equation for the 'variance of variance' follows from Eq. (8.2.2.5) at $\tau = 0$:

$$\overline{[y(t) - \bar{y}]^2} = 2 \int_0^\infty du\, e^{-u} \psi_x^2(uT). \qquad (8.2.2.6)$$

For a noise, whose spectral density is proportional to $1/f$ in a certain

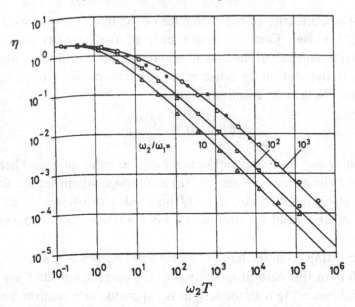

Fig. 8.9 Dependence of the dimensionless quantity η (Eq. (8.2.2.7)) on the effective averaging time T (see text). ω_1 and ω_2 are, respectively, the minimum and maximum angular frequencies in the spectrum of measured $1/f$ noise. Open triangles, boxes, and circles: measurements on a carbon film resistor, solid circles: measurements on a *n-p-n* transistor. Solid curves: calculation. From Stoisiek & Wolf (1976).

frequency range from f_1 to f_2 but zero outside this range (Eq. (8.2.1.3)), the variance $\overline{(\delta x)^2}$ and the correlation function ψ_x can be easily found (see Eqs. (8.2.1.4) and (8.2.1.5)). Using Eq. (8.2.2.6) one can calculate the dimensionless ratio:

$$\eta = \frac{\overline{(y - \bar{y})^2}}{\left[\overline{(\delta x)^2}\right]^2} \tag{8.2.2.7}$$

as a function of the quantities $\omega_2 T$ and f_2/f_1 which can be varied in the experiment. The same quantity η can be measured directly using the records of the noise $x(t)$ and Eq. (8.2.2.4), and both results can be compared. The experiments by Stoisiek & Wolf (1976) have been performed on carbon resistors and bipolar transistors. As is seen from Fig. 8.9, the agreement between calculated and measured values of η is good. It means that one of the necessary conditions for $1/f$ noise to be a Gaussian random process (Eq. (8.2.2.5)) is satisfied in the specimens used in these experiments.

Clear non-Gaussian behavior of $1/f$ noise has been found in various samples: in a carbon resistor (Restle, Weissman & Black, 1983), in small samples of silicon-on-sapphire resistors (Restle, Hamilton, Weissman & Love,

1985). But most of the resistors in these experiments have shown no non-Gaussian effects.

As was pointed out in Sec. 1.3, the noise power in a frequency range from $\bar{f} - \Delta f/2$ to $\bar{f} + \Delta f/2$, that is $[\delta x(t|\bar{f}, \Delta f)]^2$, is a random function of time which fluctuates around its mean value

$$\overline{[\delta x(t|\bar{f}, \Delta f)]^2} = \int_{\bar{f}-\Delta f/2}^{\bar{f}+\Delta f/2} df\, S_x(f). \tag{8.2.2.8}$$

The spectral density of these fluctuations, $S_x^{(2)}(f|\bar{f}, \Delta f)$ is called 'the second spectral density' (Eq. (1.3.18)). Measurements of $S_x^{(2)}$ have been suggested and performed to reveal the non-Gaussian properties of $1/f$ noise in some conductors and spin glasses (Weissman, 1988; Garfunkel, Alers & Weissman, 1990; Israeloff, Alers & Weissman, 1991; Parman, Israeloff & Kakalios, 1992).

It is convenient to measure the second spectral density in consequent octaves, i.e., in frequency ranges in which the upper frequency is twice the lower one ($\Delta f = 2\bar{f}/3$). For $1/f$ noise the mean noise powers, given by Eq. (8.2.2.8), are equal for different octaves. Parman, Israeloff & Kakalios (1992) measured the noise power $[\delta U(t|\bar{f}, 2\bar{f}/3)]^2$ of $1/f$ noise in n-type amorphous Si:H as a function of time t and found the second spectral densities $S_U^{(2)}(f|\bar{f}, 2\bar{f}/3)$ for 7 octaves : 5–10 Hz, 10–20 Hz, and so on. The results are presented in Fig. 8.10.

The second spectral densities have been normalized so that if the noise is Gaussian, then $S_U^{(2)}(f)$ is unity for all frequencies f (compare with $\tilde{S}_x^{(2)}(f)$ in Eq. (1.3.22)). The second spectral density was fitted to the expression:

$$S_U^{(2)}(f) = S^{(2)}(0) + \alpha_2(\bar{f}_l/f)^{\gamma_2}. \tag{8.2.2.9}$$

Here $\bar{f}_l = (2\sqrt{2}/3)\bar{f}$ is the geometrical (logarithmic) mean of the lower and upper frequencies of the octave, $S^{(2)}(0)$, α_2, and γ_2 are constants. One can see that at low relative frequencies $f/\bar{f}_l$ the second spectrum is of $1/f$ type ($\gamma_2 = 1.13$ at 300 K) and the deviation from Gaussian statistics is very great. It may result from the fact that in amorphous semiconductors the conductivity and resistance noise are determined by a small number of current paths (Ch. 7).

The problem of $1/f$ noise stationarity attracted much attention. Recall that if a random process $x(t)$ is stationary, its correlation function $\psi_x(t_1, t_2)$ depends only on the difference $t_1 - t_2$, and the spectral density $S_x(f)$ depends on one frequency f only and is independent of time, i.e., repeated measurements yield the same $S_x(f)$, within experimental accuracy. The last implicitly assumes that the time of averaging (time of measurement t_m) is

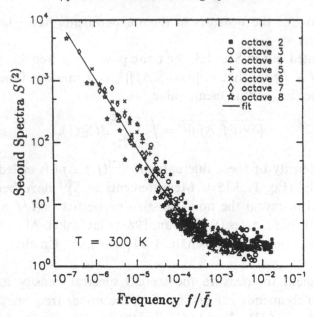

Fig. 8.10 Second spectral density of resistance fluctuations in hydrogenated amorphous silicon. $\bar{f}_l$ is the logarithmic central frequency of each octave (see text). From Parman *et al.* (1992).

many times longer than the longest relaxation time of the system. The continuous increase of the spectral density of $1/f$ noise with decreasing frequency means that relaxation times longer than any possible measurement time t_m exist. Such systems may be considered as partially nonergodic (Sec. 1.4). The long relaxation times may correspond to rare transitions of the system, overcoming high energy barriers, from one 'valley' in phase space to another one in which the spectrum of relaxation times $\tau < t_m$ is different. Such transitions result in variations of the spectrum or its 'wandering' (Sec. 8.6 and Weissman, 1993). If this effect takes place in the absence of any external perturbation or instability, it is more reasonable to attribute it not to nonstationarity of the noise but to incomplete ergodicity of the fluctuating system.

Some authors (Malakhov, 1959; Tandon & Bilger, 1976) tried to explain the paradoxical properties of $1/f$ noise, particularly the absence of a minimum frequency below which the spectral density does not rise with decreasing frequency, by its nonstationarity. A mathematical model has been proposed (Tandon & Bilger, 1976) of a nonstationary random process whose mean spectral density is proportional to $1/f$. However, one can not see any connection between this model and any physical processes.

The following idea is the basis of several experiments to reveal a possible nonstationarity of $1/f$ noise (Brophy, 1968; Greenstein & Brophy, 1969; Purcell, 1972; Moore, 1974; Stoisiek & Wolf, 1976; Restle, Weissman & Black, 1983). A large number of realizations of the noise, each over a time interval T, is recorded. One can find the variance $\overline{(\delta x)^2_T}$ for each of the obtained realizations. One can find then their mean value $\overline{(\delta x)^2}$, and also the variance of the magnitudes of variances, $\overline{[(\delta x)^2_T - \overline{(\delta x)^2}]^2}$. The nonstationarity is expected to affect the magnitude of the dimensionless ratio (compare with Eq. (8.2.2.7)):

$$\eta = \frac{\overline{[(\delta x)^2_T - \overline{(\delta x)^2}]^2}}{[\overline{(\delta x)^2}]^2}. \tag{8.2.2.10}$$

If the nonstationarity is large enough, this ratio will prove to be too large to be explained within the assumption of a stationary noise. If in the experiments by Stoisiek & Wolf (1976) the noise were nonstationary, the experimental points in Fig. 8.9 would lie appreciably higher than the calculated points. No such manifestation of nonstationarity was found.

The stationarity of the noise has also been tested by repeating the noise measurements on the same sample over a long period of time. The intensity of noise in a resistor prepared by ion implantation remained invariant, within the accuracy of measurements $\pm 10\%$, over 2.5 years. In a semiconductor stabilitron it remained invariant for 4.5 years within an accuracy $\pm 20\%$ (Tandon & Bilger, 1976).

8.2.3 *Dependence of* $1/f$ *noise on mean voltage or current*

As a rule, in uniform conductors the spectral density of $1/f$ noise, $S_I(f)$ or $S_U(f)$, like the spectral density of any modulation noise, is proportional to the square of the mean voltage U^2 or the mean current, I^2, throughout the voltage region in which Ohm's law is obeyed. Equations (6.1), $S_U \propto U^2 \propto I^2$, have been verified many times on various conductors. For continous metal films they have been verified, for instance, by Voss & Clarke (1976). These relations break down at very high currents at which either the entire specimen is overheated or the current carriers become hot. It is not surprising that the dependence of the noise on the bias voltage in nonuniform semiconductor structures, devices, etc, is more complex, because an increase in the voltage alters the barriers for the current carriers.

An Ohmic relationship between the mean current and voltage does not always correspond to a linear dependence of the noise on U^2. In island Pt

films the spectral density $S_U \propto U^\beta$, where the exponent β in different specimens varies from 1 to 4 (Williams & Stone, 1972). In granular composites
consisting of Ni particles in an Al_2O_3 matrix, the spectral density of $1/f$
noise $S_U \propto U^2$ at small U, while at higher U it is linear in U, although
the deviations from Ohm's law are inappreciable throughout the entire voltage region (Mantese, Goldburg, Darling et al., 1981). It must be recalled
that in strongly disordered conductors, such as island films and granular
composites, the conduction is confined to some very intricate paths, and the
current density and the electric field are highly nonuniform, reaching high
values in some segments of these paths. As was shown in Ch. 7, this strong
inhomogeneity affects the noise much more strongly than it does the mean
quantities, e.g., the current–voltage characteristic.

Another example is the contact noise in n-InSb specimens (Vande Voorde
& Love, 1981b). Although the resistance of these contacts is very small and
rectification is immeasurable, the contact $1/f$ noise varies from 6 to 10 times
upon reversing the direction of the current.

The problem of the dependence of S_U on U involves the question: does
the electric current cause the observed $1/f$ noise, or does it only reveal the
resistance fluctuations that occur in the absence of any mean current, i.e., in
a state of thermodynamic equilibrium? The opinion has been expressed long
ago in the literature, but not proved, that $1/f$ stems from some instability
of the gas of current carriers caused by current. This mechanism contradicts
not only the usually observed relationship $S_U \propto U^2$, but also a specially
designed experiment by Voss & Clarke (1976). The idea of this experiment
consists of the following. The spectral density of the equilibrium (Johnson–
Nyquist) voltage fluctuations depends on the resistance R of the specimen
(Eq. (2.2.4)). If the resistance fluctuates, then the equilibrium noise also
fluctuates at the same frequencies. By measuring the spectrum of low-
frequency fluctuations of the equilibrium noise ('noise of the noise'), one
can verify that it is proportional to $1/f$ at low enough frequencies, as is
implied by the results of the common method of noise measurements, i.e.,
by measurement of $S_U(f)$ under a nonzero current. The important point is
that, in contrast to the common method, no voltage was applied across the
specimen in the experiment by Voss and Clarke ($\bar{U} = 0, \bar{I} = 0$), and only
fluctuational currents are flowing.

This idea is realized as follows. The voltage fluctuations $\delta U(t)$ taken
from the conductor's electrodes are filtered with a passband of width Δf and
central frequency $\bar{f}$, i.e., from some frequency $v_1 = \bar{f} - \Delta f/2$ to $v_2 = \bar{f} + \Delta f/2$.
The signal $\delta U(t|\bar{f}, \Delta f)$, obtained after filtration, is squared. The quantity
$P(t) = [\delta U(t|\bar{f}, \Delta f)]^2$ fluctuates around its mean value given by the Nyquist

formula (Eq. (2.2.4)):

$$\bar{P} = \int_{v_1}^{v_2} dv \cdot 4k_B T \operatorname{Re} Z(v). \tag{8.2.3.1}$$

Here Z is the impedance of the sample, and

$$\operatorname{Re} Z(v) = R(1 + 4\pi^2 v^2 R^2 C^2)^{-1}, \tag{8.2.3.2}$$

where C is the capacitance of the specimen.

Experimentally, one measures the spectral density of the fluctuations $\delta P(t) = P(t) - \bar{P}$ at low frequencies $f \ll v_1$ (in Sec. 1.3 this spectral density was called the second spectral density and denoted by $S^{(2)}(f|\bar{f}, \Delta f)$). The low-frequency fluctuations of $P(t)$ arise from two different sources: (1) from fluctuations of the temperature T and resistance R, which enter explicitly into Eq. (8.2.3.1) for $\bar{P}$, and (2) from sources which can be considered as Langevin sources (Sec. 1.10), and originate from random motion and scattering of current carriers in the conductor. The second source would give rise to fluctuations of $P(t)$ (we shall denote them by $\delta P^{\text{int}}(t)$), even if T and R were not fluctuating. Thus, we have

$$\delta P(t) = \frac{\partial \bar{P}}{\partial T} \delta T(t) + \left(\frac{\partial \bar{P}}{\partial R}\right)_T \delta R(t) + \delta P^{\text{int}}(t). \tag{8.2.3.3}$$

In the second term in the r.h.s., δR is that part of the fluctuation of the resistance which does not involve the temperature fluctuation.

Since the individual terms in the r.h.s. of Eq. (8.2.3.3) are uncorrelated, the total spectral density is a sum of spectral densities corresponding to each term in the r.h.s. of Eq. (8.2.3.3):

$$S_P(f) = \left(\frac{\partial \bar{P}}{\partial T}\right)^2 S_T(f) + \left(\frac{\partial \bar{P}}{\partial R}\right)_T^2 S_R(f) + S_P^{\text{int}}(f). \tag{8.2.3.4}$$

According to Eq. (1.3.21), if the equilibrium fluctuations are Gaussian for fixed T and R, the second spectral density S_P^{int} at frequencies $f \ll v_1$ equals:

$$S_P^{\text{int}} = 2 \int_{v_1}^{v_2} dv S_U^2(v). \tag{8.2.3.5}$$

It does not depend on frequency f at low f. Therefore, if $S_T(f)$ or $S_R(f)$ contains a component proportional to $1/f$, it becomes predominant in the measured spectral density $S_P(f)$ at low enough frequencies.

Voss & Clarke (1976) performed these measurements on an InSb specimen and a Nb film, while Beck & Spruit (1978), using the same method, made the experiments on carbon resistors. As one can see in Fig. 8.11, actually $S_P \propto 1/f$ at low frequencies. The magnitude of S_R found from these

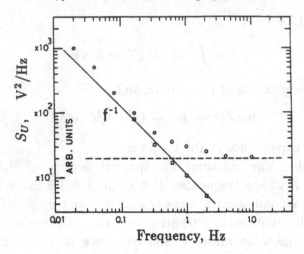

Fig. 8.11 Spectral density of low-frequency fluctuations of the equilibrium (Johnson–Nyquist) voltage noise in carbon resistors. Open circles: observed noise; dashed line: white background noise from a metal-film resistor as a reference; open boxes: $1/f$ spectrum corrected for the white background. From Beck & Spruit (1978).

measurements by using Eq. (8.2.3.4) agrees with the magnitude of S_R found by the common method, i.e., from measurements of the current noise under a nonzero bias. An analysis of the experiments of Voss & Clarke (1976) and Beck & Spruit (1978) was given by Tremblay & Nelkin (1981).

Thus, the measurements of the low-frequency fluctuations of the equilibrium noise in the absence of mean current confirm that $1/f$ noise is produced by fluctuations of the conductor's resistance, and the current usually serves only to detect these fluctuations.

8.2.4 Anisotropy of the conductivity fluctuations

In spatially uniform isotropic media and crystals of cubic point symmetry, in the absence of a magnetic field, the tensor of electrical conductivity is $\sigma_{ij} = \sigma \delta_{ij}$, that is, it reduces to a scalar quantity σ. However, a fluctuation of the local electrical conductivity $\delta\sigma_{ij}(\mathbf{r}, t)$ may be of any tensor symmetry, even in an isotropic medium. It reduces to a scalar quantity, for instance, when it is produced by fluctuations of temperature δT or density δn of the charge carriers. If the noise is produced by temperature fluctuations,

$$\delta\sigma_{ij} = \frac{\partial\sigma_{ij}}{\partial T}\delta T = \delta_{ij}\frac{\partial\sigma}{\partial T}\delta T. \tag{8.2.4.1}$$

One should expect that spontaneous displacements of defects in the crystal are a source of anisotropic conductivity fluctuations (Secs. 6.1 and 6.2). In order to elucidate the mechanism of $1/f$ noise, it is useful to determine whether the corresponding electrical conductivity fluctuations are isotropic or anisotropic (Weissman, 1980).

The correlation radius of the electrical conductivity fluctuations is often many times smaller than the dimensions of the region of the specimen that is controlling the current fluctuations in the experiment (for spatial correlation of conductivity fluctuations see Sec. 8.2.5). In this case, one can consider the correlation function of the electrical conductivity fluctuations at two points $\mathbf{r}_1$ and $\mathbf{r}_2$ as proportional to $\delta(\mathbf{r}_1 - \mathbf{r}_2)$:

$$\langle \delta\sigma_{ij}(\mathbf{r}_1, t_1)\delta\sigma_{kl}(\mathbf{r}_2, t_2) \rangle = \psi_{\sigma,ijkl}(t_1 - t_2)\delta(\mathbf{r}_1 - \mathbf{r}_2). \tag{8.2.4.2}$$

The symmetry of the tensor $\psi_{\sigma,ijkl}(t)$ introduced here reflects the anisotropy of the conductivity fluctuations. In the case of isotropic fluctuations $\psi_{\sigma,ijkl}(t) = \psi_\sigma(t)\delta_{ij}\delta_{kl}$, where $\psi_\sigma(t)$ is a certain scalar function.

In a general case, one can represent the conductivity fluctuation tensor as a sum of three terms:

$$\delta\sigma_{ij} = \delta_{ij}\delta\sigma + \delta\sigma_{ij}^{(s)} + \delta\sigma_{ij}^{(a)},$$

$$\delta\sigma = \frac{1}{3}\mathrm{Tr}\,\widehat{\delta\sigma}, \quad \delta\sigma_{ij}^{(s)} = \frac{1}{2}(\delta\sigma_{ij} + \delta\sigma_{ji}) - \delta_{ij}\delta\sigma, \quad \delta\sigma_{ij}^{(a)} = \frac{1}{2}(\delta\sigma_{ij} - \delta\sigma_{ji}). \tag{8.2.4.3}$$

The first term in the r.h.s. of the first Eq. (8.2.4.3) is the isotropic component of the fluctuation, the second forms an irreducible second-order tensor, the last is the antisymmetric part of the fluctuation. The same representation is used for the fluctuations of dielectric permittivity in the theory of electromagnetic waves' scattering (Landau & Lifshitz, 1984).

In an isotropic medium, only fluctuations of the same tensor symmetry are correlated, and therefore

$$\psi_{\sigma,ijkl}(t) = \psi_{\sigma,0}(t)\delta_{ij}\delta_{kl} + \psi_{\sigma,s}(t)\left(\delta_{ik}\delta_{jl} + \delta_{il}\delta_{jk} - \frac{2}{3}\delta_{ij}\delta_{kl}\right)$$
$$+ \psi_{\sigma,a}(t)(\delta_{ik}\delta_{jl} - \delta_{il}\delta_{jk}). \tag{8.2.4.4}$$

In studying $1/f$ noise, the frequencies of interest are many times smaller than the inverse kinetic relaxation times of the current carriers and phonons, and the lengths exceed the kinetic lengths. One should expect that the conductor with a conductivity fluctuation is in a quasi-equilibrium state, and therefore the fluctuation satisfies Onsager's symmetry principle $\delta\sigma_{ij}(\mathbf{r}, t; \mathbf{B}) = \delta\sigma_{ji}(\mathbf{r}, t; -\mathbf{B})$, where $\mathbf{B}$ is the magnetic field (Sec. 1.2). It means that at $\mathbf{B} - 0$ the antisymmetric part of the fluctuation $\delta\sigma_{ij}^{(a)} = 0$ and $\psi_{\sigma,a}(t) = 0$. The tensor $\psi_{\sigma,ijkl}(t)$ is then symmetric not only with respect to transposition of the first and second pairs of indices, but also with respect to transposition

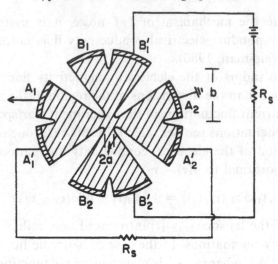

Fig. 8.12 Configuration of a four-probe resistor with separate current and voltage contacts used in studies of the conductivity fluctuations' anisotropy. $2a$ is the dimension of the constriction. From Weissman, Black & Snow (1982).

of indices within each pair. In an isotropic medium, the tensor is determined by two nonzero quantities, $\psi_{\sigma,0}$ and $\psi_{\sigma,s}$, in cubic crystals by three nonzero quantities. The magnitude of the ratio of Fourier transforms $\psi_{\sigma,s}(\omega)/\psi_{\sigma,0}(\omega)$ is a measure of the anisotropy of the electric conductivity fluctuations at a given frequency $f = \omega/2\pi$.

Weissman (1980) has suggested the use, when studying the anisotropy of conductivity fluctuations, of specimens in the form of a 'Maltese cross' (Fig. 8.12) with four arms. Each arm has two separated contacts: one current contact, the other a potentiometric one. The voltage fluctuations at the potentiometric contacts are determined by fluctuations of the electric conductivity in the central narrow region of the cross. Under the conditions that the currents I_A, between the arms A_1 and A_2, and I_B, between B_1 and B_2, are fixed, the corresponding voltage fluctuations equal:

$$\delta U_A = I_A \delta R_{AA} + I_B \delta R_{AB}, \quad \delta U_B = I_A \delta R_{AB} + I_B \delta R_{BB}. \qquad (8.2.4.5)$$

According to Eq. (7.2.9),

$$\delta R_{AB}(t) = \int d\mathbf{r} e_{Ai}(\mathbf{r}) \delta\sigma_{ij}(\mathbf{r}, t) e_{Bj}(\mathbf{r}). \qquad (8.2.4.6)$$

Here, as in Sec. 7.2, vector $\mathbf{e}_A(\mathbf{r})$ is the ratio of the Ohmic field $\mathbf{F}(\mathbf{r})$ at the point $\mathbf{r}$, caused by the passage of the current I_A, and the magnitude of this current. One can write similar expressions for δR_{AA} and δR_{BB}. Equations (8.2.4.2) and (8.2.4.6) imply that the spectral density of the fluctuations δR_{AB}

equals:

$$S_{R_{AB}R_{AB}}(f) = 2\langle\delta R_{AB}(t_1)\delta R_{AB}(t_2)\rangle_\omega = 2\psi_{\sigma,ijkl}(\omega)\int d\mathbf{r}\, e_{Ai}e_{Bj}e_{Ak}e_{Bl}. \quad (8.2.4.7)$$

By measuring the spectral densities of the voltage fluctuations S_{UA} and S_{UB} and the cross-density S_{UAB} for various currents I_A and I_B, one can find the spectral densities of resistance fluctuations, and obtain the components of the tensor $\psi_{\sigma,ijkl}$ by using Eq. (8.2.4.7). The distributions $e_A(\mathbf{r})$ and $e_B(\mathbf{r})$ can be either calculated or measured on an analogue model.

Upon substitution of Eq. (8.2.4.4) for the tensor $\psi_{\sigma,ijkl}$ of an isotropic medium (with $\psi_{\sigma,a} = 0$, see above) into Eq. (8.2.4.7) and into a similar expression for the spectral density $S_{R_{AA}R_{BB}}$, one can find that

$$\frac{\psi_{\sigma,s}}{\psi_{\sigma,0}} = \frac{G+1-Q}{(G+\frac{4}{3})Q - \frac{2}{3}(2-G)}. \quad (8.2.4.8)$$

Here

$$Q = \frac{\langle\delta R_{AA}(t_1)\delta R_{BB}(t_2)\rangle_\omega}{\langle\delta R_{AB}(t_1)\delta R_{AB}(t_2)\rangle_\omega}, \qquad G = \frac{\int d\mathbf{r}([e_A \times e_B])^2}{\int d\mathbf{r}(e_A \cdot e_B)^2}. \quad (8.2.4.9)$$

The specimens studied by Weissman, Black & Snow (1982) had $G = 1.35$. If the values of Q measured by them are substituted into Eq. (8.2.4.8), one obtains for two carbon films $\psi_{\sigma,s}/\psi_{\sigma,0} = 0.43$ and 0.56, for an Au film 0.65, and for two Cr films 1.18 and 3.42 (see also Black, Snow & Weissman, 1982). An even greater anisotropy of conductivity fluctuations has been found in Bi (Black, Restle & Weissman, 1983).

The considerable anisotropy of conductivity fluctuations in metal films indicates that they do not arise from temperature fluctuations. It can be explained by the motion of scattering centers in the metal (Secs. 6.1 and 6.2). Particularly, the hopping of defects, whose symmetry is lower than the point-group symmetry of the crystal, between several positions with equal free energies but different orientation in the lattice, gives $\mathrm{Tr}\,\widehat{\delta\sigma} = 0$, $\psi_{\sigma,0} = 0$. It means that in this case the conductivity fluctuations are purely anisotropic (Kogan & Nagaev, 1984).

8.2.5 Correlation length of conductivity fluctuations and dependence of noise on the dimensions of the conductor

The correlation function $\psi_x(\mathbf{r}_1 t_1; \mathbf{r}_2, t_2)$ of fluctuations $\delta x(\mathbf{r}, t)$ measured at different instants of time, t_1 and t_2, and at different points, $\mathbf{r}_1$ and $\mathbf{r}_2$, depends not only on the time difference $t_1 - t_2$ but also on the distance between these two points. As ψ_x falls off with increasing $|t_1 - t_2|$, it falls off as well with increasing $|r_1 - r_2|$ due to the same relaxation processes (the steepness of decline may be different in different directions). If this decrease is exponential, one can define the correlation length L_c as the length at which the correlation function drops by $e = 2.718\ldots$ times (Sec. 1.2).

Of course, the correlation length of electrical conductivity fluctuations depends on the dominant mechanism of kinetics (relaxation) in the system, on temperature, electric and magnetic fields, etc. Usually this length is the free-path length or the energy relaxation length of the charge carriers (Sec. 5.4).

A number of attempts to measure the correlation length of $1/f$ noise in various conductors have been performed, mainly to verify the model according to which $1/f$ noise is produced by temperature fluctuations (Secs. 6.4 and 8.7). Scofield, Darling & Webb (1981) found that the fluctuations with an $1/f$ spectrum in two Au films isolated from one another galvanically, but not thermally, are uncorrelated. There is no correlation between fluctuations in two adjacent regions of a thin and narrow metal film lying at a distance $\simeq 1$ mm (Black, Weissman & Fliegel, 1981). No correlation was found between current fluctuations in two adjacent transistors in an integrated circuit, despite the fact that the transistors were separated by a distance of several tens of μm only (Kilmer, Chenette, van Vliet & Handel, 1982). No correlation was found between conductivity fluctuations in different sections of an inversion channel on p-Si (Adkins & Koch, 1982). The same result was obtained for two different segments of a p-Si strip separated by a distance 63 μm (Çelik-Butler & Hsiang, 1988).

If (1) the conductor is on average uniform, (2) the conductivity fluctuations are of bulk type (rather than determined by the surface or contacts), and (3) the correlation length of the fluctuations is small as compared with the dimensions of the conductor, then the spectral density of relative (fractional) voltage or current fluctuations is inversely proportional to the volume V of the conductor, i.e., $S_U/U^2 \propto V^{-1}$ (Sec. 1.2). Actually, according to Eqs. (7.1.6), (7.2.11), and (8.2.4.2),

$$\frac{S_U(f)}{U^2} = \frac{S_R(f)}{R^2} = 2\psi_{\sigma,ijkl}(\omega)\frac{\int d\mathbf{r} e_i e_j e_k e_l}{(\sigma \int d\mathbf{r} |\mathbf{e}(\mathbf{r})|^2)^2}. \tag{8.2.5.1}$$

If the conductor is uniform and $\mathbf{e}(\mathbf{r}) = \text{const.}$, the expression in the r.h.s. of Eq. (8.2.5.1) varies as V^{-1}. The same happens when the conductor is nonhomogeneous but its dimensions exceed the correlation length of the conductivity fluctuations. If the macroscopic conductor is two-dimensional, the spectral density of relative fluctuations is proportional to the inverse area A^{-1} of the conductor.

De Felice (1976) measured the spectral density of $1/f$ noise in complex resistors put together by connecting m nominally identical carbon resistors, in series and parallel, such that the total resistance was equal to the resistance of a single resistor. He found that S_U/U^2 is proportional to $1/m$, i.e., to the inverse volume of the complex resistor. Fleetwood, Masden & Giordano (1983) found that in thin Pt films and filaments the spectral density of noise obeys the law $S_U f/U^2 \propto 1/N$, where N is the number of Pt atoms in the specimen, over a vast interval of variation of N from $\simeq 10^7$ to $\simeq 10^{14}$.

If the correlation radius of the conductivity fluctuations were larger than the dimensions of the specimen, the spectral density of relative voltage fluctuations with a current flowing may not vary as V^{-1} upon changing these dimensions. For example, this would happen if $1/f$ noise were caused by some geophysical or cosmophysical phenomenon (such hypotheses have also been speculated). The experimentally established small correlation lengths of the current $1/f$ noise enables one to rule out such explanations of this noise.

8.2.6 Empirical Hooge relationship

Often the spectral density of $1/f$ noise follows a power law $f^{-\gamma}$, with the exponent γ very close to unity, and is proportional to U^2. In this case, one can represent the spectral density of fractional fluctuations in the form $S_U/U^2 = C/f$, where C is a dimensionless coefficient. Hooge (1969) correlated a large bulk of experimental data obtained by different authors on the magnitude of $1/f$ noise in semiconductor and metal films. He plotted these data on the $C - N_c$ plane, where $N_c = nV$ is the total number of free charge carriers in the specimen, n is their density, and V is the volume of the specimen. Each point corresponds to a certain semiconductor or metal. The distribution of the points revealed a qualitative rule, which Hooge approximated as $C = \alpha/N_c$. He assumed that $\alpha = 2 \cdot 10^{-3}$ is a universal coefficient. The following empirical relationship has become termed in the literature as the Hooge relationship:

$$\frac{S_U(f)}{U^2} = \frac{\alpha}{N_c f}. \qquad (8.2.6.1)$$

The coefficient α was called the Hooge constant and was denoted by α_H.

The qualitative tendency toward decrease in the $1/f$ noise with decreasing resistivity ρ of the specimen (while $\rho \propto 1/n$) was noted as early as the 1950s in Ge samples (Montgomery, 1952). It was well known also that $1/f$ noise in metals is smaller by orders of magnitude than in semiconductors. Hooge was the first to try to impart the form of a quantitative relationship to this qualitative tendency.

If the Hooge relationship were exact, and this implies universal, it would be a convincing argument in favor of one universal mechanism of $1/f$ noise. However, even a cursory examination of experimental data shows that this relationship is not exact, and hence cannot be universal.

For example, the noise in metallic films is known to depend strongly (varying by two orders of magnitude or even more) on the temperature and the type of substrate, although the number of free charge carriers does not vary. In Bi samples, $1/f$ noise is of the same order as in metals, despite the fact that the density of free carriers in the semimetal Bi is smaller by several orders of magnitude. As a rule, films made of the same metal, prepared by the very same technique and having approximately the same resistivity ρ, strongly differ in the magnitude of $1/f$ noise: the difference can exceed an order of magnitude (see, for instance, Fleetwood & Giordano, 1983).

Hooge (1969) made, in fact, a conclusion that the mechanism of $1/f$ noise is always of bulk type. This conclusion contradicts the known effects of surface treatment and ambient atmosphere on $1/f$ noise in semiconductors (Sec. 8.5.1). The decrease in noise with increasing n in this case can be explained within the framework of a surface source of noise: as n increases the thickness of the space charge region, where the conductivity is altered by the charge fluctuations in surface states, decreases.

Table 8.2 shows the values of the dimensionless quantity

$$\alpha = \frac{S_U(f)fnV}{U^2} \qquad (8.2.6.2)$$

for a number of semiconductors and metals. It is interesting to compare them with the Hooge constant $\alpha_H = 2 \cdot 10^{-3}$. If the experimental values of α were always greater than α_H, one could assume an universal noise described by Hooge relationship (Eq. (8.2.6.1)) ('α-noise'), accompanied by some other noise with the same spectrum. However, values of α which are many times less than α_H are often observed. For instance, in some samples of n-InSb, whose spectra are shown in Fig. 8.4, α is smaller than α_H by two orders of magnitude. However, there are samples of the same material with α an order of magnitude higher than α_H.

At the same time, the Hooge relationship cannot be undervalued. It is the only relationship that allows one, however crudely, to estimate the magnitude of the spectral density of $1/f$ noise in uniform conductors with an accuracy that most often is no poorer than one or two orders of magnitude. In strongly disordered and inhomogeneous conductors, the measured values of

Table 8.2. Values of $\alpha \equiv f n V S_U / U^2$ in various materials

Material	T, K	n, cm^{-3}	V, cm^3	α	References
n-Ge	300	$0.8 \cdot 10^{15}$	$2.1 \cdot 10^{-3}$	$1 \cdot 10^{-3}$	Brophy (1956)
n-Ge	300	$1.5 \cdot 10^{14}$	$1.8 \cdot 10^{-3}$	$3.5 \cdot 10^{-3}$	Montgomery (1952)
n-Ge	300	$1.3 \cdot 10^{14}$	$1.8 \cdot 10^{-3}$	$1.8 \cdot 10^{-4}$	Montgomery (1952)
n-Ge	300	$5 \cdot 10^{13}$	$1.8 \cdot 10^{-5}$	$6 \cdot 10^{-4}$	Noble & Thomas (1961)
a-Si:H	400	–	–	~ 1	Parman et al. (1991)
p-Ge	300	$8.7 \cdot 10^{15}$	$1.8 \cdot 10^{-3}$	$5 \cdot 10^{-3}$	Montgomery (1952)
InSb	300	$2.8 \cdot 10^{16}$	$2.7 \cdot 10^{-5}$	$5.1 \cdot 10^{-2}$	Epstein (1965)
InSb	300	$4.6 \cdot 10^{16}$	$2.3 \cdot 10^{-5}$	$5.5 \cdot 10^{-2}$	Epstein (1965)
InAs	300	$1.2 \cdot 10^{18}$	$2.5 \cdot 10^{-6}$	$1.8 \cdot 10^{-3}$	Epstein (1965)
n-InSb	76	$1 \cdot 10^{14}$	$2.1 \cdot 10^{-4}$	$4 \cdot 10^{-4}$	[a]
n-InSb	76	$1 \cdot 10^{14}$	$1.1 \cdot 10^{-3}$	$3.4 \cdot 10^{-5}$	[a]
n-InSb	77	$1.6 \cdot 10^{14}$	$3.2 \cdot 10^{-5}$	$1 \cdot 10^{-3}$	Vandamme (1974)
n-InSb	295	$1.6 \cdot 10^{16}$	$3.2 \cdot 10^{-5}$	$4 \cdot 10^{-3}$	Vandamme (1974)
n-InSb	77	$1.6 \cdot 10^{14}$	$1.1 \cdot 10^{-4}$	$1.6 \cdot 10^{-3}$	Vandamme (1974)
n-InSb	295	$1.6 \cdot 10^{16}$	$1.1 \cdot 10^{-4}$	$2.7 \cdot 10^{-3}$	Vandamme (1974)
p-InSb	77	$1.2 \cdot 10^{16}$	$1.4 \cdot 10^{-4}$	$7 \cdot 10^{-3}$	Vandamme (1974)
p-InSb	295	$1.6 \cdot 10^{16}$	$1.4 \cdot 10^{-4}$	$3.7 \cdot 10^{-3}$	Vandamme (1974)
n-GaAs	295	$2.3 \cdot 10^{16}$	$8 \cdot 10^{-4}$	$6 \cdot 10^{-3}$	Vandamme (1974)
p-GaAs	295	$2.3 \cdot 10^{16}$	$1 \cdot 10^{-3}$	$1.7 \cdot 10^{-3}$	Vandamme (1974)
p-GaAs	295	$2.3 \cdot 10^{16}$	$5.1 \cdot 10^{-4}$	$5 \cdot 10^{-3}$	Vandamme (1974)
n-GaP	295	$2.9 \cdot 10^{16}$	$1.1 \cdot 10^{-4}$	$9 \cdot 10^{-3}$	Vandamme (1974)
n-GaAs	60	$6 \cdot 10^{17}$	–	$1 \cdot 10^{-8}$	Tacano et al. (1992)[b]
n-InP	50	$1.2 \cdot 10^{18}$	–	$1 \cdot 10^{-7}$	Tacano et al. (1992)[b]
Au	300	$5.9 \cdot 10^{22}$	–	$2.4 \cdot 10^{-3}$	Hooge et al. (1969)
Au	300	$5.9 \cdot 10^{22}$	–	$4 \cdot 10^{-3}$ [c]	[d]
Cu	300	$8.4 \cdot 10^{22}$	–	$4.8 \cdot 10^{-2}$ [e]	[d]
Ag	300	$5.8 \cdot 10^{22}$	–	$6.8 \cdot 10^{-2}$	[d]
Bi	300	$2.8 \cdot 10^{18}$	$1.1 \cdot 10^{-10}$	$4.3 \cdot 10^{-6}$	Voss & Clarke (1976)
Sn	300	–	–	$0.7 \cdot 10^{-3}$	[f]
Sn	300	–	–	$5.4 \cdot 10^{-3}$	[f]

[a] Vande Voorde & Love (1979)
[b] Focused ion beam implanted quarter-μm filaments
[c] α increases by a factor of 50 with increase of T from 100 to 500 K
[d] Eberhard & Horn (1978)
[e] α varies by almost 3 orders of magnitude with increase of T from 100 to 410 K
[f] Fleetwood & Giordano (1982)

α are usually many orders higher than α_H. Apart from other mechanisms, this may result from the fact that in these conductors the current paths occupy only a small part of the total volume V which enters in Eq. (8.2.6.2) (see Ch. 7). Parameter α, being dimensionless, is widely used to compare the $1/f$ noise magnitude in different samples, materials, etc.

8.3 Basic models of $1/f$ noise

8.3.1 The concept of nonexponential kinetics

In the simplest case, when the kinetics of the fluctuations is characterized by a single relaxation time τ, the correlation function of the fluctuating quantity $x(t)$ is purely exponential, i.e., $\psi_x(t) = \overline{(\delta x)^2} \exp(-|t|/\tau)$, while the spectral density, according to the Wiener–Khintchine theorem (Eq. (1.3.8)), is a Lorentzian function of frequency:

$$S_x(f) = 4 \int_0^\infty dt \psi_x(t) \cos(\omega t) = \overline{(\delta x)^2} \frac{4\tau}{1 + \omega^2 \tau^2}. \qquad (8.3.1.1)$$

In more complex cases, the kinetics of the quantity $x(t)$ is a superposition of several, or even many, relaxation processes with different, but definite, relaxation times. In general, a continuous distribution of relaxation times may exist. The spectral density given by Eq. (8.3.1.1) has to be averaged over this distribution with a weight function $p_x(\tau)$ which includes the number of subsystems with relaxation time τ and the corresponding variance of fluctuations. Of course, the kinetics in this case is nonexponential, that is, the correlation function falls off with time $|t|$ not exponentially, and the spectral density equals:

$$S_x(f) = \int_0^\infty d\tau p_x(\tau) \frac{4\tau}{1 + \omega^2 \tau^2}. \qquad (8.3.1.2)$$

Since the total variance of the fluctuations is

$$\overline{(\delta x)^2} = \int_0^\infty df \, S_x(f) = \int_0^\infty d\tau \, p_x(\tau), \qquad (8.3.1.3)$$

the quantity $p_x(\tau)d\tau$ is the contribution to the variance of processes whose relaxation times lie in the interval from τ to $\tau + d\tau$.

If $p_x(\tau) \propto 1/\tau$ in some interval from τ_1 to $\tau_2 \gg \tau_1$, but is zero outside this interval, then, according to Eq. (8.3.1.2), one has $S_x(f) \propto 1/f$ in the frequency range $\tau_2^{-1} \ll f \ll \tau_1^{-1}$. This idea was suggested by Surdin (1939, 1951).

The next step in the interpretation of $1/f$ noise spectrum was made by Du Pré (1950) and Van der Ziel (1950). They assumed that $1/f$ noise is a result of superposition of activation kinetic processes with different relaxation times, each of which depends exponentially on the inverse temperature: $\tau = \tau_0 \exp(E/k_B T)$, where E is the activation energy, and τ_0^{-1} is the frequency of attempts to surmount the activation barrier E. The distribution in activation energies $F_x(E)$ was assumed to be almost constant in a broad enough interval (this was the main idea). Then the distribution of relaxation times $p_x(\tau)$ has the required form. In fact, because $p_x(\tau)d\tau = F_x(E)dE$, the function $p_x(\tau) = F_x(E)/(d\tau/dE) = k_B T F_x(E)/\tau$. When $F_x(E) = $ const., $p(\tau) \propto 1/\tau$.

If the kinetics of the fluctuating quantity is controlled not by activation but by tunneling processes, the relaxation times depend approximately exponentially on the width and the height of the tunnel barriers. If the distribution of these parameters of the tunnel barriers is almost constant in a wide interval, the resulting distribution of relaxation times also has the required form, $p_x(\tau) \propto 1/\tau$. It was first shown by McWhorter (1957) in his model of $1/f$ noise in semiconductors (see below Sec. 8.3.2).

Let us study more accurately the noise arising from activation processes. The weighting function of activation energies $F_x(E, T)$ may depend on temperature T, for instance, when the mean concentration of defects, whose transitions between different states create the noise, is not fixed, but increases with T, or the fluctuating quantity is temperature dependent. The spectral density of noise is then:

$$
\begin{aligned}
S_x(f) &= \int_0^\infty dE\, F_x(E, T) \frac{4\tau_0 e^{E/k_B T}}{1 + \omega^2 \tau_0^2 e^{2E/k_B T}} \\
&= \frac{1}{\pi f} \int_0^\infty dE \frac{F_x(E, T)}{\cosh[(E - E_\omega)/k_B T]}.
\end{aligned}
\tag{8.3.1.4}
$$

In the last integral the notation $E_\omega = k_B T \ln(\omega\tau_0)^{-1}$ has been introduced. In condensed media usually $\tau_0 \sim 10^{-14} - 10^{-11}$ sec. At those frequencies at which $1/f$ noise is usually observed, the magnitude of $(\omega\tau_0)^{-1}$ is $\gg 1$, and even $\ln(\omega\tau_0)^{-1} \gg 1$. Therefore the energy E_ω may be of the same order as ordinary activation energies ($\sim 0.1 - 1$ eV) (Dutta, Dimon & Horn, 1979).

The function $1/\cosh[(E - E_\omega)/k_B T]$ of activation energy E in the r.h.s. of Eq. (8.3.1.4) constitutes a narrow peak of width $\sim k_B T$ with a maximum at $E = E_\omega$. As E moves away from E_ω on either side by several $k_B T$, this function falls off exponentially. If the width of the weighting function of activation energies $F_x(E)$ is considerably smaller than $k_B T$, the noise spectral density, obviously, reduces to the Lorentzian function (Eq. (8.3.1.1)). However, if one takes into account that the thermal energy $k_B T$ is small in comparison with ordinary activation energies, even when T is several hundred Kelvins, the opposite case seems real, in which the width of the weighting function $F_x(E)$ is far greater than $k_B T$. Then one can remove $F_x(E, T)$ for $E = E_\omega$ from within the integral in Eq. (8.3.1.4). The spectral density equals (Du Pré, 1950; Dutta, Dimon & Horn, 1979):

$$
S_x(f) = k_B T F_x(E_\omega, T) \frac{1}{f}.
\tag{8.3.1.5}
$$

Since E_ω depends on the frequency f only logarithmically, any not too strong energy dependence of $F(E, T)$, e.g., not an exponential decline, results in only a small deviation of $S_x(f)$ from the $1/f$ law. The spectrum remains of $1/f$ type. Let us recall (Sec. 8.2.1) that the experimental noise spectrum is close to a power law $f^{-\gamma}$, and γ differs from unity only within the range $\pm \sim 0.2$. Since $F_x(E, T)$ can either increase or decrease with increasing E near $E = E_\omega$, the spectral density $S(f)$ may deviate from $1/f$ either toward steeper or gentler decline with increasing f.

If the weight function $F_x(E, T)$ does not depend on T explicitly, i.e., depends on T only through the temperature dependence of E_ω, both the deviation of the temperature dependence of noise from a linear one and the deviation of its frequency dependence from purely $1/f$ law stem from the function $F_x(E_\omega)$, and thus are connected by a simple relation (Dutta, Dimon & Horn, 1979):

$$\gamma - 1 = [\ln(1/\omega\tau_0)]^{-1} \left\{ \frac{\partial \ln S_x(f)}{\partial \ln T} - 1 \right\}, \qquad (8.3.1.6)$$

where $\gamma = -\partial \ln S_x(f)/\partial \ln f$.

According to Eq. (8.3.1.6), the closer the exponent γ is to 1, the closer the temperature dependence of noise is to the linear one. The deviations from Eq. (8.3.1.6) may be indicative of the explicit dependence of $F_x(E, T)$ not only on E but also on T.

The fact that the $1/f$ spectrum originates from the superposition of random processes having different relaxation times has been demonstrated many times. In the one of the first experiments of this kind (Ralls, Sckocpol, Jackel *et al.*, 1984), fluctuations of current through the inversion channel of Si MOSFETs having extremely small dimensions $(1 \times 0.1\,\mu\text{m}^2)$ have been measured. In each such transistor the fluctuations amount to a random sequence of switchings between two states having resistances that differ by tenths of percent, but are definite under given conditions. The spectrum of this random telegraph noise is Lorentzian (Eq. (1.7.10) and Sec. 8.8). In certain transistors the noise is a superposition of two types of switching with different characteristic times, and the spectrum is a superposition of two Lorentzians. Interestingly enough, in transistors of larger dimensions $(10 \times 20\,\mu\text{m}^2)$, grown in a single process with submicron ones, the individual switchings could not been resolved and only an ordinary $1/f$ noise was seen.

These results are interpreted by Ralls *et al.* (1984) as follows. The fluctuations of the channel's resistance are caused by electron capture into and emission from traps that lie in the oxide layer. In the submicron specimens, owing to their small area, the noise is determined by one or two traps, while in larger specimens it is determined by a superposition of effects from a large number of traps with various activation energies and relaxation times, which leads to $1/f$ noise.

Thus, according to the model presented in this section, the $1/f$ spectrum is a result of an exponentially broad distribution of relaxation times τ that encompasses many orders of magnitude. Such broad distributions of τ are characteristic of disordered systems. The almost universal character of $1/f$ noise is due not to the existence of some (unknown) universal mechanism of low-frequency fluctuations but to definite universal features of low-frequency kinetics of disordered systems (Secs. 8.4–8.6).

It is worth noting that, in any physical model of broad distribution of relaxation times, a maximum relaxation time τ_{max} appears because the heights and thicknesses of the barriers are finite (in a sample of finite dimensions). For instance, one should expect that the function $F_x(E, T)$ starts to drop exponentially with E at some high activation energy (high E are rare). At $f < \tau_{max}^{-1}$ the spectral density stops increasing with decreasing f and reaches a finite value. However, the time τ_{max} may be greater than the age of the Universe. It may exceed all accessible times of measurement t_m.

Since in systems with $1/f$ noise the time of measurement t_m is always somewhere within the spectrum of relaxation times ($t_m < \tau_{max}$), these systems are always partly nonergodic (Sec. 1.4). Specifically, they are nonergodic with respect to the relaxation processes with $\tau > t_m$. Strictly speaking, these systems, being nonergodic, are also not completely equilibrium ones: the system, within the time t_m, does not penetrate into some regions of its phase space, i.e., the space of its quantum states. The measured spectral density reflects the motion of the system over a part of its states only. It may result in a dependence of the data on the specific 'valley' of the phase space into which the system happened to be placed by the initial preparation procedure. The difference between experimental data measured in different experimental runs is usually within statistical dispersion of data. Perhaps, in these systems each 'valley' is large enough and the measurements are insensitive to the specific 'valley' in which the system is moving (Sec. 8.6). However, the transitions from one 'valley' to another one with different spectrum of relaxation times was observed in spin-glasses (see the review by Weissman, 1993).

The kinetics of fluctuations can be represented by a superposition of independent processes with different but definite relaxation times only in simple systems such as, for instance, systems having a variety of *noninteracting* traps. In general, due to many-body effects, it becomes impossible to separate out individual relaxation modes in a disordered system. However, these complications can not change the qualitative conclusions regarding the origin of the $1/f$ spectrum.

Most often $1/f$ noise can not be associated with other phenomena in the same systems. Moreover, for several decades no connections between $1/f$ noise and other kinetic phenomena in solids have been established and, hence, $1/f$ noise was considered as an isolated and even 'mysterious' phenomenon. One can explain this with the fact that measurement of the noise spectrum (fluctuation spectroscopy) is far more sensitive than other methods of investigation of low-frequency kinetics of disordered physical systems, so that relaxation phenomena observable from the noise spectrum can not be simultaneously observed on the same specimens from other physical effects. The noise measurements, especially in metals, are performed on specimens having very small dimensions (e.g., thin films). It is difficult to measure other low-frequency phenomena on the same specimens. However, now such connections of $1/f$ noise with other low-frequency phenomena are firmly established in some systems (Secs. 8.4 and 8.6).

The exponentially broad distribution of relaxation times or, equivalently, smooth distribution of activation energies or tunneling parameters, by which one explains the $1/f$ noise spectrum, enables one also to explain a number of other phenomena in real, disordered solids. These phenomena are dielectric relaxation in disordered dielectrics, magnetic relaxation in spin-glasses, amorphous ferromagnets and type II superconductors, the anomalous low-temperature heat capacity, heat conduction, and acoustic attenuation in amorphous solids (Sec. 8.3.3), and the flat spectrum of internal friction (Sec. 8.4.3). It means that this picture of the low-frequency kinetics of real solids can not be considered merely as a hypothesis devised to explain a single phenomenon, the $1/f$ noise.

8.3.2 The McWhorter model

In this model, suggested by McWhorter (1957), fluctuations in the number of charge carriers in the surface layer of a semiconductor and, consequently, conductivity fluctuations, arise from exchange of electrons between the surface layer and the traps lying in the oxide layer covering the surface, or on the outer surface of the oxide. Transfer of an electron into or from the trap occurs by tunneling. Therefore, the characteristic inverse relaxation time of this process falls off exponentially with the distance x between the surface and the trap: $\tau^{-1} = \tau_0^{-1} \exp(-x/\lambda)$. Here $\lambda \sim 0.1$ nm, while τ_0^{-1} is a pre-exponential factor that depends only weakly on x. Since the distances to the traps vary, with a scatter $\gg \lambda$, the distribution of relaxation times is exponentially wide and encompasses many orders of magnitude.

One should expect that the exponential increase of τ with x extends only to

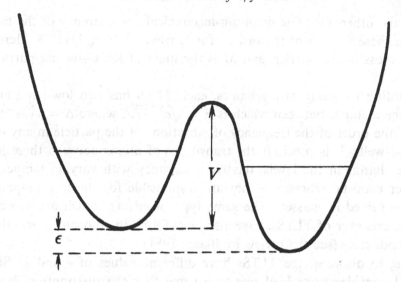

Fig. 8.13 Schematic diagram of the potential energy of a two-level tunneling system versus coordinate. ϵ is the asymmetry of the two wells, V is the height of the barrier. Thin lines are the two energy levels of the TLTS.

distances x of the order of the mean distance between traps in the dielectric: exchange with remote traps occurs by tunneling first to a trap closer to the surface; the electron then reaches the remote traps by hopping motion via the traps.

This model has been used many times to interpret $1/f$ current noise in MOSFETs.

8.3.3 *Two-level tunneling systems*

A number of disordered systems, first of all dielectric glasses, exhibit an anomalous temperature dependence of the heat capacity and the heat conductivity at low temperatures, as well as specific phenomena in the absorption of sound (see reviews by Black, 1981, and Smolyakov & Khaimovich, 1982). All these phenomena have been successfully explained on the basis of a model of two-level tunneling systems (TLTS) (Anderson, Halperin & Varma, 1972; Phillips, 1972). According to this model, in disordered, especially amorphous, structures atoms or groups of atoms exist that can occupy two positions, so one can represent their energy as a function of configuration in the form of two potential wells separated by a barrier of some height V (Fig. 8.13). In general, the wells are asymmetric and the energies of their minima differ by some energy ϵ. Atoms or groups of atoms can tunnel from one well to the other one. The quantum-mechanical transparency of the barrier between these wells is of the order of $e^{-\lambda}$, where $\lambda = d\sqrt{2MV}/\hbar$. Here d is the thickness of the barrier, and M is the mass of the tunneling particle or particles.

According to quantum mechanics, each TLTS has two low-lying energy levels, the distance between which is $E = \sqrt{\epsilon^2 + \Delta^2}$, where $\Delta = \hbar\omega_0 e^{-\lambda}$, and ω_0 is of the order of the frequency of vibrations of the particle in any of the potential wells. It is precisely the transitions of atoms between these levels, and the change in the levels' relative occupancy with varying temperature or under acoustic vibrations, that are responsible for the low-temperature effects observed in glasses. The same type of effects, which are associated with the existence of TLTSs, have also been found in amorphous metals and ionic conductors (see the review by Black, 1981).

Owing to disorder, the TLTSs have different values of ϵ and λ. Simple physical considerations lead one to assume that the distribution function $P(\epsilon, \lambda)$ of TLTSs in ϵ and λ is almost constant at those values of these parameters ($\lambda \gg 1$, $\epsilon \ll \hbar\omega_0$) that control the effects observed in experiments: in this region one has $P(\epsilon, \lambda) \simeq \bar{P}$, where $\bar{P}$ is a constant. The magnitude of $\bar{P}$ can be found by comparing the theory with experiment.

The rate of transitions between the two levels of a TLTS is determined by its interaction with phonons (in insulating solids) or electrons (in metals). The corresponding inverse relaxation times for the deviation of the occupancy numbers of the levels from the equilibrium ones equal:

$$\tau_{ph}^{-1} = aE\Delta^2 \coth\left(\frac{E}{2k_B T}\right), \qquad \tau_e^{-1} = b\frac{\Delta^2}{\hbar E} \coth\left(\frac{E}{2k_B T}\right). \tag{8.3.3.1}$$

Here a is a coefficient that depends on the parameters of the interaction of the TLTS with acoustic phonons and on the sound speed, and b is a dimensionless coefficient that depends on the interaction of the TLTS with electrons of the metal and on the density of electron states at the Fermi surface (b can be ~ 1). Equations (8.3.3.1) imply that the distribution in relaxation times τ and energies E equals $p(\tau, E) = (E/2\epsilon\tau)P(\lambda, \epsilon)$. Larger values of τ exist in systems having $\Delta \ll \epsilon \simeq E$. Thus, for large τ the function $p(\tau, E) = P(\lambda, \epsilon)/2\tau \simeq \bar{P}/2\tau$. Owing to the exponential dependence of τ on the tunneling parameter ($\tau \sim e^{2\lambda}$) and to the approximately uniform distribution in λ, the distribution with respect to τ is inversely proportional to τ in an exponentially broad interval, as is characteristic of systems exhibiting $1/f$ noise (Sec. 8.3.1).

Spontaneous transitions between the levels of the TLTS can lead to fluctuations of macroscopic quantities, e.g., resistance of disordered metals

(Kogan & Nagaev, 1984a; Ludviksson, Kree & Schmid, 1984), density of electron states in semiconductors and MOS structures, tunneling current through dielectric layers between metals (Kogan & Nagaev, 1984b), etc. This noise has a $1/f$ spectrum.

Let $\delta x = x_2 - x_1$ be the variation of the quantity x (e.g., resistance) when a TLTS goes from the lower to the upper level. According to Eq. (1.7.10), the spectral density of x in this two-state system equals:

$$S_x^{(i)}(f) = \overline{(\delta x)^2} \frac{4w_1 w_2 \tau}{1 + \omega^2 \tau^2}, \tag{8.3.3.2}$$

where the upper index i indicates that this spectral density is related to a definite (i-th) TLTS. Since the ratio of the probabilities of the TLTS to be in the upper (w_2) and lower (w_1) state is $w_2/w_1 = \exp(-E/k_B T)$, and $w_1 + w_2 = 1$, the product

$$4w_1 w_2 = [\cosh(E/2k_B T)]^{-2}.$$

If the transitions in different TLTSs are uncorrelated, the average spectral density of the fluctuations of the quantity x equals:

$$S_x(f) = V \int d\tau dE \overline{(\delta x)^2_{\lambda, \epsilon}} \frac{p(\tau, E)}{\cosh^2(E/2k_B T)} \frac{\tau}{1 + \omega^2 \tau^2}. \tag{8.3.3.3}$$

Here $\overline{(\delta x)^2_{\lambda, \epsilon}}$ is the square of δx averaged over the TLTSs having similar values of λ and ϵ, and V is the volume over which the fluctuation δx is averaged.

At low temperatures, when $k_B T$ is smaller than the range E_{max} of energies E, the main contribution to $S_x(f)$ comes from the TLTS with $E <$ or $\simeq k_B T$. The number of these TLTSs increases with temperature and one may expect that the noise is proportional to T (apart from the temperature dependence of $\overline{(\delta x)^2_{\lambda, \epsilon}}$). Let τ_T be the relaxation time of those TLTSs for which $\Delta \simeq k_B T$, $E \simeq k_B T$, and τ_{max} is the relaxation time corresponding to the maximum λ at which the distribution function $P(\lambda, \epsilon)$ is still approximately constant and equal to $\bar{P}$ (this value of λ is denoted below as λ_{max}). In the frequency range $\tau_{max}^{-1} \ll f \ll \tau_T^{-1}$, the main contribution to the spectral density (Eq. (8.3.3.3)) comes from values of τ and E for which $p(\tau, \epsilon) \simeq \bar{P}/2\tau$. The quantity $\overline{(\delta x)^2_{\lambda, \epsilon}}$ is expected not to depend strongly on λ and ϵ. The spectral density (Eq. (8.3.3.3)) acquires the form:

$$S_x(f) \simeq \frac{1}{4} V \overline{(\delta x)^2} \bar{P} k_B T \frac{1}{f}, \qquad \tau_{max}^{-1} \ll f \ll \tau_T^{-1}, \quad k_B T \ll E_{max} \ll \hbar\omega_0. \tag{8.3.3.4}$$

In the opposite case of high temperatures, $k_B T \gg E_{max}$, one may approximate $\cosh(E/2k_B T) \simeq 1$. The spectral density then equals:

$$S_x(f) \approx \frac{V \bar{P} E_{max}}{8} \overline{\delta x^2} \frac{1}{f} \qquad E_{max} \ll k_B T \ll \hbar\omega_0. \qquad (8.3.3.5)$$

It may depend on temperature only through $\overline{\delta x^2}$.

The lower frequency bound of the $1/f$ spectrum is determined either by λ_{max} or by the rate of activation processes (transitions over the barrier) that can be significant at higher temperatures. According to the estimates by Karpov, Klinger & Ignat'ev (1983), $\lambda_{max} \sim 10^2$. The corresponding value of τ_{max}^{-1} (see Eq. (8.3.3.1)) is in any case smaller than any frequency that is technically accessible for noise measurements. The inverse relaxation time for activation processes is $\sim \omega_0 \exp(-\hbar\omega_0/k_B T)$, and is also very small at low temperatures $T \ll \hbar\omega_0/k_B$.

Thus, owing to the exponential dependence of the relaxation time of the TLTSs on the tunneling parameter λ and the uniform distribution of λ, spontaneous transitions in the TLTSs produce fluctuations in physical quantities, which are sensitive to these transitions, with a $1/f$ spectrum over an extremely broad range of f (at low temperatures). The lowest frequency of this range $\sim \tau_{max}^{-1}$ may be by many orders of magnitude smaller than the inverse time of the existence of the Universe. It means that the spectral density of noise follows the $1/f$ dependence down to all accessible frequencies and at the same time it does not diverge at $f \to 0$. The paradox noted in Sec. 8.1 and 8.2.1 does not arise.

It is interesting that in the TLTS model, owing to the uniform distribution of the TLTSs' excitation energies E, the spectral density of $1/f$ noise increases linearly with the temperature (at low temperatures), just like the contribution of the TLTSs to the heat capacity.

8.3.4 Other models

Halford (1968) suggested a general model of noise produced by random time-dependent perturbations (pulses). The author called this model 'mechanical'. In particular, at definite assumptions about the characteristics of the random pulses, the spectral density of this noise has a $1/f$ spectrum in an assumed frequency interval. Ultimately, this model seems to be equivalent to the model of a $1/\tau$ distribution of processes with various relaxation times τ in an assumed broad interval (see above Sec. 8.3.1).

Marinari, Parisi, Ruelle & Windey (1983) have suggested a model of $1/f$ noise called by the authors 'Random walk in a random environment'.

The random quantity $x(t)$ obeys a first-order stochastic equation: $\dot{x} = F(x) + \eta(x,t)$, where the random drift $F(x)$ and the Langevin source of random walk $\eta(x,t)$ are assumed to be uncorrelated, and each of them to be delta-correlated, i.e., the correlation function $\langle F(x_1)F(x_2)\rangle = \delta(x_1 - x_2)$ and that of the Langevin source is proportional to $\delta(x_1 - x_2)\delta(t_1 - t_2)$. The randomness of the 'environment' bridges this model with that of disordered systems with a wide spectrum of activation energies or tunneling constants (Sec. 8.3.1). The spectral density in this model, $S_x(f) \propto (\ln f)^4/f$. It diverges at $f \to 0$ even stronger than $1/f$. The integral over frequency diverges at the lower limit as $(\ln f)^5$ instead of $\ln f$ in the case of $1/f$ noise.

The idea of the model of 'Self-organized criticality' (Bak, Tang & Wiesenfeld, 1987) is that many dynamical systems with spatial degrees of freedom organize themselves in a dynamical state which is critical in the sense that no characteristic time or length scale exist. The spectra found numerically for some such systems are of $1/f$ type.

A lattice gas model was suggested by Jensen (1990). Particles occupy the sites of a lattice, and the fluctuating quantity is the number of particles in the lattice. Each site can be empty or occupied by no more than one particle. The particles interact with each other, pinning centers and a driving force acting on all particles. The particles obey definite rules of entering and leaving the lattice. Such a model exhibits a power-law spectrum of fluctuations.

A number of $1/f$ noise models can be viewed as mathematical ones (Lowen & Teich, 1993; Frieden & Hughes, 1994, and references therein).

8.4 1/f noise in metals

8.4.1 Direct connection between 1/f noise and defects' motion

Many experiments show that $1/f$ noise in metals is produced by the motion of defects or impurities.

Dutta, Eberhard & Horn (1978), Eberhard & Horn (1977, 1978) , and Dutta, Dimon & Horn (1979) measured the spectral density of $1/f$ noise in a variety of metal films in a broad temperature range up to 600 K (see review by Dutta & Horn, 1981). They found a satisfactory agreement between their data and the model of activation energies' distribution (Sec. 8.3.1). Equation (8.3.1.6), which connects the deviation of the exponent γ from unity with the deviation of noise temperature dependence from the linear one, was tested quantitatively (Eberhard & Horn, 1978). The distribution of activation energies E was obtained (of course, within an unknown factor). For Au films this distribution peaks around $E \approx 1$ eV. The activation energies were

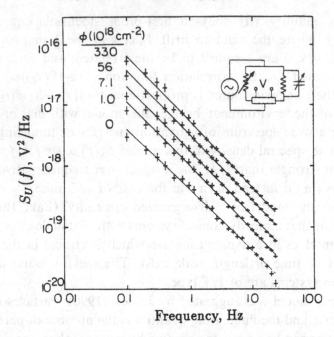

Fig. 8.14 Spectra of excess voltage noise in copper films irradiated by 0.5 MeV electrons for different electron dose ϕ. The dose is shown at the left side of each spectrum. From Pelz & Clarke (1985).

found to be of the same order as the known activation energies of defects' generation and motion.

Pelz & Clarke (1985) (see also Pelz, Clarke & King, 1988) measured the resistivity ρ and spectral density of voltage noise $S_U(f)$ in polycrystalline copper films versus the dose ϕ of 500-keV electron irradiation which induced defects in the films. The measurements of both ρ and $S_U(f)$ have been performed at 90 K. The increase of resistivity, $\Delta\rho$, scaled with the dose as $\phi^{1/2}$. The maximum relative increase of the resistivity was $\sim 10\%$. At the same time the dimensionless noise parameter α (Eq. (8.2.6.2)), which before the irradiation was $\sim 5.5 \cdot 10^{-4}$, increased to $\sim 6 \cdot 10^{-3}$ at the maximum dose, i.e., by an order of magnitude. The increment of the noise parameter, $\Delta\alpha$, increased as $\sim \rho^{0.6}$ (Fig. 8.14). If, as it is usually assumed, $\Delta\rho$ is proportional to the added defect concentration n_d, the noise increment is proportional to $n_d^{0.6}$. The exponent is less than unity: perhaps, the noise is produced only by mobile defects that are only a fraction of those defects that affect the resistivity, and the dependence of the number of mobile defects on ϕ is different (presumably, all defects induced by irradiation contribute to $\Delta\rho$).

Annealing experiments support the idea that noise is produced only by a part of the total number of defects. Annealing of the samples reduces the

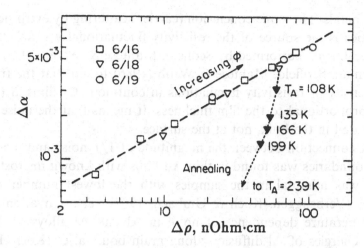

Fig. 8.15 Change in the dimensionless magnitude of $1/f$ noise, $\Delta\alpha$, due to irradiation by 0.5 MeV electrons vs change of sample resistivity $\Delta\rho$ in Cu films. Open squares, triangles, and circles correspond to different dates of measurement. Solid triangles along the dotted line correspond to annealing at progressively higher temperatures. At $T_A = 239$ K $\Delta\rho \approx 11.6$ nOhm·cm, $\Delta\alpha \approx 7 \times 10^{-5}$. From Pelz & Clarke (1985).

noise much more rapidly than the resistivity (Fig. 8.15). For instance, after annealing at 239 K $\Delta\rho$ is reduced by several times while $\Delta\alpha$ drops by two orders of magnitude down to $\sim 7 \cdot 10^{-5}$.

It is very difficult to identify the defects that produce resistivity fluctuations. Moreover, the presence of such defects is often due to some details of the procedure of the samples' preparation that are not under direct experimental control. Fleetwood & Giordano (1983a) found for a great number of metals that the magnitude of $1/f$ noise of nominally identical samples of one and the same metal, at room temperature, varies by as much as an order of magnitude, and these variations are much greater than the variations of the samples' resistivity. At the same time, the sample-to-sample (and even metal-to-metal) variations of the exponent γ in the frequency dependence $f^{-\gamma}$ of the spectral density are small. The values of γ are all close to 1.15 (in the case of Au films $\gamma \approx 1.08$).

The same authors (1983b) studying strained metal films, found that the magnitude of the $1/f$ noise decays with time as the strain relaxes. It proves that the processes of resistance noise and strain relaxation are linked.

Fleetwood & Giordano (1985) found that annealing of an AuPd film drastically reduces $1/f$ noise. They found the distribution $F(E)$ of activation energies E of processes responsible for $1/f$ noise. It is significantly lowered by annealing. Their main conclusion is that a link exists between $1/f$ noise and defects in metals. The same conclusion (carrier scattering by extrinsic defects or impurities is the source of the resistivity fluctuations) was substantiated by the experiments performed by Scofield, Mantese & Webb (1985).

Zimmerman, Scofield, Mantese & Webb (1986) found that the fractional spectral density of resistivity fluctuations in continous Cr films, $S_\rho(f)/\rho^2$ is inversely proportional to the film thickness. It means that the noise sources are distributed in the bulk, not at the surface.

A direct connection between the magnitude of $1/f$ noise and the number of grain boundaries was found in thin Au films with known microstructure: the noise was minimal in the samples with the lowest number of grain boundaries (Verbruggen, Koch & Umbach, 1987). The activation energies of the temperature dependence of noise in Al and Al alloys are close to activation energies of Al diffusion along grain boundaries (Koch, Lloyd & Cronin, 1985).

Zimmerman & Webb (1988) studied the resistance noise in Pd with various concentrations of dissolved hydrogen. They found that the noise in the frequency range 0.001–100 Hz is of $1/f$ type at temperatures below ~ 150 K. At higher temperatures the noise in same frequency range is produced by diffusion of hydrogen into and from the contacts (Sec. 6.3). The spectral density of $1/f$ resistance fluctuations is proportional to the concentration of hydrogen (in the absence of hydrogen $1/f$ noise is about three orders lower). This dependence and other experimental findings prove that $1/f$ fluctuations in this metallic system are produced by the motions of protons on an atomic scale.

The same authors (1990) studied the temperature dependence of $1/f$ resistance noise in amorphous PdSi metal films at various dissolved hydrogen contents in the temperature range $\sim 40 - 200$ K. The noise produced by hydrogen was dominant at these temperatures. The spectral density $S_R(f, T)$ exhibits two thermally activated peaks, one at about 80 K and one shifting with dissolved hydrogen content from ≈ 160 K down to ≈ 130 K. The second peak closely matches the peak of the internal friction temperature dependence which shifts with hydrogen content in the same way as the $S_R(f, T)$ peak. Some mechanisms of impurity motion contribute simultaneously to low-frequency resistance noise and to internal friction (Kogan & Nagaev, 1982, 1984a) (Sec. 8.4.3).

Keener & Weissman (1991) proved that the origin of $1/f$ noise in Bi is also consistent with defects' motion, despite some anomaly found in earlier experiments: the noise magnitude was greater in thick films than in thin films or films with more disorder.

Zhigal'skii (1991) measured on the same Mo and Cr films (1) the spectral density of $1/f$ noise S_U/U^2 and (2) the cubic nonlinearity of the current–voltage characteristic. Both quantities depend exponentially on the inverse temperature $1/T$ and on internal stress σ. The activation energies of temperature dependencies of both effects, as well as the activation volumes (for dependence on σ) have been found to be very close, indicating that both effects have a common mechanism. The author argues that temperature and stress affect the formation of vacancies that are responsible for both $1/f$ noise and nonlinearity.

The main conclusion of this section is: $1/f$ noise in metals is produced mainly by mobile defects.

8.4.2 *The mechanisms of* $1/f$ *noise in metals*

Two main mechanisms of resistance fluctuations caused by defects' motion are known: the change of the anisotropic scattering cross-section of an impurity, whose symmetry is lower than the point-group symmetry of the host metal (Sec. 6.1), and the change of the interference pattern created by multiple scattering of electrons at the impurities (Sec. 6.2).

Let us assume, in accordance with the general model of $1/f$ noise presented in Sec. 8.3.1, a smooth distribution of the defects' activation energies E. In the case of incoherent scattering at mobile defects (Eq. (6.1.14)) the spectral density of fractional resistance fluctuations can be represented in the form:

$$\frac{S_R(f)}{R^2} = \frac{k_B T l^2}{fV} F(E_\omega, T). \tag{8.4.2.1}$$

This equation is similar to Eq. (8.3.1.5). The function $F(E, T)$ is the distribution function of the product of concentration of defects with given E and their scattering cross-section squared, σ_s^2.

The temperature dependence of this noise is controlled mainly by the product $T l^2$. The free-path length l is determined by all scattering processes, including phonon scattering. It may increase as the temperature is lowered, and, if the increase of l^2 exceeds the decrease of T, the noise grows with decreasing temperature. However, such unusual temperature dependence can be observed, within the mechanism of incoherent scattering, only in the temperature region where the phonon scattering dominates over or is comparable with the impurity one. At sufficiently low temperatures and in 'dirty' metals, when the resistivity is controlled by impurity scattering only, the noise is expected to be approximately linear in temperature T.

When the temperature is low and the concentration of scattering centers is high, the physics of resistance fluctuations is different (Sec. 6.2). In this case the coherence length L_ϕ (Eq. 6.2.1) is greater than the elastic scattering free-path length l. The electron wave functions create very intricate interference patterns, which are coherent over distances $\sim L_\phi$. The hop of only one impurity changes the interference pattern within the same distance L_ϕ around this impurity. If at least one of the conductor's dimensions is greater than L_ϕ, the spectral density of fractional conductance fluctuations for one kind of mobile impurities is given by Eq. (6.2.20). However, at low temperatures the most probable candidates for mobile impurities are the so called two-level tunneling systems (TLTS) (Sec. 8.3.3). Owing to the exponential dependence of the tunneling probability on the tunneling parameter λ, the TLTS relaxation times τ vary over a vast range. Comparing Eqs. (8.3.3.3) and (6.2.20), one can see that the spectral density of fractional conductance fluctuations can be represented in the form:

$$\frac{S_G(f)}{G^2} = \frac{1}{V} \int d\tau \, dE \left(\frac{3\pi^2 L_\phi^2}{(e^2/\hbar)k_F^2 l}\right)^2 \overline{[\delta G_1(L_\phi)]^2} \frac{p(\tau,E)}{\cosh^2(E/2k_B T)} \frac{\tau}{1+\omega^2\tau^2}.$$

$$(8.4.2.2)$$

Here the notations are the same as in Secs. 6.2 and 8.3. The overline means averaging over TLTSs with comparable parameters λ and ϵ.

At low temperatures, when $k_B T \ll E_{max}$ (E_{max} is the range of gaps E between the two TLTS's levels), the spectral density of fractional conductance fluctuations equals:

$$\frac{S_G(f)}{G^2} = \frac{1}{fV} k_B T \bar{P} \left(\frac{3\pi^2 L_\phi^2}{2(e^2/\hbar)k_F^2 l}\right)^2 \overline{[\delta G_1(L_\phi)]^2}, \qquad k_B T \ll E_{max} \ll \hbar\omega_0.$$

$$(8.4.2.3)$$

At higher temperatures $k_B T \gg E_{max}$, the thermal energy $k_B T$ in the r.h.s. must be substituted by $E_{max}/2$ (Eq. (8.3.3.5)). The factor $k_B T$ appears also in the case of activation hopping of impurities (Eq. (8.3.1.5)).

The coherence length L_ϕ, which is usually determined by inelastic (phonon) scattering of electrons, increases as the temperature is lowered, approximately as a power function T^{-p}, where the value of p is not universal. Since the noise is proportional to a high power of L_ϕ (fourth power in the case of a thin film when $\delta G_1(L_\phi)$ is independent of L_ϕ), it is expected to *increase* as the temperature is reduced.

The unusual temperature dependence of the noise was found in amorphous Ni$-$Zr alloys at $T <\simeq 100$ K (Alers, Weissman, Averback & Shyu, 1989). This interesting phenomenon was clearly observed also in Bi films below ~ 70 K (Birge, Golding & Haemmerle, 1989, 1990; McConville & Birge, 1993). The noise approached a power law $T^{-0.9}$ at low temperatures (Fig. 8.16). This result was interpreted as follows. At low temperatures in Bi

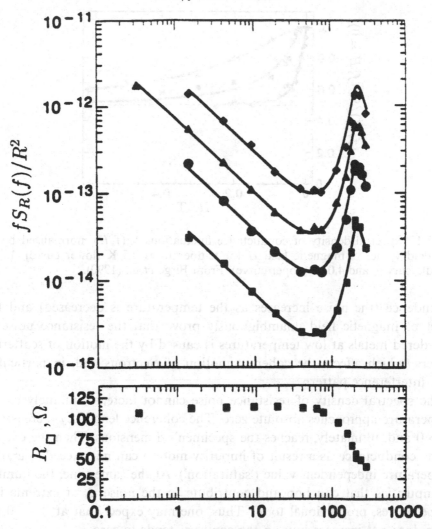

Fig. 8.16 Temperature dependence of the normalized spectral density of resistance fluctuations, $fS_R(f)/R^2$, in Bi films. Their dimensions are (from the lowest to the upper curve), in μm: 10×100; 4×40; 2×20; 1.3×10. The lower figure shows the temperature dependence of the sheet resistance. From Birge *et al.* (1990).

both $L_T = \sqrt{\hbar D/k_B T}$ and $L_{Din} = \sqrt{D\tau_{in}}$ vary as $T^{-1/2}$. Thus, the factor $L_\phi^2 L_{Din}^2 \propto T^{-2}$. The number of impurities that are able to hop, by tunneling or by surmounting some activation barriers, is proportional to T. The product of these two factors is proportional to T^{-1}.

Birge, Golding & Haemmerle (1990) and McConville & Birge (1993) found also that the magnitude of $1/f$ noise is reduced precisely by a factor of 2 by a magnetic field of $\simeq \Phi_0/L_\phi^2$ (Fig. 8.17). The unusual temperature

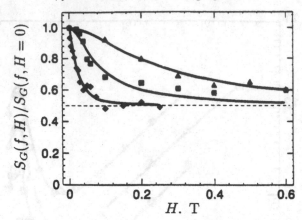

Fig. 8.17 Spectral density of conductance fluctuations $S_G(f, H)$, normalized by its zero-field value, vs magnetic field H for temperatures 0.5 K (lower curve), 1.5 K (middle curve), and 4.0 K (upper curve). From Birge *et al.* (1990).

dependence (the noise increases as the temperature is decreased) and the effect of magnetic field unambiguously prove that the resistance noise in disordered metals at low temperatures is caused by the motion of scattering centers, which affects the coherent motion of electrons and, in particular, their interference pattern.

The spectral density of resistance noise cannot increase infinitely as the temperature approaches absolute zero. The coherence length L_ϕ increases as $T \to 0$ and, ultimately, reaches the specimen's dimensions. Then the change of the conductance as a result of impurity motion can not exceed $\sim e^2/h$, a temperature independent value ('saturation'). At the same time, the number of impurities that can hop and contribute to the noise is, at extreme low temperatures, proportional to T. Thus, one may expect that at $T \to 0$ the noise does not increase but, on the contrary, tends to zero.

8.4.3 *$1/f$ noise, internal friction, and anelastic piezoresistance in metals*

For many years, before the 1980s, $1/f$ noise as a phenomenon seemed to be isolated and not linked to other physical phenomena. This hampered its interpretation. The idea seemed rather natural of associating the low-frequency current noise with other low-frequency relaxation phenomena with reciprocal relaxation times of the same order as the frequencies at which $1/f$ noise is measured. One of these phenomena is internal friction, the mechanisms of which are well studied (see Nowick & Berry, 1972). At low frequencies it stems from various motions of defects – hopping, reorientation, etc. The simplest model of the defect's contribution to the internal friction is the Snoeck's model, which was, in fact, presented in Sec. 6.1. If the symmetry of a defect is lower than the point-group symmetry of the host crystal, it can exist in several positions whose free energies are identical. Elastic stresses σ_{ij} make the free energies, corresponding to different positions of the defect, different. Hence, these positions become not equally probable. If the frequency ω of deformation is not too low and not too high in comparison with the reciprocal relaxation time τ^{-1} of the defect's hopping between its positions, a phase difference arises between $\sigma_{ij}(\omega)$ and the strain (deformation) $u_{ij}(\omega)$. It results in dissipation of the elastic energy. The reciprocal Q-factor, which is the characteristic of the internal friction, can be written in the form (Nowick & Berry, 1972):

$$Q^{-1} = B \frac{n_d}{k_B T} \frac{\omega\tau}{1 + \omega^2\tau^2}, \tag{8.4.3.1}$$

where B is a coefficient of order $10^{-41} - 10^{-40}$ J·cm³. Q^{-1}, as a function of frequency, has a maximum at $\omega = \tau^{-1}$. If the hopping of the defects is an activation process, i.e., $\tau = \tau_0 \exp(E/k_B T)$, with an activation energy E, the internal friction as a function of temperature has a peak, which is called the Debye peak, at $k_B T = E / \ln(\omega\tau_0)^{-1}$.

Hopping of defects with symmetry lower than the point-group symmetry of the crystal is not only a source of internal friction but also a source of resistance fluctuations (Kogan & Nagaev, 1982) (Sec. 6.1). The spectral density of fractional resistivity tensor fluctuations, $\delta\rho_{ij}(t)$, averaged over a volume V, equals:

$$\frac{S_{\rho,ijkl}(f)}{\rho^2} = \frac{4n_d\tau}{sV(1 + \omega^2\tau^2)} \sum_{\alpha\beta} \frac{1}{\rho^2} \frac{\partial\rho_{ij}}{\partial n_\alpha} \frac{\partial\rho_{kl}}{\partial n_\beta} \left(\delta_{\alpha\beta} - \frac{1}{s}\right). \tag{8.4.3.2}$$

Here n_α is the concentration of defects located in the position α, s is the total number of defect's positions.

Let us assume that the hopping of defects is an activation process and the distribution of the activation energies, $F(E, T)$, is broad and flat (Sec. 8.3.1). Integrating $S_{\rho,ijkl}(f)$ with $F(E, T)$ over E, one finds that the spectral density of resistivity fluctuations is close to $1/f$. Taking into account the estimate (6.1.13)–(6.1.14), one obtains an estimate for the dimensionless parameter $\alpha = fnVS_\rho/\rho^2$ (Eq. (8.2.6.1)) of $1/f$ noise:

$$\alpha \simeq nk_B T F(E_\omega, T)(l\sigma_s)^2. \tag{8.4.3.3}$$

Here, as in Sec. 8.3.1, $E_\omega = k_B T \ln(\omega\tau_0)^{-1}$, n is the electron density, l and σ_s are the free-path length and the scattering cross-section of the impurities.

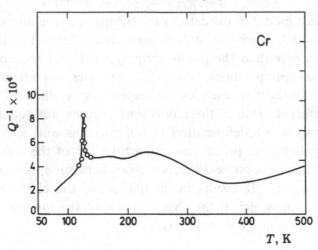

Fig. 8.18 Spectrum of internal friction in Cr: reciprocal Q-factor versus temperature. Frequency $f \simeq 1.7$ Hz. A Debye peak at 126 K and the flat background of internal friction are seen. From Weller & Moser (1981).

In the same material with a broad distribution of activation energies, the internal friction is given by

$$Q_{\text{flat}}^{-1} \approx \frac{\pi}{2}F(E_\omega, T)B(E_\omega). \qquad (8.4.3.4)$$

This internal friction, unlike the case of a single activation energy (see Eq. (8.4.3.1)), depends only weakly on the frequency f and temperature T. Its spectrum is featureless and almost flat. Flat spectra of internal friction are often measured in metals (see, for instance, Weller & Moser, 1981; Harangozo & Kedves, 1984; Fantozzi & Vincent, 1983). The typical values of Q_{flat}^{-1} are $10^{-4} - 10^{-3}$ (Fig. 8.18).

This means that both phenomena, the $1/f$ spectrum of resistance fluctuations and the flat spectrum of internal friction, stem from the motion of lower-symmetry defects with a broad and flat distribution of the defects' hopping activation energies. One can eliminate $F(E_\omega, T)$ from Eqs. (8.4.3.3) and (8.4.3.4) and express the dimensionless noise parameter α in terms of Q_{flat}^{-1}:

$$\alpha \approx [nk_B T(l\sigma_s)^2/B(E_\omega)]Q_{\text{flat}}^{-1}. \qquad (8.4.3.5)$$

Substitution of typical values of parameters in the r.h.s. yields for α values of order $10^{-3} - 10^{-2}$ that are of the same order as usually measured in metals.

Zimmerman & Webb (1990) found that the peak of the resistance $1/f$ noise, $S_R(f, T)$, in amorphous films of PdSi alloy matches the peak of the internal friction $Q^{-1}(T)$ of the same alloy at the same temperatures. The shifts, with hydrogen content, of both peaks, that of the noise and that of the internal friction, from ≈ 160 K to ≈ 130 K are also very similar.

The mobile defects and their relaxation are responsible also for another effect, namely, for the out-of-phase, or imaginary part, or anelastic, piezoresistance (Alers & Weissman, 1991):

$$m''(f) = \frac{1}{u_{zz}(f)} \mathrm{Re} \left(\frac{\Delta\rho(f)}{\rho} \right). \tag{8.4.3.6}$$

Here $\Delta\rho(f)/\rho$ is the fractional response of the conductor's resistivity to an ac mechanical strain with dimensionless amplitude $u_{zz}(f)$ along the direction of the applied elastic stress σ_e with frequency f. The strain $u_{zz}(f) = \sigma_e/E_Y$, where E_Y is the Young's modulus. The main contribution to $m''(f)$ comes from defects with inverse relaxation time $\tau^{-1} \sim f$. In principle, the same defects that determine the resistance noise $S_R(f)$ may contribute also to the internal friction $Q^{-1}(f)$ and to $m''(f)$.

Alers & Weissman measured both $m''(f)$ and $S_R(f)/R^2$ (1/f noise) in the same film samples of three metals. In Bi below 200 K both effects could be attributed to one class of mobile defects. In Nb both effects grow with the concentration of hydrogen impurity. However, their temperature dependence is different: a peak in noise vs temperature at $\simeq 110$ K was found, and no peak in $m''(f, T)$. The noise in Fe with 10 at. % Mn and 0.5 at. % C exhibits a peak at $\simeq 225$ K, unlike the anelastic piezoresistance which is a monotonically increasing function of temperature. In more clean Fe films with only $\simeq 0.03$ at. % C, the noise increases with temperature without any clear peak, and $m''(f)$ is almost independent of temperature.

Two conclusions can be made: (1) 1/f noise in metals is produced by mobile defects, and (2) 1/f noise is related to other low-frequency kinetic phenomena in metals with defects.

8.5 Low-frequency noise in semiconductors

Intensive studies of the low-frequency noise in semiconductors and semiconductor devices have been conducted from the beginning of the 1950s, i.e., from the onset of vigorous growth of the physics of semiconductors and semiconductor electronics, which followed the creation of pure semiconductor materials technology and the invention of the transistor. Since then an enormous amount of work has been done. Of course, the main goal was to find how to eliminate this noise. The solution of this problem requires, at least, the knowledge of the noise sources, in particular, whether they are at the surface or in the bulk. There is much evidence of surface mechanisms of low-frequency noise, as well as some evidence of the presence of noise sources in the bulk.

8.5.1 Surface noise

The most obvious evidence of surface sources of $1/f$ noise is the strong effects of purely surface treatment on the intensity and the spectrum of the noise. As is well known, the conductance (inverse resistance) of the near-surface space-charge layer of a semiconductor depends on the magnitude of the charge in surface states. Fluctuations of this charge must lead to fluctuations of the near-surface region conductance and, hence, of the entire specimen. One can change the properties of the surface and space-charge layer without any change of the bulk properties of the semiconductor: (1) by application of an electric field perpendicular to the surface through a dielectric or vacuum layer (field effect), and (2) by chemisorption or desorption of various substances by changing the ambient atmosphere in which the specimen is placed.

The effect of an electric field perpendicular to the surface on $1/f$ noise has been studied on n- and p-type Ge specimens 50–100 μm thick (McRae & Levinshtein, 1960). The noise increased as the type of conduction in the space-charge region became opposite to the type of bulk conduction, i.e., as the inversion layer at the surface increased. The greatest increase in noise was approximately by $5-6$ times. The noise increased only when the surface potential ϕ_s in p- and n-type specimens increased in the positive and in the negative direction, respectively. Such variation of ϕ_s, which increased the number of majority carriers in the space-charge layer, caused no appreciable effect on the noise. It is interesting and important for the interpretation of the mechanism of noise that upon changing the electric field the noise relaxes following a very slow (logarithmic) law, unlike the conductivity (field effect) which relaxes far more rapidly.

The effect of gaseous ambient on $1/f$ noise has been observed in n- and p-type Ge specimens with a transverse cross-section of 0.5×0.5 mm^2 (Mapple, Bess & Gebbie, 1955). The relative change in the spectral density of noise in going from dry to moist nitrogen was, in different specimens, from -0.5 to $+5.5$. A ten-fold increase in the noise was also observed upon varying the ambient atmosphere from dry nitrogen to liquid CCl$_4$. This effect was accompanied by a change of the exponent γ in the frequency dependence of the spectral density ($f^{-\gamma}$) from 1 to 1.2. In the experiments by Noble & Thomas (1961) with rather thin n-Ge specimens the composition of the ambient atmosphere was cyclically changed from wet nitrogen to ozone. It resulted in a cyclical change of the surface potential from positive values

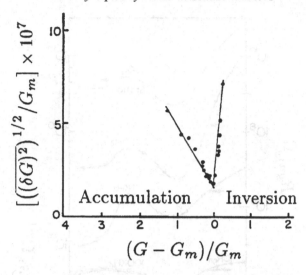

Fig. 8.19 Dependence of the spectral density of fractional conductance fluctuations, S_G/G^2, in a n-Ge specimen on the relative variation of conductance in the cycle of changing the ambient atmosphere from wet nitrogen to ozone. Thickness of the sample was 6 μm, $\rho = 32$ Ohm cm. From Noble & Thomas (1961).

(the near-surface region enriched in majority carriers) to negative values that cause the type of conduction in the near-surface region to invert. As one should expect, the conductance of the specimens passed through a minimum G_m (Garrett & Brattain, 1955). The spectral density of noise was directly measured as a function of $(G - G_m)/G_m$ (Fig. 8.19). Remarkably, the noise is minimal at the same composition of the ambient atmosphere as the conductance. At the extreme points of the cycle, the noise is higher than at the minimum by a factor of 25 to 40. In contrast with the experiments by McRae & Levinshtein (1960), an increase of noise was found not only upon varying the potential ϕ_s in the direction of inversion layer formation but also in the opposite direction.

One of the most important devices in semiconductor electronics, the metal–oxide–semiconductor transistor, often exhibits considerable $1/f$ fluctuations of the current in the channel parallel to the dielectric layer. This noise has been studied by many authors. Sah & Hielscher (1966) compared two dependencies measured on the same structure: the spectral density of noise $S_U(f; V_G)$ and the surface states' density $\rho_{SS}(V_G)$ versus the gate voltage V_G across the oxide layer between the metal and semiconductor. These two relationships proved similar at low frequencies $f < 1$ kHz (Fig. 8.20). The relationships $S_U(V_G)$ and the transverse conductivity $G_{SS}(V_G)$, which involves transfer of charge into the surface states also proved similar. A

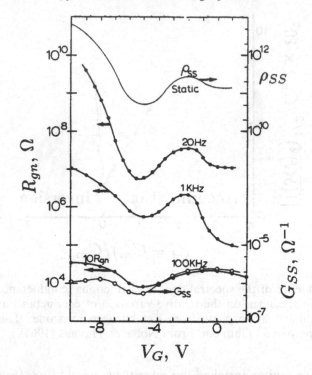

Fig. 8.20 Equivalent noise resistance of a n-channel MOS field-effect transistor structure (left-hand scale), density of surface states ρ_{SS} in states/(eV·cm²) (upper right-hand scale), and the reduced real part of the input admittance of the gate G_{SS} (lower right-hand scale) versus gate voltage. From Sah & Hielscher (1966).

similar comparison of S_U and ρ_{SS} has been made also by Abowitz, Arnold & Leventhal (1967).

If the conductance fluctuations of the near-surface region stem from the fluctuations of the density of current carriers, and, ultimately, from fluctuations of the surface potential ϕ_s, then the spectral density of voltage fluctuations equals:

$$\frac{S_U(f)}{U^2} = \left(\frac{dG_s}{d\phi_s}\right)^2 \frac{S_\phi(f)}{(G_V + G_s)^2}. \qquad (8.5.1.1)$$

Here G_V is the conductance of the specimen in the absence of any surface charge, G_s is the conductance of the near-surface region, and $S_\phi(f)$ is the spectral density of surface potential fluctuations.

Within this model, the surface noise must vanish at the same surface potential at which the surface conductance G_s is minimal. It agrees with experiment (Fig. 8.19) but only in the sense that the noise passes through a minimum simultaneously with G. The noise does not vanish at $G = G_m$, although it is relatively small. This nonzero minimum noise can be explained by inhomogeneity of the surface and its potential, or by the presence of some bulk noise.

The dependence of the measured noise on the dimensions of the specimens, i.e., the surface area A and thickness d, can also be used to distinguish between the surface and bulk noise (Mircea, Roussel & Mittonneau, 1972). If the surface is uniform and the correlation length of the charge fluctuations in the surface states is many times smaller than the linear dimensions of the surface, the spectral density of purely surface relative fluctuations is inversely proportional to the surface area, i.e., $S_U/U^2 \propto A^{-1}$. Indeed, in Eq. (8.5.1.1) $S_\phi \propto A^{-1}$, $G_s \propto w/L$, $G_V \propto wd/L$, where d, w and L are the thickness of the specimen, its width, and length, respectively. When $G_V \gg G_s$, the spectral density of relative fluctuations, $S_U/U^2 \propto (Ad^2)^{-1}$. In the opposite case when the surface conductance is dominant, $S_U/U^2 \propto A^{-1}$. However, if the noise is of purely bulk origin $S_U/U^2 \propto V^{-1}$ (Sec. 8.2.5). Experiments on $n^+\text{-}n\text{-}n^+$ GaAs resistors confirmed the dependence $S_U/U^2 \propto (Ad^2)^{-1}$ (Mircea *et al.*, 1972), and enabled the authors to deduce a surface mechanism of noise in these resistors.

It is interesting that in rather typical conditions the spectral density of relative fluctuations S_U/U^2 produced by surface sources is approximately inversely proportional to the density of the charge carriers in the bulk, i.e., to n^{-1} (see the review by Kogan, 1985). The proportionality of the spectral density of noise to the bulk resistivity of the material has been found by Montgomery (1952) in experiments on n- and p-type Ge. This empirical rule corresponds to the empirical Hooge relationship (Sec. 8.2.6) which, however, was considered by Hooge as favoring a bulk mechanism of $1/f$ noise.

8.5.2 Bulk $1/f$ noise

As was mentioned above, the problem of separating the bulk noise from the surface one is rather difficult. One of the ways to observe a purely bulk noise is to confine the charge carriers to a channel in the bulk which does not touch the specimen's surface (see below).

The measured intensity of $1/f$ noise estimated by the value of the dimensionless Hooge constant α varies in an extremely wide interval $\sim 10^{-7} - \simeq 1$ (see Table 8.2). The noise intensity is different not only for different materials and devices but also for a given specimen in different conditions. It depends on the technology of the specimen's preparation.

Many experiments have been made to understand what fluctuates: the number or the mobility of charge carriers. The idea of an elegant experiment by Levinshtein & Rumyantsev (1983) was the following. If a specimen, the length L of which is much shorter than its width w, is placed in a magnetic field B, the Hall current (the current perpendicular to the electric field) is not zero, and the Hall field is small (unlike specimens with opposite relation between L and w). The same takes place in a Corbino disk, i.e., a circular specimen in which the current flows in radial directions between the central contact and the peripheral one. As is well known, under these conditions the measured mobility $\mu(B)$ of the charge carriers depends on the magnetic field B:

$$\mu = \frac{\mu_0}{1 + (\mu_0 B/c)^2},\tag{8.5.2.1}$$

where μ_0 is the mobility at $B = 0$, c is the speed of light (we are using the Gaussian system of units).

If $1/f$ fluctuations are produced entirely by mobility fluctuations, the spectral density of relative mobility and, hence, resistance fluctuations in a magnetic field should be:

$$\frac{S_\mu(f;B)}{\mu^2} = \left[\frac{1 - (\mu_0 B/c)^2}{1 + (\mu_0 B/c)^2}\right]^2 \frac{S_\mu(f;0)}{\mu_0^2}.\tag{8.5.2.2}$$

One should expect that in the case of mobility fluctuations, the spectral density of noise as a function of magnetic field should drop to zero at $\mu_0 B/c = 1$ and then increase with magnetic field up to a level that is independent of B. Experiments on GaAs have shown, however, that the spectral density of noise is independent of the product $\mu_0 B$, i.e., of the magnetic field, and no drop of the noise at $\mu_0 B/c = 1$ was ever observed. These results have been confirmed later (Song & Min, 1985; Song, Birbas, van der Ziel & van Rheenen, 1988). Thus, in these specimens of n-GaAs the noise more probably originates from the fluctuations of the number of charge carriers.

This conclusion is substantiated by another observation. Levinshtein and his group found a strong effect of illumination by light with quanta $h\nu$ close to the band gap E_g on $1/f$ noise in Si and epitaxial n-GaAs (Vainstein, Levinshtein & Rumyantsev, 1987; Guk, Dyakonova & Levinshtein, 1988; Dyakonova, Levinshtein & Rumyantsev, 1988; Dyakonova & Levinshtein, 1988). The effect dropped steeply as the photon energy $h\nu$ became smaller than the gap E_g at the temperature of measurement, but it dropped also when the quanta $h\nu$ were increased in excess of E_g. The noise spectral density decreased at low frequencies $f < \sim 1$ kHz, and increased at higher frequencies. The absolute magnitude of the effect of illumination on noise ranged up to several decibels (several times) despite the change of conductance, i.e., the photoconductivity, being not greater than $0.1 - 1\%$. The effect became more pronounced as the temperature was lowered. The authors interpret these phenomena by photogeneration of holes. The concentration of these nonequilibrium holes was no more than 1% of that of electrons. The holes are trapped by the defects, the levels of which are deep in the forbidden energy gap. In the absence of illumination, these levels, being lower than the n-type specimen, are permanently occupied by electrons and do not participate in the generation–recombination (G–R) processes. After trapping holes, such levels can trap conduction electrons. These processes give rise to a G–R noise. As the spectrum of the deep levels is broad and the trapping time depends on the energy of the level, this noise has, in the opinion of the authors, a $1/f$ type spectrum in a very broad frequency range.

Lukyanchikova (1993) examined extensive experimental data on $1/f$ noise and the life-time of charge carriers τ in the same samples. She was able to show that a relation between $1/f$ noise and recombination of charge carriers holds: $\alpha = \beta\tau$. The 'universal' coefficient $\beta \simeq 3 \cdot 10^2 \ \text{sec}^{-1}$. This empirical relation points out that some connection between $1/f$ noise and recombination of charge carriers exists.

Examination of Table 8.2 shows that the smallest dimensionless $1/f$ noise intensity parameter $\alpha \sim 10^{-7}$ occurs in the case when the charge carriers are confined to a channel that does not touch the specimen's surface (Tacano, Tanoue & Sugiyama, 1992). This trend was found also by some other authors. One may speculate that the bulk $1/f$ noise, unlike the surface noise, can reach very low values.

8.6 Magnetic noise in spin glasses and some other magnetic systems

Spin glasses are magnetic systems in which the interaction energy between the elementary magnets (spin magnetic moments) takes randomly different signs due to some frozen-in structural disorder. Therefore the magnetic moments cannot establish any order of ferromagnetic or antiferromagnetic type. Each such system exhibits at some temperature T_G a 'freezing transition' to a state in which the magnetic moments are aligned in random but correlated directions. In the absence of any external magnetic field before and after this freezing transition, the net magnetic moment is zero. The randomness of the magnetic moments' alignment and the freezing transition resemble in some sense an ordinary ('structural') glass, that is why these magnetic spin systems are called spin glasses. There are several excellent reviews on the physics of spin glasses (Binder & Young, 1986; Dotsenko, Feigelman & Ioffe, 1990; Fisher & Hertz, 1991).

A large number of spin glasses are known at the present time. One of the most investigated class of spin glasses is dilute solutions of transition metals (Mn, Fe) in noble metals (Au, Ag, Cu). The spin magnetic moments of the transition metal ions interact with each other via the degenerate electron gas. The energy of this indirect exchange interaction between two spin moments (it is called the Ruderman-Kittel-Kasuya-Yosida, or RKKY, interaction) equals:

$$J(\mathbf{R}) = J_0 \frac{\cos(2k_F R + \phi_0)}{(k_F R)^3}. \tag{8.6.1}$$

Here $\mathbf{R}$ is the radius-vector that connects the two interacting magnetic ions in the metal, k_F is the wave-vector of electrons at the Fermi surface, J_0 and ϕ_0 are constants. Equation (8.6.1) is valid at $k_F R \gg 1$. As in metals $k_F \sim a_0^{-1}$, where a_0 is the lattice parameter, this inequality is satisfied if the distance $R \gg a_0$. The last condition is met in dilute alloys.

Since the Fermi wave-vector $k_F \sim a_0^{-1}$, a change of the distance R by one lattice parameter only leads to a drastic change of the argument of the cosine function in the r.h.s. of Eq. (8.6.1) and, possibly, to the change of the very sign of the RKKY interaction. Because the transition metal ions are distributed at random, the signs and absolute values of the interactions between their spin magnetic moments vary at random.

Let the interaction of two definite spin moments be positive (ferromagnetic type of interaction) and they tend to be aligned parallel. However, the interaction of any of these two magnetic moments with other magnetic moments may occasionally be stronger, and the actual alignment of the two spins can become antiparallel. It means that at temperatures $T < T_G$ the ferromagnetic and antiferromagnetic interactions ('bonds') between the magnetic moments in a spin glass compete and are 'in conflict' with each other.

At temperatures higher than T_G the spin glass is in a paramagnetic state. The correlation between the spins is small and each spin frequently changes its direction. Its time-average is zero. Therefore both the static magnetic susceptibility $\chi(0)$ and the real part of the frequency-dependent susceptibility $\chi'(f)$ grow as the temperature is lowered. In a narrow temperature range around T_G the magnetic moments of the spin glass become frozen. Although below T_G the directions of the magnetic moments are random, most of the time the direction of each individual magnetic moment remains fixed. Thus, the spin glass state is a 'frozen' state. The alignment of the magnetic moments under a small magnetic field is hindered, hence the susceptibilities $\chi(0)$ and

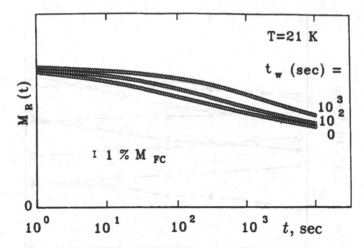

Fig. 8.21 Relaxation of the remanent magnetization $M_R(t)$ in Cu–5 at. % Mn spin glass. The sample was cooled in a magnetic field 2 G from a temperature higher than $T_G = 28$ K to the temperature $T = 21$ K. The magnetic field was switched off after a wait time t_w at constant temperature. 1 % of the field cooled magnetization is shown. The curves correspond to different t_w. From Nordblad *et al.* (1986).

$\chi'(f)$ decrease with decreasing temperature. If the spin glass is cooled from above T_G to a temperature below T_G in a strong enough magnetic field, and then the magnetic field is switched off, the remanent magnetization of the specimen decays with time so slowly that this decay is observed during macroscopic times. At low temperatures these times become huge (Fig. 8.21).

The relaxation of the magnetic moment of a spin glass specimen is obviously not exponential. It is approximated by some authors by a logarithmic law (other representations of the experimental relaxation curves are also used):

$$M(t) = M_1 - m\ln(t/t_1). \qquad (8.6.2)$$

Here t_1 is an instant of time at which the logarithmic decay starts, M_1 is the magnetic moment at this instant, and m is a constant that strongly depends on temperature. The logarithmic law Eq. (8.6.2) can hold only up to some maximum time t_{max} that does not exceed $t_1 \exp(M_1/m)$. It may be anticipated that at such long times the logarithmic law gives way to an exponential relaxation.

The 'cusp' in the temperature dependence of $\chi'(f)$ at $T = T_G$ is observed at all frequencies in a wide frequency range in which the measurements are usually made, i.e., from 10^{-3} Hz up to 1 MHz (these frequencies are smaller by several orders of magnitude than the inverse microscopic time of spin-flip, $\sim 10^{12}$ Hz).

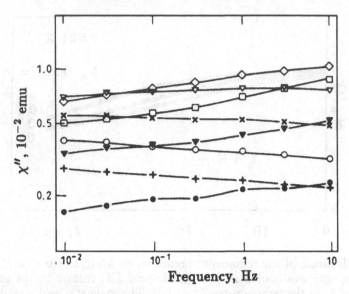

Fig. 8.22 Imaginary part of the magnetic susceptibility, $\chi''(f)$, of spin glass $CsNiFeF_6$ vs frequency f for different temperatures: $+$, 4.0 K; $\circ$, 4.2 K; $\times$, 4.4 K; open ∇, 4.6 K; $\diamond$, 4.8 K; open squares, 5.0 K; solid ∇, 5.2 K; $\bullet$, 5.4 K. Below 4.6 K the logarithmic slopes of χ'' become negative. From Refregier, Alba, et al. (1987).

Another striking feature of spin glasses is the frequency dependence of the imaginary (out-of-phase, dissipative) magnetic susceptibility $\chi''(f)$. Unlike ordinary magnets with exponential relaxation in which $\chi''(f) \to 0$ as $f \to 0$, $\chi''(f)$ of a spin glass is nonzero and almost independent of f in a very broad frequency range down to extremely small frequencies f (Fig. 8.22).

The logarithmic decay of the remanent magnetization, the independence of the drastic change of the temperature dependence of $\chi'(f)$ on frequency f, and the nonzero value of $\chi''(f \to 0)$ mean that the effect of freezing of spin magnetic moments cannot be reduced to an increase of a single relaxation time: it can be interpreted only as a vast broadening of the relaxation times' spectrum.

The simplest way to estimate the spectral density of magnetic moment fluctuations of a spin glass is to assume that both the state of the specimen before the magnetic field is switched off and its final state after the relaxation is completed are equilibrium states, and to use the fluctuation–dissipation relation (FDR) in the forms (2.1.19) and (2.1.43), taking into account that in this case $hf \ll k_B T$:

$$S_M(f) = V\frac{2k_B T}{\pi f}\chi''(f) = \frac{2k_B T}{\pi f}\int_0^\infty dt\, \sin(2\pi f t)\left[-\frac{d}{dt}\frac{M(t)}{H_0}\right]. \qquad (8.6.3)$$

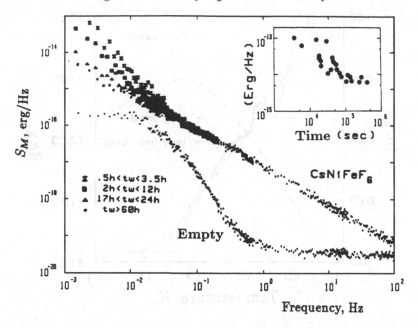

Fig. 8.23 Spectral density of magnetic noise in CsNiFeF$_6$ spin glass vs frequency f (upper curve) after subtracting the noise of the system in the absence of the sample (lower curve). t_w is the waiting time after cooling the sample. In the insert, time evolution of the noise at $f = 3 \cdot 10^{-3}$ Hz. $T = 4.2$ K. $T_G = 5.4$ K. From Ocio *et al.* (1986).

Here M is the magnetic moment of the entire specimen of volume V, H_0 is the initial magnetizing field which is supposed to be small enough (linear approximation). Substitution of the logarithmic law (Eq. (8.6.2)) into Eq. (8.6.3) allows one to conclude that the spectrum of magnetic fluctuations in a spin glass is of $1/f$ type (Kogan, 1981). The same result follows from the nonzero value of $\chi''(f \to 0) > 0$:

$$S_M(f) = \frac{2 k_B T V \chi''(f \to 0)}{\pi} \frac{1}{f} = \frac{k_B T m}{H_0} \frac{1}{f}. \qquad t_{\max}^{-1} \ll 2\pi f \ll t_1^{-1}. \quad (8.6.4)$$

The spectra of magnetic fluctuations in several dielectric spin glasses have been measured and found to be of $1/f$-type at low frequencies (Ocio, Bouchiat & Monod, 1985, 1986; Reim, Koch, Malozemoff *et al.*, 1986; Refregier, Alba, Hammann & Ocio, 1987; Alba, Hammann, Ocio *et al.*, 1987; Refregier, Ocio & Bouchiat, 1987; Refregier, Ocio, Hammann & Vincent, 1988; Gunnarson, Svedlindh, Nordblad *et al.*, 1989). A typical spectrum of magnetic fluctuations in spin glasses is presented in Fig. 8.23.

Slow kinetic phenomena, including $1/f$ magnetic noise, and theoretical studies (see below) show that the spectrum of relaxation times in spin glasses

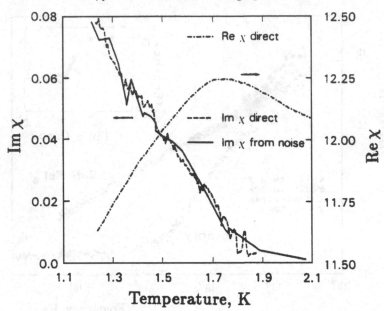

Fig. 8.24 Comparison of temperature dependencies of (1) the imaginary part of magnetic susceptibility Im $\chi \equiv \chi''(f)$ calculated from the measured value of magnetic noise (solid curve), and (2) directly measured ac $\chi''(f)$ (dashed curve). In both cases $f = 50$ Hz. Dash-dot curve: Re $\chi \equiv \chi'$ (right scale). Spin glass $Eu_{0.4}Sr_{0.6}S$ with freezing temperature $T_G = 1.53$ K. From Reim et al. (1986).

is extremely broad. Such systems are not ergodic (see the discussion in Sec. 1.4), and the measured values of physical quantities averaged over the time of measurement t_m may differ from the values obtained by averaging over the Gibbs ensemble. Hence, the fluctuation–dissipation relation (Eq. (8.6.3)) may not hold. It was interesting to verify experimentally the validity of FDR. Such verification was accomplished by Reim, Koch, Malozemoff et al. (1986), by Refregier, Ocio, Hammann & Vincent (1988) (see also Hammann, Ocio & Vincent, 1990), and by Gunnarson, Svedlindh, Nordblad et al. (1989). In Fig. 8.24 the imaginary part of the magnetic susceptibility $\chi_N''(f)$ calculated using measured values of the spectral density of magnetic noise $S_M(f)$ and Eq. (8.6.3) is compared with directly measured ac $\chi''(f)$. One can see that both susceptibilities are close to one another in the whole temperature range. Refregier et al. (1988) also found that these two susceptibilities are similar and close to the susceptibility $\chi''(f)$ calculated from the relaxation of the magnetization using the second Eq. (8.6.3). One can see that, despite spin glasses being nonergodic, all three susceptibilities appear to be consistent, and the FDR are obeyed within a satisfactory accuracy. This result is important for the kinetics of spin glasses. It means that their magnetic noise is close to an equilibrium one. It is important also for the problem of $1/f$ noise as a whole: $1/f$ noise in spin glasses does not appear as an isolated phenomenon, its connection with other kinetic phenomena in the same systems is firmly established.

Electrical resistance noise of $1/f$-type in thin films of $Cu_{1-x}Mn_x$ spin glasses at temperatures around the spin glass freezing temperature T_G was found and studied by Israeloff, Weissman, Nieuwenhuys & Kosiorowska (1989). Several experimental facts prove that the resistance fluctuations are produced by the fluctuations of the spin system. The most striking is a dramatic increase of the resistance noise by almost two orders of magnitude when the temperature crosses a narrow temperature interval around T_G. The temperature of maximum temperature derivative of the resistance noise agrees very well, at all values of the Mn content x, with the value of T_G determined from magnetic susceptibility measurements. Moreover, cooling of the spin glass from above T_G in a magnetic field ≈ 2.1 T strongly affects the spectrum of the resistance noise, specifically, the exponent γ in the frequency dependence $S_R(f) \propto f^{-\gamma}$. Magnetic fields of this magnitude can affect, in this metallic alloy, only the dynamics of the spin system.

The spectral density of magnetic noise $S_M(f)$ in the films was determined as proportional to $T\chi''(f)$ (Eq. (8.6.3)). Below T_G the behavior of the resistance noise S_R, on one hand, and S_M, on the other hand, are different. As temperature decreases, the resistance noise increases, while the magnetic noise drops. The growth of the resistance noise indicates that it results from fluctuations of the electron interference pattern (Secs. 6.2 and 8.4.2).

The behavior of spin glasses is determined both by many-body competing interactions and disorder. Hence, the problem of kinetics of the spin glasses is extremely complicated and is not yet fully solved. Two models are discussed: the model of droplet-like excitations and the model of hierarchy of metastable states (for a comprehensive review see Weissman, 1993).

The droplet model (Fisher & Huse, 1988a, 1988b) takes into account that the interaction between spins is short-range. According to this model, any degeneracy of the ground state of the spin glass can result only from the symmetry of the Hamiltonian, e.g., with respect to the reversal of all spins. The excitations of the spin-glass are droplet-like compact clusters of spins that flip (turn over). The excitation energies are determined by the interactions of the droplet's spins with the surrounding spins across the interface. These energies grow with the linear dimension, L, of the droplet according to a power law. Due to the randomness of the signs of the interaction energies the exponent of this power law is substantionally lower than it would be in an ordered magnet. More important is that, owing to the same randomness, the excitation energy is not a definite function of L: at each L there is a distribution of energies, including small ones. The statistical probability of a thermal excitation with a dimension L at a given temperature T falls off with increasing L not exponentially but more smoothly. One should expect that the activation energy of a droplet flip, which determines the relaxation time τ, increases with L. Therefore the spectrum of τ is very broad, the long τ corresponding to large L, the short τ to small L. It results in a nonexponential kinetics. The above mentioned smooth fall of the statistical probability of low-energy excitations with increasing L and, hence, increasing τ, means also a smooth fall of the probability of long-τ fluctuations and, particularly, an increase of the spectral density of noise with decreasing frequency f at low f.

According to the second model, the spin glasses have a specific, hierarchical, structure of their states. Each such thermodynamic state is a set of microscopic states, the transitions between which are frequent on the scale of some characteristic time (for instance, the time of experiment) and the relative probabilities of finding the system in each of these microscopic states correspond to the Boltzmann formula. Each such thermodynamic state is a metastable state, and may be viewed as a valley in the phase space, i.e., in the space of microstates. These metastable states differ by orientation of all or some of the spins. They are separated by energy activation barriers. Two or more shallow valleys form a more deep valley of the next level. The metastable states of this level are separated by higher barriers, and so on. If these states strongly differ (a large fraction of spins has to be overturned to convert one state into another), the necessary activation energy is high and the relaxation time is long. Thus, such a system has a hierarchy of relaxation times, and its kinetics is obviously nonexponential.

In the Sherrington–Kirkpatrick model, from which the hierarchical model originates, the interactions between spins have been assumed random but independent of the distance between spins (infinite radius). In this model there are an infinite number of low-lying states with infinite barriers between them. One can guess that in a more realistic model with finite radius of interaction there is a single ground state (the number of ground states may be doubled by the symmetry of the Hamiltonian) but in any vicinity of the ground level there are metastable states with a hierarchical structure.

Thus, both known models of spin glasses show nonexponential kinetics. However, the statistics of fluctuations in these models is in some respects different. Weissman and his group suggested and performed experiments which can distinguish between the two models (Weissman, Israeloff & Alers, 1992; Weissman, 1993). The idea was to measure the electrical resistance fluctuations in such small specimens of spin glasses that the deviations of the statistics of fluctuations from the Gaussian one may become distinct (the fluctuations of the magnetic moment of these small specimens can not be measured by modern quantum magnetometers). The resistance fluctuations are generated by spin flips, which affect the electron interference pattern in the metal (Sec. 6.2).

In a small specimen only a small number of large droplets can be effective. The fluctuations then are expected to be a superposition of several Lorentzian functions, each corresponding to flips of a droplet between two of its states. It is very important that within the model of noninteracting droplets this spectrum remains fixed in time. According to the hierarchical model of kinetics, on the contrary, the systems of spins performs thermal transitions within one deep valley between more shallow valleys, but sometimes undergoes transitions to another deep valley with a different spectrum of relaxation times of thermal motion between shallow valleys. This 'wandering' of the low-frequency noise is a distinctive feature of hierarchical kinetics. The wandering of the spectrum of relaxation times can be detected by observing the fluctuations in the time domain. A quantitative characterization of this effect can be achieved by measuring the second spectral density, $S^{(2)}(f|\bar{f}, \Delta f)$ (Sec. 1.3). Actually, this quantity is the spectral density of fluctuations of the noise power in the frequency range Δf around the frequency $\bar{f}$. The magnitude of the relative second spectral density (divided by the first spectral density squared) is a quantitative characteristic of the intensity of wandering of the first spectrum, and its dependence on f reveals the characteristic relaxation times of this wandering.

The experiments on Cu:Mn spin glasses by Weissman, Israeloff & Alers have shown that the effect of wandering of the $1/f$ noise spectrum is very significant. These results rule out the droplet model in its simplest version and support the hierarchical model of spin glass kinetics.

Noise in ferromagnets is usually measured using an alternating magnetic field or switching the field. Lebed', Marchik, Noskin & Shirko (1982) found noise with a $1/f^\gamma$ spectrum in various ferrites (Mn–Zn, Ni–Zn, and others) placed in a *static* magnetic field and in the absence of any external magnetic field. Three methods have been used: inductive transducer, measurement of resistance noise, and measurements of magneto-optical fluctuations. The spectrum of resistance noise was $\propto 1/f$ ($\gamma = 1$) in the absence of a magnetic field, and was $\propto 1/f^2$ at $H = 0.03$ T. The magnetic noise had $\gamma = 2$, i.e., it was not $1/f$ noise. Possibly the sources of the resistance and magnetic noises were different.

A clearly magnetic noise with $1/f$-type spectrum was found in soft amorphous ferromagnets (mainly Co-based alloys) at temperatures $T \leq 4.2$ K (Prodi, Vitale, Cerdonio & Falferi, 1989; Vitale, Tommasini, Cerdonio *et al.*, 1992; Durin, Falferi, Cerdonio *et al.*, 1993) and in the soft Ultraperm alloy $Fe_{12}Ni_{77}Cu_5Mo_4$ (Vitale, Cavalleri, Cerdonio, Maraner & Prodi, 1994). The noise was measured using SQUID magnetometers in the absence of any external field. These measurements have been complemented by measurements of low-frequency complex susceptibility $\chi(f)$. As in spin glasses (see above), the dissipative part $\chi''(f)$ tends to a nonzero value at low frequencies. The measured spectral density of magnetic noise and the dissipative part of the susceptibility satisfy the fluctuation–dissipation relationship (Eq. (8.6.3)). It means that within satisfactory accuracy the measured $1/f$ magnetic noise is a thermal and equilibrium one. The mechanism of the magnetic fluctuations is not yet clear: the transitions of, say, domain walls over energy barriers, tunnel transitions of magnetization, or hopping of various structural defects which result in variations of magnetization.

In metallic superlattices consisting of alternating transition metal and nonmagnetic layers, each of several lattice parameters thick, the magnetizations of the transition metal layers in zero external magnetic field align ferromagnetically (the magnetic moments of all these layers are parallel) or antiferromagnetically (antiparallel alignment of neighbouring layers). The alignment in zero magnetic field $H = 0$ depends, in particular, on the thickness of the nonmagnetic layers (spacers). If at $H = 0$ the alignment is antiferromagnetic, an applied magnetic field forces the magnetizations of all layers to align in parallel. The resistance R of the superlattice depends on the magnetic state: whether it is antiparallel (R_{AP}) or parallel (R_P). Therefore, the superlattice exhibits a giant magnetoresistance (GMR) which comprises several tens of percents (Baibich, Broto, Fert *et al.*, 1988; Binach, Grunberg, Sauerenbach & Zinn, 1989). The resistance and magnetization are connected by an 'interpolation' equation (Zhang, Levy & Fert, 1991):

$$R = R_{AP} - (M/M_s)^2 \Delta R, \tag{8.6.5}$$

where M_s is the saturation ferromagnetic magnetization, $\Delta R = R_{AP} - R_P$.

Since the resistance and magnetization of the superlattice are strongly coupled, the fluctuations of the specific magnetization M result in a resistance noise:

$$S_R(f) = \left(\frac{dR}{dM}\right)^2 S_M(f). \tag{8.6.6}$$

The resistance noise in Co/Cu superlattices was measured by Hardner, Weissman, Salamon & Parkin (1993) (see also Hardner, Parkin, Weissman *et al.*, 1994). The spectra of this noise have been found to be of $1/f$ type.

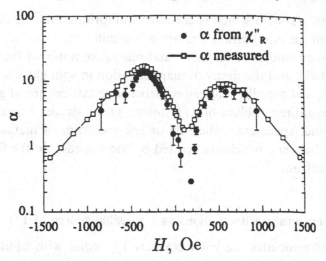

Fig. 8.25 Comparison of the directly measured dimensionless $1/f$ noise parameter $\alpha(H)$ and noise parameter $\alpha(H)$ calculated from the χ_R'' data using Eq. (8.6.3) in a metallic superlattice with giant magnetoresistance. From Hardner et al. (1993).

The dimensionless noise magnitude

$$\alpha(H) \equiv \frac{fS_R(f)N}{R^2} = \frac{fS_M(f)N}{R^2}\left(\frac{dR}{dM}\right)^2, \tag{8.6.7}$$

where N is the number of atoms in the sample.

In order to find whether the measured noise satisfies the fluctuation–dissipation relation (Eq. (8.6.3)), the authors measured the linear ac response of the resistance $\delta R(t)$ to ac magnetic field $\delta H(t)$ and obtained the out-of-phase susceptibility:

$$|\chi_R''(f)| = \left|\text{Im}\frac{\delta R(f)}{\delta H(f)}\right| = \chi''(f)\left|\frac{dR}{dM}\right|, \tag{8.6.8}$$

where $\chi''(f)$ is the magnetic out-of-phase susceptibility.

If the magnetic noise is an equilibrium one, its spectral density is given by Eq. (8.6.3), divided by V^2 because in Eq. (8.6.8) M is the specific magnetization. Using Eqs. (8.6.5)–(8.6.8) one obtains:

$$\alpha(H) = \frac{4k_BT}{\pi\mu_sR^2}[\Delta R(R_{AP} - R(H))]^{1/2}|\chi_R''(f)|, \tag{8.6.9}$$

where $\mu_s = M_sV/N$ is the saturation magnetic moment per atom.

The comparison of directly measured $\alpha(H)$ with $\alpha(H)$ found from χ_R'' (Eq. (8.6.9)) is presented in Fig. 8.25. The agreement between the two quantities is satisfactory, except at small fields. It proves that the measured $1/f$ noise is produced in a system which is close to equilibrium, and the deviations of the noise from the equilibrium one are not significant.

The combined studies of magnetic and resistance noise, of the response to ac magnetic field, and the decay of magnetization in spin glasses, amorphous ferromagnets, and metallic magnetoresistive superlattices are of great importance for the entire problem of $1/f$ noise. These studies revealed the role of disorder and, specifically, the role of the spectrum of metastable states, separated by barriers, which are created by the disorder, in the formation of $1/f$ noise spectrum.

8.7 Temperature fluctuations as a possible source of $1/f$ noise

A model that associates the low-frequency $1/f$ noise with equilibrium fluctuations of the resistor's temperature (Sec. 6.4) was suggested by Voss & Clarke (1976). According to this model the resistance fluctuations stem from the temperature dependence of the resistance: $\delta R(t) = (dR/dT)\delta T(t)$. The spectral density of the noise then equals:

$$S_R(f) = \left(\frac{dR}{dT}\right)^2 S_T(f). \tag{8.7.1}$$

As the theory of temperature fluctuations (Sec. 6.4 and below) gives no indications for $S_T(f)$ to have a $1/f$ spectrum in any significant frequency range, the authors postulated that for a specimen with dimensions $L_1 \gg L_2 \gg L_3$ the spectral density $S_T(f) \propto 1/f$ in the range from $f_1 = D/\pi L_1^2$ to $f_2 = D/\pi L_2^2$, where D is the heat diffusion coefficient. The spectral density at $f < f_1$ was assumed to be constant, at $f > f_2$ as declining $\propto f^{-3/2}$ (Secs. 6.3 and 6.4). The condition of $S_T(f)$ continuity and the condition imposed on the integral of $S_T(f)$ over all frequencies (it must be equal to the known variance $\langle (\delta T)^2 \rangle$, given by Eq. (6.4.1)) determined the model $S_T(f)$ uniquely. In the range where a $1/f$ spectrum was assumed,

$$\frac{S_R(f)}{R^2} = \frac{k_B T^2 \beta^2}{cV[3 + \ln(f_2/f_1)]}\frac{1}{f}, \qquad f_1 \leq f \leq f_2, \tag{8.7.2}$$

where V is the specimen's volume, c is the specific heat capacity, and $\beta = (dR/dT)/R$ is the temperature coefficient of resistance.

Several arguments have been considered as substantiating this model. Voss & Clarke have measured the voltage noise across thin (25–200 nm) continuous films of Au, Ag, Cu, Sn, Bi, and the alloy manganin at room temperature. In all cases but one, they found a significant noise with a spectrum which was close to $1/f$. The remarkable exception was manganin, whose temperature coefficient of resistance is anomalously small at room temperatures: $|\beta| < 10^{-4}$ K^{-1}. The magnitude of the noise measured in all metals was close to the value given by Eq. (8.7.2) (it was the second argument).

Voss & Clarke measured also the correlation of voltage fluctuations $\delta U_1(t)$ and $\delta U_2(t)$ across the ends of two segments of the same film lying at a distance L from one another. If the voltage fluctuations are caused by temperature fluctuations, they must be correlated at low frequencies $f < D/\pi L^2$, when the temperature fluctuations in both portions of the film are equal, $\delta T_1 = \delta T_2$. They are uncorrelated at high frequencies $f \gg D/\pi L^2$. An experiment performed on two Bi films found that the correlation actually decreases with f and vanishes approximately at $f \approx D/\pi L^2$. However, in one of these specimens the measured correlation passed through zero and became negative. The authors explained this by experimental errors and considered the results as confirming the temperature mechanism of $1/f$ noise.

In subsequent experiments (Eberhard & Horn, 1977, 1978, see also the review by Dutta & Horn, 1981) the spectral density $S_U(f)$ was measured in a broad temperature range $100 - 600$ K. Although at room temperature S_U is close to the value measured by Voss & Clarke and hence to the value given by Eq. (8.7.2), at other temperatures $S_U(f)$ was found to be very far from this value. Moreover, the temperature dependence of the noise is absolutely different from that of the model Eq. (8.7.2). The conclusion was made that the agreement between the experiments by Voss & Clarke and their temperature model was fortuitous.

The idea, suggested by Voss & Clarke, of measuring the correlation of voltage fluctuations in different but thermally coupled specimens has proved very fruitful. A number of experiments have been performed (see below), but in contrast to experiments by Voss & Clarke, no spatial correlation was found.

Scofield, Darling & Webb (1981) studied a system of two Au films (of thickness $\simeq 60$ nm) galvanically isolated from one another by a SiO layer $\sim 0.6\,\mu$m thick (Fig. 8.26). Two types of measurements were performed. First, the spectral densities of temperature fluctuations in the first and second films, $\delta T_1(t)$ and $\delta T_2(t)$, have been measured using pulse Joule heating of one film and measuring the temperature response in the same and in the second film (Sec. 6.4 and Eq. (6.4.9)). Thus, spectral densities $S_{T11}(f)$, $S_{T22}(f)$ and the cross spectral density $S_{T12}(f)$ have been found. The quantity

$$\gamma_T(f) = \frac{S_{T12}(f)}{\sqrt{S_{T11}(f)S_{T22}(f)}} \tag{8.7.3}$$

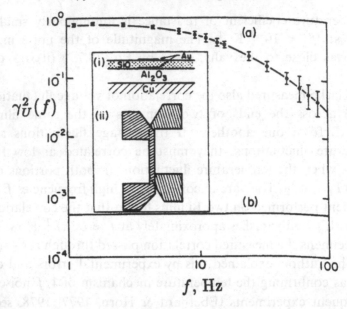

Fig. 8.26 Experimental study of coherence of $1/f$ fluctuations in two Au films separated by a thin insulating film of SiO. Curve (a): computed coherence γ_T assuming that the measured $1/f$ noise is due to temperature fluctuations; (b): the upper limit of directly measured coherence γ_U. Insert: Schematic cross section (i) and plan view (ii) of the specimen. From Scofield et al. (1981).

is a measure of the correlation of temperature fluctuations in the two films. At frequencies lower than the reciprocal time of heat diffusion from one film to another, their temperatures fluctuate 'synchronously', and $\gamma_T(f) \approx 1$. The measurements confirmed this.

On the other hand, the correlation of the voltage fluctuations $\delta U_1(t)$ and $\delta U_2(t)$ across the first and second films, i.e., the spectral densities $S_{U11}(f)$, $S_{U22}(f)$ and the cross spectral density $S_{U12}(f)$ have been also measured. The correlation coefficient $\gamma_U(f)$ analogous to $\gamma_T(f)$ (Eq. (8.7.3)) was calculated using these spectral densities.

If the resistance fluctuations of the films were completely associated with the temperature fluctuations, the following equation would hold:

$$\gamma_U^2 = \gamma_T^2 \{[1 + S_{B1}/S_{U11}(f)][1 + S_{B2}/S_{U22}(f)]\}^{-1}. \qquad (8.7.4)$$

Here S_{B1} and S_{B2} are the background noise spectral densities, which in the two films are known to be uncorrelated (hence a correction was introduced in Eq. (8.7.4)). The frequency dependence of $\gamma_U^2(f)$ calculated from Eq. (8.7.4) using measured $\gamma_T^2(f)$ is shown in Fig. 8.26 (curve a).

However, $\gamma_U^2(f)$ determined by direct measurements of $S_{U12}(f)$, S_{U11} (f) and $S_{U22}(f)$ has shown no correlation of the voltage fluctuations: to an accuracy of $5 \cdot 10^{-4}$ the dimensionless quantity γ_U^2 was found to be zero (line *b* in Fig. 8.26).

Kilmer, Chenette, van Vliet & Handel (1982) found no correlation of voltage fluctuations across adjacent transistors in a Si-integrated circuit within the accuracy $\sim 2\%$, although the temperature fluctuations in the same transistors at the same frequencies are completely correlated due to small distances. Black, Weissman & Fliegel (1981) have found no correlation of voltage fluctuations across the ends of two halves of a metal film ~ 1 mm long. Çelik-Butler & Hsiang (1988) have found no correlation between fluctuations in two different segments of a *p*-Si stripe separated by a distance of 63 μm.

The problem of temperature fluctuations as a possible mechanism of $1/f$ noise stimulated theoretical studies of the spectra of temperature fluctuations in various systems. If a specimen of finite dimensions is in contact with a three-dimensional infinite or semi-infinite medium, then at sufficiently low frequencies f the heat propagation from the specimen to the medium is three-dimensional in character and $S_T(f) \to$ const. as $f \to 0$. The frequency below which S_T does not vary with f depends on the dimensions of the specimen, the parameters of heat conduction in the specimen and in the medium, and on the thermal contact between them.

The most interesting situation from the standpoint of the mechanism of low-frequency noise arises when the heat diffusion may be considered one-dimensional. Van Vliet, van der Ziel & Schmidt (1980) studied temperature fluctuations in a film on a substrate of the same area. They found that a spectrum $S_T(f)$ of $1/f$ type can arise whenever there is a purely surface fluctuational energy flux, e.g., owing to fluctuations of thermal radiation. However, this can happen only in a narrow frequency range because a $f^{-1/2}$ spectrum dominates.

Vainshtein (1982) studied temperature fluctuations in a specimen in contact with an uniform wire or external circuit. When the length of the external circuit is infinitely large the spectrum at $f \to 0$ is $S_T(f) \propto f^{-1/2}$. If this length is finite the spectral density tends to a finite value at sufficiently low f.

In conclusion, experiments and theoretical studies prove that temperature fluctuations are unlike to be the source of $1/f$ noise. However, the paper by Voss and Clarke (1976), in which the temperature model was suggested, triggered numerous and elegant experiments and theoretical studies on $1/f$ noise and brought this field to a much higher level.

8.8 Random telegraph noise

8.8.1 Introduction

In many systems that comprise such semiconductor devices as p-n junctions, metal–oxide–semiconductor field-effect transistors (MOSFETs), metal–insulator–metal (MIM) tunnel junctions, small semiconductor resistors, or small metallic samples, the resistance and, consequently, the current or the voltage, switches at random between two (or several) discrete values (Fig. 1.5). The time intervals between switchings are random, but the two values of the fluctuating quantity are time-independent. This kind of noise is now usually called 'random telegraph noise' (RTN). It was observed beginning from the 1950s in reverse-biased p-n junctions. Then it was called 'burst noise', or 'popcorn noise' and most frequently interpreted as some local breakdown, accompanied by formation of microplasmas in the region of high electric field, or by switching on and off of some surface conduction channel, or by other mechanisms (for a detailed review see Buckingham, 1983). In any case, these phenomena have not been interpreted as events or transitions involving only a single electron or a single hole. Later, in the early 1980s, resistance switchings caused definitely by transitions of a *single* electron or a *single* defect were discovered in various conductors and devices. Just these kinds of RTN are discussed below in this section.

All such RTNs have some common features and, consequently, common statistical properties. First of all, the times spent by the system in each of the two or several states are much longer than the microscopic relaxation times (time of energy relaxation in the electron trapping or detrapping processes, time of momentum relaxation in metals, and so on) which play the role of the times of forgetting the previous state of the system. Therefore these random processes can be considered as Markov processes, specifically, discrete Markov processes (Sec. 1.7). The main quantities that determine the statistical properties of these processes are (for two state systems): (1) the equilibrium probabilities w_1 and w_2 of finding the system in the first or in the second state, respectively; (2) the probabilities of transition per unit time from one state to another: $W(1 \to 2)$ and $W(2 \to 1)$.

The states, between which the system is switching, may have different charge, i.e., different number of electrons. In thermodynamic equilibrium the probabilities w_i (i numbers the states of the system) are given by the Gibbs statistics for systems with variable number of particles (see, for instance,

Landau & Lifshitz, 1985):

$$w_i = g_i \frac{\exp[(E_F n_i - E(i))/k_B T]}{\sum_j g_j \exp[(E_F n_j - E(j))/k_B T]}. \tag{8.8.1.1}$$

Here $E(i)$ and n_i are the energy and number of electrons in the i-th state of the system, respectively, g_i is the degeneracy, i.e., the number of quantum states that have the same $E(i)$ and n_i and are considered as one 'state'. E_F is the Fermi energy, or Fermi level, of electrons in the system, that is, their chemical potential. The sum in the denominator is the partition function.

If the system can be only in two states, the partition function consists of only two terms. In this case:

$$w_1 = \left\{ \frac{g_2}{g_1} \exp\left[\frac{E(1) - E(2) - (n_1 - n_2)E_F}{k_B T} \right] + 1 \right\}^{-1}, \quad w_2 = 1 - w_1,$$

$$\frac{w_1}{w_2} = \frac{g_1}{g_2} \exp\left[\frac{(n_1 - n_2)E_F - (E(1) - E(2))}{k_B T} \right].$$

$$\tag{8.8.1.2}$$

For definiteness, let $n_1 - n_2 = 1$. It means that the microscopic system (it may be called an electron trap) is negatively charged in the state 1 (charge $-e$) and neutral in the state 2, or it is positively charged in state 2 and neutral in state 1. Then the energy difference $E(1) - E(2) \equiv E_t$ is just what in solid state physics is called the electron level of the system (or of the trap). When the Fermi level $E_F = E_t - k_B T \ln(g_1/g_2)$, the probabilities of the system being occupied by an electron or being empty are equal, $w_1 = w_2 = 1/2$. When the electron level is much higher than the Fermi level, i.e., $E_t - E_F \gg k_B T$, the system is very rarely occupied by an electron and the fluctuations of its occupancy are improbable. The fluctuations are improbable also in the opposite case, when E_t is lower than E_F by many $k_B T$.

The probabilities of transition, per unit time, between the two states, $W(2 \to 1)$ and $W(1 \to 2)$, on one hand, and the occupancy probabilities w_1 and w_2, on the other hand, are not independent. Under stationary conditions, the number of transitions per unit time from the first state to the second one must be equal to the number of reverse transitions: $w_1 W(1 \to 2) = w_2 W(2 \to 1)$. It is equivalent to the equation:

$$\frac{W(2 \to 1)}{W(1 \to 2)} = \frac{w_1}{w_2}. \tag{8.8.1.3}$$

Since the sum $w_1 + w_2 = 1$, one obtains (Eq. 1.7.7):

$$w_1 = \frac{W(2 \to 1)}{W(1 \to 2) + W(2 \to 1)}, \quad w_2 = \frac{W(1 \to 2)}{W(1 \to 2) + W(2 \to 1)}. \tag{8.8.1.4}$$

Let us find the probability density function, $F_i(t)$, of the time t spent in a definite state i before any transition to another state occurs. This problem is similar to the following one. Imagine a great number of identical systems which, starting at the initial instant $t = 0$, disappear with a probability per unit time W. Then $F(t)$ is the fraction of the initial number of systems which remains by the time $t > 0$. Its decay rate at this time t is proportional to $WF(t)$. Therefore

$$\frac{dF_1(t)}{dt} = -W(1 \to 2)F_1(t), \qquad \frac{dF_2(t)}{dt} = -W(2 \to 1)F_2(t). \qquad (8.8.1.5)$$

The solution must satisfy the condition: the total probability for all times t up to infinity must be unity. The solutions to Eq. (8.8.1.5) which satisfy this normalization condition equal:

$$\begin{aligned} F_1(t) &= W(1 \to 2)\exp[-W(1 \to 2)t], \\ F_2(t) &= W(2 \to 1)\exp[-W(2 \to 1)t]. \end{aligned} \qquad (8.8.1.6)$$

It means that the average time spent in the first state before making a transition to the second one, and the mean time spent in the second state before making a transition to the first one equal:

$$\tau_1 = \int_0^\infty dt\, t F_1(t) = 1/W(1 \to 2), \quad \tau_2 = \int_0^\infty dt\, t F_2(t) = 1/W(2 \to 1). \qquad (8.8.1.7)$$

The ratio of these two times is, obviously, equal to the ratio of the probabilities, w_1/w_2 (Eq. (8.8.1.1)), of finding the system in the first or the second state, respectively.

The time spent in each state before making a transition to another state fluctuates around its mean value given by Eq. (8.8.1.7). The variance (mean fluctuation squared) can be easily calculated by averaging the fluctuation squared with the known functions $F_i(t)$:

$$\langle (\delta \tau_i)^2 \rangle = \tau_i^2. \qquad (8.8.1.8)$$

The spectral density of fluctuations of a quantity $x(t)$ which randomly switches between two values x_1 and x_2 was derived in Sec. 1.7 (Eq. (1.7.10)). We rewrite this equation in the notations used in this section:

$$S_x(f) = \frac{4w_1 w_2 (x_1 - x_2)^2 \tau}{1 + \omega^2 \tau^2}, \qquad (8.8.1.9)$$

where the inverse relaxation time

$$\tau^{-1} = W(1 \to 2) + W(2 \to 1) \qquad (8.8.1.10)$$

is the total rate of transitions, back and forth, in the system.

The mean value of $x(t)$ is, obviously, $\langle x \rangle = w_1 x_1 + w_2 x_2$. Integrating the spectral density over all frequencies, one obtains the variance of the fluctuations of $x(t)$ around this mean value:

$$\langle (\delta x)^2 \rangle = \langle x^2 \rangle - \langle x \rangle^2 = w_1 w_2 (x_1 - x_2)^2$$

$$= (x_1 - x_2)^2 \left\{ [1 + W(1 \to 2)/W(2 \to 1)][1 + W(2 \to 1)/W(1 \to 2)] \right\}^{-1}.$$

$$(8.8.1.11)$$

One can easily verify that in systems with fluctuating electron occupancy ($n_1 \neq n_2$) the product $w_1 w_2$ is maximum and equal to 1/4 when the electron level E_t (see above) coincides with the Fermi level (for simplicity we neglect the energy $k_B T \ln(g_1/g_2)$). As the difference $|E_t - E_F|$ increases, the product $w_1 w_2$ decreases. The contribution to the noise of microscopic systems with electron levels far, on the scale $k_B T$, from the Fermi level is exponentially small.

The quantities measured in experiments are: the mean times τ_1 and τ_2 (equivalently, the transition rates $W(1 \to 2)$ and $W(2 \to 1)$) vs temperature T and, in semiconductors and insulators, also their dependence on bias or gate voltage (see below); the amplitude of resistance fluctuations $|R_1 - R_2|$; the spectral density $S_R(f)$ and the integrated spectral density, i.e., the variance $\langle (\delta R)^2 \rangle$.

The dependence of the measured mean time spent by the system in each state is found in experiments to be, at not too low temperatures, an exponential function of the inverse temperature $1/T$:

$$\tau_1 = v_1^{-1} \exp(E_1/k_B T), \quad \tau_2 = v_2^{-1} \exp(E_2/k_B T), \quad (8.8.1.12)$$

where E_1 and E_2 are some energies called activation energies, v_1 and v_2 are the corresponding frequencies of attempts to surmount the barriers. The activation energies vary from several tens of meV up to several eV, the attempt frequencies v are of the order of frequencies of atomic vibrations and usually vary from $\sim 10^{11}$ s^{-1} to $\sim 10^{14}$ s^{-1}.

As follows from Eqs. (8.8.1.2), (8.8.1.3), and (8.8.1.7), the difference of activation energies can be expressed in terms of the energies and charges of the two states, and the Fermi energy (if the charges are different):

$$E(2) - E(1) - k_B T \ln(g_2/g_1) - (n_1 - n_2) E_F = E_1 - E_2 - k_B T \ln(v_1/v_2). \quad (8.8.1.13)$$

Let us consider a simple example of a system with $n_1 = n_2$, $g_1 = g_2$. Then $E_1 - E_2 - k_B T \ln(v_1/v_2) = E(2) - E(1)$. If, say, $E_1 > E_2$, the energy $E(2)$ of the second state is higher than the energy $E(1)$ of the first one. The activation energy E_2 is then the activation barrier height which has to be overcome for transition from the higher energy state to the lower energy state (in our example, from the second state to the first one).

At a given temperature T and given frequency range of the experimental setup, only a few of the transitions in a given specimen manifest themselves as resistance switchings. At low temperatures, only the transitions with low enough activation energies can be detected. In the same specimen and at the same low temperature, the transitions with high activation energies can not be detected because the time spent in the low-energy state becomes immeasurably long. On the contrary, at high temperatures the transitions with low activation energies are too frequent to be detected. It means that at each temperature only resistance switchings having activation energies in a definite narrow range can be effectively measured, and this range shifts with temperature towards higher energies. One has to take into account that a change of temperature results not only in the change of kinetics of the transitions. It may also change the energies of states and the activation energies for transitions in the fluctuators.

It is instructive to find how the noise generated by random transitions in a two-state system varies as the difference between the energies of the two states is increased. Let $E(1) = E - \Delta E$, $E(2) = E + \Delta E$ be the energies, $2\Delta E$ is their difference. If the numbers of electrons, degeneracy numbers, and attempt frequencies are the same in both states, the activation energies for the probabilities of transitions per unit time differ by the same amount (Eq. (8.8.1.13)):

$$E_1 = E_a^{(0)} + \Delta E, \qquad E_2 = E_a^{(0)} - \Delta E.$$

Here $E_a^{(0)}$ is the activation energy at equal energies of the two states. Substituting this equation into Eq. (8.8.1.9) and using Eqs. (8.8.1.2), (8.8.1.10) and (8.8.1.12), one obtains:

$$S_x(f) = \frac{S_x^{(0)}(0)}{\cosh(\Delta E/k_B T)\left[\cosh^2(\Delta E/k_B T) + \omega^2 \tau_0^2\right]}, \qquad (8.8.1.14)$$

where τ_0 and $S_x^{(0)}(0)$ are, respectively, the relaxation time (Eq. (8.8.1.10)) and zero-frequency spectral density at $\Delta E = 0$. Obviously, the spectral density of noise decreases with increasing $|\Delta E|$.

At low enough temperatures, the kinetics is dominated by tunneling of electrons or atoms and the transition rates weakly vary with temperature.

The outlined theory is restricted to two-state fluctuators only. In this simplest case the detailed balance (the rates of transitions between any two states in both directions are equal) is always obeyed under stationary, even nonequilibrium, conditions (Eq. (8.8.1.3)). In multistate fluctuators (number of states greater than two) the detailed balance is obeyed only under true equilibrium conditions. If the fluctuator switches between three states under stationary but nonequilibrium conditions, the transition imbalance between two states is compensated by imbalances between each of these states and the third state. Deviations from detailed balance, which can be observed in the RTN kinetics, can give an insight into the effect of current, bias, etc. on the microscopic fluctuators. Such effects have been found in small samples of NbSe₃ (Bloom, Marley & Weissman, 1993, 1994).

8.8.2 RTN in metals

Random resistance switching was found in metallic nanobridges, i.e., tiny contacts 3–40 nm wide between bulk metals (Ralls & Buhrman, 1988). The volume of these metallic point contacts was less than 8000 nm^3. The main idea of this experiment is obvious: in such small volumes at low temperatures (< 150 K) the expected mean number of mobile defects, whose transitions from one state to another with different scattering cross-section generate abrupt changes of the contact's resistance, is no more than unity and, consequently, in some such contacts these changes will be seen as random resistance switchings. In fact, usually at low temperatures no noise in excess of Nyquist–Johnson noise was seen. But, in some ranges of temperature, typical random switchings of the point contact resistance have been clearly detected (Fig. 8.27).

Ralls & Buhrman have shown that at low temperatures the nanobridges are close to the ballistic regime (Ch. 5), i.e., the size of the point contact is less than the mean free-path for elastic scattering. Since the electron Fermi wavelength in metals is smaller than the size of the point contact, the ballistic conductance $G = R^{-1} \propto A$, where A is the cross-section area of the point contact. By the same reasons, the change of the conductance is proportional to the change of the scattering cross-section of the defect, $\Delta\sigma_s$. Thus, the amplitude of the resistance switchings which is proportional to the relative resistance fluctuation is, in the ballistic regime, proportional to the resistance R. The experiments have shown that this is just the case, and the cross-sections of the defects that generate the resistance switchings are of atomic dimensions ($\simeq 0.01 - 0.1$ nm^2).

The mean times spent in high-resistance and low-resistance states depend on the inverse temperature exponentially (Eq. (8.8.1.12)). The activation energies vary from $\simeq 30$ to $\simeq 300$ meV, and, on average, they increase with temperature (see the discussion in Sec. 8.8.1). The attempt frequencies vary in a wide range from $\sim 10^{11}$ s^{-1} to $\sim 10^{15}$ s^{-1}, grouping around 10^{13} s^{-1}.

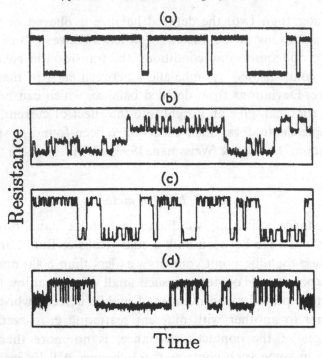

Fig. 8.27 Random resistance fluctuations in Cu nanobridges (volume 40–8000 nm³) at temperatures $T < 150$ K. (a) a single two-level fluctuator (TLF); (b) two uncorrelated TLFs; (c) correlated (interacting) TLFs: the amplitude of the small TLF is higher when the large TLF is in the state with lower resistance than when it is in the state with higher resistance; (d) correlated (interacting) TLFs: one TLF modulates the frequency of another TLF. From Ralls & Buhrman (1988).

One of the most striking results found by Ralls & Buhrman is the discovery of interaction between the switching fluctuators. In some specimens and at some temperatures two superimposed switchings are seen, as in Fig. 8.27 (b,c,d). They may be independent, i.e., uncorrelated, as in Fig. 8.27 (b). In this case, the parameters of switching of one of the fluctuators is independent of the state of the second fluctuator: whether it is in the high- or low-resistance state. However, interacting fluctuators are also observed (Fig. 8.27 (c,d)): the parameters of switching of one fluctuator (amplitude, mean times τ_1 and τ_2) are modulated by a second fluctuator. As shown in Fig. 8.27 (c), the amplitude of the resistance change generated by the fluctuator with higher frequency of switching is higher when another fluctuator is in its low-resistance state than when it is in its high-resistance state. The interaction between the fluctuators may manifest itself also in the modulation of switching frequency (Fig. 8.27 (d)). Since the parameters of switching of each fluctuator are determined by the corresponding activation energy

E_i and attempt frequency v_i, the modulation effects can be attributed to the modulation of E and v. It means that the activation energies, E_1, E_2, and attempt frequencies, v_1, v_2, of one of the two fluctuators are quite different for the high- and low-resistance states of the second fluctuator. The interpretation of these phenomena is quite obvious. The change in the configuration of one of these defects is accompanied by a change of the fields (in metals it is, most probably, the strain field) acting on the second defect.

At temperatures higher than ≈ 150 K, discrete resistance switching in copper nanobridges is not observed. Instead a continous $1/f$ noise spectrum is measured. Two possible interpretations have been suggested. According to the first one, the number of active defects capable of switching in the given experimental frequency range grows. The superposition of a great number of two-level switchings with various relaxation times τ (Eq. (8.8.1.9) and Sec. 8.3.1) is seen as a $1/f$ noise. The effect of temperature is in this case similar to the effect of the sample's dimensions: as they become larger the number of fluctuators becomes greater and discrete switchings become indistinguishable (Rogers & Buhrman, 1984).

A different interpretation was suggested by Ralls & Buhrman (1991). Let us assume that at each instant there is only one two-state fluctuator in the nanobridge whose switching frequencies fit the experimental frequency bandwidth. At high temperatures other defects are also moving. Due to interactions between the defects, the parameters of the fluctuator's switchings, first of all its activation energy, are not constant but vary in time. This interaction results in deviation from a simple Lorentzian spectrum. Moreover, the system of interacting defects may pass to a different metastable state in which a different defect plays the role of an active fluctuator. This interpretation is substantiated by the measured resistance of the nanobridge in the time domain at 300 K. The record $R(t)$ is still composed of only one or a few discrete fluctuations (as at $T < 150$ K). However, their amplitudes and switching times change randomly in time. The system of interacting defects is similar to a glass, particularly, to a spin glass (Sec. 8.6) with a great number of metastable states between which it is incessantly wandering.

8.8.3 RTN in tunnel junctions

Rogers & Buhrman (1984, 1985) studied the resistance fluctuations in metal–insulator–metal (Nb–Nb$_2$O$_5$–PbBi) tunnel junctions of various area A. They found that in large devices the spectral density of noise is roughly $\propto 1/f$ but in small devices ($0.03\ \mu m^2 < A < 2\ \mu m^2$) and at low temperatures < 100

K the noise spectrum is uniquely resolved into a sum of only a few distinct Lorentzian spectra (Eq. (8.8.1.9)), each corresponding to a discrete resistance fluctuation. As the area of the junction becomes larger, the number of these discrete fluctuations increases and the resulting noise spectrum gradually transforms into a common $1/f$ spectrum. It is a very important experimental result in favor of the interpretation of the $1/f$ spectrum as a superposition of individual fluctuation phenomena with various relaxation times (Sec. 8.3.1).

These experiments proved that the noise measurements are a powerful tool for investigations of individual microscopic fluctuators. The measured activation energy for the relaxation time τ (Eq. (8.8.1.10)) was found to be much greater than the activation energy for the variance of resistance fluctuations (Eq. (8.8.1.11)). This result demonstrated that the fluctuator is switching between two states with almost equal energies separated by a barrier. In fact, the variance is expressed in terms of the ratios of transition rates which, according to Eq. (8.8.1.2), (8.8.1.3), and (8.8.1.13), depend on the difference of the two states' energies. If these energies are equal, the strong, exponential, temperature dependence drops out. Unlike the variance, τ depends on the sum of the transition rates.

The time τ increased exponentially with inverse temperature only down to ~ 15 K. At lower temperatures the kinetics was almost temperature independent. Presumably it was dominated by tunneling.

The rates of transitions between the states of fluctuators in the metal–insulator–metal (MIM) tunnel junctions are usually enhanced by bias voltage (Rogers & Buhrman, 1985; Wakai & Van Harlingen, 1987). It is interpreted by Rogers & Buhrman as due to thinning or lowering of the barrier for ionic reconfiguration in the fluctuator. According to Wakai & Van Harlingen, who measured the RTN in Josephson junctions PbInAu–In$_2$O$_3$–Pb of very small area ($< 0.05\,\mu m^2$), RTN is generated by electron transitions from the metallic electrodes to electron traps in the oxide layer and back. The electric field of a charged trap alters the junction tunnel conduction by blocking some paths for tunneling. The bias voltage U changes the energy level position of the trap in the oxide relative to the Fermi levels of the electrodes (they differ by eU). At very low temperatures $k_B T \ll eU$, only the tunneling processes can be taken into account, and only the traps with levels between the Fermi levels of the electrodes, i.e., in the energy range of eU, exchange electrons with the electrodes. The rates of tunneling transitions increase with U, regardless of polarity.

Some difficult problems in this field are: what the microscopic structure of the fluctuators is, what the states between which they are switching are, and how these transitions affect the tunnel resistance. The trapping and detrapping of electrons seems plausible. The Coulomb field of a charged trap alters the tunnel barrier within an area of radius $\sim d$ around the trap (d is the thickness of the oxide layer), because at greater distances this field is screened by the electrons in metals. The relative effect $|R_1 - R_2|/R_1$ is expected to be of the order of d^2/A, where A is the total junction area. If $A \approx 0.03 \, \mu m^2$ and $d \approx 1.5$ nm, the ratio $|\Delta U|/U = |R_1 - R_2|/R \sim 10^{-4}$. According to calculations by Schmidlin (1966), the conductance of the tunnel barrier area $\sim d^2$ around the charged defect in the insulator is changed by an order of magnitude. It means that $|R_1 - R_2|/R$ may be $\sim 10 \quad d^2/A$. The experimental values are close to these estimates (Wakai & Van Harlingen, 1987).

The amplitude of resistance switchings in metal–oxide–semiconductor tunnel junctions of $1 \, \mu m^2$ area was found to vary from less than 0.1 % to more than 10 % (Farmer, Rogers & Buhrman, 1987). If the fluctuators are not charged and are of microscopic ($<\sim 1$ nm) dimensions and if their effect is simply determined by their cross-sections ($< nm^2$), the relative amplitude of resistance switching should be $\sim 10^{-6}$, several orders less than the measured value.

RTN can be measured even in tunnel junctions of large area if the tunnel current density is strongly nonuniform and the major part of the current flows through a small area where the tunnel barrier is lowered, for instance, by a dislocation propagating through the tunnel barrier. Campbell, Snow, Moore *et al.* (1991) observed light-activated switching between discrete resistance states in GaAs–14 nm GaAlAs–GaAs tunnel structures $\simeq 0.1$ mm across. The RTN was observed only in samples with anomalous current–voltage characteristic in which the current in a wide range of bias voltage was not exponential in voltage, unlike normal samples, and was by orders higher. Moreover, many devices which exhibited RTN have shown switching only when illuminated. The rate of transitions from the low-current state was found to be proportional to the intensity of the light with photons exceeding the band gap. It means that the transition from the low-current state is a single-photon event. The most plausible mechanism of light-activated switching is capture of a single photogenerated hole by a single hole trap.

8.8.4 RTN in MOSFETs

The metal–oxide–semiconductor field-effect transistor (MOSFET) is one of the basic semiconductor electron devices and may be considered as the most important. It consists of a semiconductor and a metal separated by an insulating oxide layer. Application of a voltage between the semiconductor and the metal across the insulating layer (gate voltage) results in the bending of the edges of the conduction and valence (hole) bands in the near-surface region of the semiconductor. If, for instance, the positive potential is on the metal and the gate voltage is high enough, a channel with high electron density and high conduction is created. The resistance of this channel in the direction parallel to the surface and of the entire device (between the source and drain electrodes) can be controlled by the gate voltage. Thus, the MOSFET works like a valve.

Kandiah & Whiting (1978), studying the resistance noise in large area (gate width 0.62−13.6 mm) junction field-effect transistors (JFETs), observed sharp peaks of the spectral density of noise as a function of the bias voltage V_{ss} on the substrate. Later the authors also observed the RTN. They have shown that the peaks of noise and the RTN are both manifestations of random electron transitions to and from *individual* impurity (defect) centers, i.e., of random charging and discharging of *single* such defects in the transition region between the conduction channel and the region fully depleted of charge carriers. This generation–recombination noise (Ch. 4) sharply increased when the Fermi level, controlled by V_{ss}, crossed the electron level of the center.

Ralls, Sckocpol, Jackel *et al.* (1984) made a very important step in this field. They studied resistance noise in very small Si MOSFETs (width 0.1 μm, length 1.0 μm) at temperatures in the range 4.2 − 111 K and were the first to observe RTN, i.e., discrete resistance switching events in these systems. Thanks to the small area of the devices, only a small number of active fluctuators were present and individual switching processes could be resolved. Moreover, the background noise is in this case many times smaller than the magnitude of RTN.

The authors have found that the mean times τ_1 and τ_2 (Eqs. (8.8.1.7) and (8.8.1.12)) depend exponentially not only on the inverse temperature $1/k_B T$ but also on the gate voltage V_G. These data can be used to obtain the energy level of the trap and its location in the oxide layer.

The applied gate voltage V_G partially drops in the near-surface region of the semiconductor. Let ϕ_s be the potential at the semiconductor–oxide interface, with respect to the bulk of the semiconductor, due to the applied gate voltage. The difference $V_G - \phi_s$ is then the potential drop across the oxide layer. Let the electron trap be located inside the oxide at a distance x from the semiconductor surface. The difference $E_t - E_F$ between the trap energy level and the Fermi level in the semiconductor is decreased as V_G is increased. This decrease consists of two parts. The first is the shift of the conduction band edge at the semiconductor's surface $-e\phi_s$, the second equals $-e(V_G - \phi_s)x/d$, where d is the thickness of the oxide layer. The same decrease of the difference $E_t - E_F$ can be found using the measured mean times: (1) time τ_c of an electron staying in the channel (time of electron capture by the trap) and (2) time τ_e of an electron residing in the trap (time of electron emission into the channel). According to Eq. (8.8.1.2),

$$\frac{\tau_c}{\tau_e} = g \exp[(E_t - E_F)/k_B T], \qquad (8.8.4.1)$$

where g is the ratio of the numbers of effective states for the electron in the channel and in the trapped state.

Differentiating with respect to V_G, one obtains:

$$\frac{d(E_t - E_F)}{dV_G} = k_B T \frac{d}{dV_G} \ln\left(\frac{\tau_c}{\tau_e}\right) = -e \frac{d\phi_s}{dV_G} - e \frac{x}{d} \frac{d}{dV_G}(V_G - \phi_s)$$

$$= -e \frac{d\phi_s}{dV_G} - e \frac{x}{d}\left(1 - \frac{d\phi_s}{dV_G}\right). \qquad (8.8.4.2)$$

The distance of the trap from the surface equals:

$$x = d\left(-\frac{1}{e} \frac{d(E_t - E_F)}{dV_G} - \frac{d\phi_s}{dV_G}\right)\left(1 - \frac{d\phi_s}{dV_G}\right)^{-1}. \qquad (8.8.4.3)$$

Usually the derivative $d\phi_s/dV_G \ll 1$. The trap was assumed to be located in the oxide layer, $x > 0$. Hence Eq. (8.8.4.3) can be applied only if the r.h.s. is positive.

Using this method, Ralls *et al.* were able to find that the traps that manifest themselves in RTN are located ~ 2 nm from the surface. It is a reasonable value: much greater distance would result in very long times τ_c and τ_e. For a discussion of this method and of the data that can be extracted from the gate voltage dependence of the times τ_c and τ_e, see the comprehensive review by Kirton & Uren (1989).

The magnitude of the resistance jump due to a transition of a single electron from the channel into a trap in the oxide layer is determined by two mechanisms. The first one is the decrease of the total number N of conduction electrons in the channel by one. The relative change of the resistance is $|\Delta R|/R \simeq 1/N$. The value measured by Ralls *et al.* (several tenths of a percent) is of the same order. The second mechanism is the change of the mobility of electrons in the channel by the change of electrostatic fields in the channel accompanying the trapping of one electron. The magnitude of this effect is more difficult to estimate.

The discovery of RTN generated by *single* electron transitions or *single* defects and the application of this method to various devices can be compared with the discoveries of such devices that give the opportunity to visualize individual atoms or molecules, e.g., the scanning tunnel microscope by Binnig and Rohrer in 1986.

8.9 Fluctuations with $1/f$ spectrum in other systems

In some materials (NbSe$_3$, TaS$_3$, quasi-one-dimensional organic crystals), the conductivity of which is metallic at high temperatures, a charge density wave (CDW) is established at a critical temperature T_P. If the period of the CDW is not commensurate with the period of the crystal, it can, in principle, slide freely through the lattice. In real materials with CDW, the latter is pinned by impurities and defects, each of which affects the local phase and amplitude of the CDW and its motion. The CDW is brought into sliding motion by an electric field which must exceed some threshold value that depends on the quality of the crystal (for a review see Grüner, 1988). A very intensive noise with a $1/f^\gamma$ spectrum ($0.5 < \gamma < 1.0$) is observed in conductors with CDW (Richard, Monceau, Papoular & Renard, 1982; Bhattacharya, Stokes, Robbins & Klemm, 1985). Owing to random pinning by defects, the charge density wave has a great number of different metastable configurations (states) separated by barriers. Much experimental evidence has been found in favor of transitions between the metastable states as the source of the broad-band resistance fluctuations (Bhattacharya et al., 1985).

In small samples of conductors with CDW, random telegraph noise (RTN) is observed in the pinned state of the CDW (Zaitsev-Zotov & Pokrovskiĭ, 1989; Pokrovskiĭ & Zaitsev-Zotov, 1990) and in the sliding state (Bloom, Marley & Weissman, 1993, 1994). Bloom et al. found that some fluctuators that produce the observed RTN are significantly influenced by dc current, their kinetic properties depend on the bias. The kinetics of multistate fluctuators, that are switching between more than two states, does not satisfy the detailed balance, i.e., it is nonequilibrium.

The frequency v of the signal generated by frequency standards undergoes slow small deviations from the central frequency v_0 along with a systematic 'drift'. The spectral density of the relative fluctuations of the frequency $S_y(f) = v_0^{-2} S_v(f)$ contains not only a component independent of frequency ('white noise') but also a component inversely proportional to the frequency f. One can write it in the form C/f, where C is a dimensionless coefficient. The commonly used measure of the spread (inaccuracy) of the frequency v is $\sigma^2(t_m)$, the mean square of the relative deviations of v during the time of measurement t_m. One can show that the contribution of the 'white' noise to $\sigma^2(t_m)$ is proportional to t_m^{-1}, while the contribution of $1/f$ noise is independent of t_m and equals $2C \ln 2$ (Vessot, 1974). The following conclusion is very important: $1/f$ noise imposes a limit on the attainable accuracy of measurement of frequency and time.

Measurements on many quartz frequency standards have shown that the magnitude of $1/f$ noise in the spectrum of frequency fluctuations, i.e., the coefficient C, depends on the Q-factor of the quartz resonator which, in its turn, is determined usually by the internal friction in the crystal: $C = 62Q^{-4.3}$ (Gagnepain & Uebersfeld, 1977). This equation may be viewed as a specific empirical analogue of the fluctuation–dissipation relationship.

Noise with $1/f$ spectrum is found in biological systems, in particular, in biological membranes (see reviews by DeFelice, 1981, and Bezrukov & Vodyanoy, 1994). The spectral density of the voltage fluctuations across the membrane of a living, i.e., nonequilibrium, but resting nerve follows a $1/f$ law (Verveen & Derksen, 1968).

The amount of insulin required by a diabetic patient to maintain a constant amount of sugar in the blood fluctuates (with unvarying diet). Correlation analysis of the records for more than 8 years have shown that the spectral density follows a $1/f$ law over a broad frequency range (Campbell & Jones, 1972).

Some geophysical processes, e.g., velocities of currents, frequencies of river floods, etc., as well as astrophysical processes (sunspot activity, intensities of various radiation sources) fluctuate, and often the spectral density is of $1/f$ type (see review by Press, 1978).

The interpretation and discussion of $1/f$ noise in biological and geophysical systems is beyond the scope of this book.

8.10 General conclusions on $1/f$ noise

The data on $1/f$ noise accumulated by the present time owing to a huge amount of experimental and theoretical work are extremely extensive. They provide the possibility of making some definite general conclusions and answering the questions stated in Sec. 8.1. First a summary of direct conclusions drawn from experiments is presented.

1. Numerous experiments prove that no $1/f$ universal spectrum exists. The spectra of $1/f$ noise in different systems are clearly different, they vary even from sample to sample of one and the same material, and depend on temperature and other conditions.
2. Experiments have found no universal relationship between $1/f$ noise and any general parameter of the conductors such as, for instance, the number of charge carriers. They have found no finite minimum intensity of $1/f$ noise.

3. As the dimensions of a conductor grow and so does the number of individual fluctuators producing random telegraph noise, the noise spectrum varies from one Lorentzian or a superposition of few Lorentzians to a continuous $1/f$ noise.

4. The current and/or voltage $1/f$ noise is produced by resistance fluctuations and is not a result of some instability generated by the current in the conductor.

5. No clear evidence was found in favor of temperature fluctuations as a source of $1/f$ noise.

6. According to a number of experiments, the correlation length of $1/f$ noise is usually of microscopic dimension.

7. The spectral density of magnetic noise in spin glasses and amorphous ferromagnets is of $1/f$ type. The experimental spectral density of the magnetic noise $S_M(f)$, the dissipative part of the magnetic susceptibility $\chi''(f)$, and the magnetic moment relaxation $M(t)$ after switching off the magnetic field are found to satisfy, within experimental accuracy, the general relations for equilibrium systems, namely, the fluctuation–dissipation relations. It means that $1/f$ noise in spin glasses and amorphous ferromagnets is directly related to other low-frequency characteristics of these systems and is close to an equilibrium noise.

8. Many experiments prove that in metals the $1/f$ resistance noise is produced by mobile defects.

We are able now to answer the questions stated in Sec. 8.1.

All experimental data and theoretical calculations produce evidence in favor of disorder complemented in some systems by many-body interactions as the major source of $1/f$-type noise spectra. In order to answer the main question 'why $1/f$ noise is so universal and is found in systems with very different other properties?', let us compare the magnetic noise in spin glasses below the freezing temperature T_G (this noise was measured on dielectric spin glasses) and the electric and magnetic noise in type II superconductors below the critical temperature T_c. The spin magnetic moments in spin glasses form a disordered system with strong many-body interactions. Their thermal motion, which manifests itself in $1/f$ magnetic noise, can be envisioned as transitions between different metastable states by surmounting barriers of various heights, or between 'valleys' in the phase space (Sec. 8.6). It is important that the spectrum of the metastable states' energies in a macroscopic sample is continuous, without any gap between these states and the ground state. The vortices in type II superconductors also form a disordered system, due to pinning centers, with strong many-body interactions (see below Sec. 9.2). The last system is considered in some theories as a 'vortex glass'. Obviously, the low-frequency kinetic properties of both systems are similar despite all their differences (the first are insulators, the second are superconductors). In other words, many quite different physical mechanisms of noise have similar $1/f$ noise spectra owing to definite common kinetic properties of disordered systems.

As we have seen in this chapter, the presence of a variety of defects with a broad spectrum of activation energies and/or tunneling parameters results in $1/f$ type of noise in metals, semiconductors, and dielectrics. Their only common property is that the distribution of these activation energies and tunneling parameters is more or less flat. Even if this distribution follows a power law, the resulting noise spectrum is of $1/f$ type. $1/f$ noise gives way to constant spectral density only when the distribution drops exponentially or stronger. That is why the conditions required for the noise spectrum to be of $1/f$ type are met so often.

Is there a minimum frequency f_{min} at which the spectral density stops growing? Such a minimum frequency does exist in any physical model which takes into account that the barriers in any solid of finite dimensions are finite. For those solids for which qualitative or even quantitative theory exists (e.g., spin glasses) f_{min} is extremely small (even smaller than the inverse age of the Universe) making meaningless the experiments aiming to measure f_{min}. As was mentioned in this chapter, measurements always find $1/f$ noise down to the lowest accessible frequency. Since, in principle, a minimum frequency of $1/f$ noise exists, the variance is also finite. The measured variance is always lower than this 'theoretical' value because, in practice, f_{min} is never reached.

Experiments do not support the idea of the existence of some universal $1/f$ noise inherent to all systems. No properly justified theory of such a noise was ever suggested.

$1/f$ noise is at present not an isolated phenomenon with no connections (links) with other low-frequency kinetic phenomena. Besides relations between the magnetic noise in spin glasses, on one hand, and their relaxation function and the dissipative part of the magnetic susceptibility, on the other hand, one can mention also the relation between the low-frequency resistance noise, on one hand, and internal friction and anelastic piezoresistance, on the other hand, in metals.

Does $1/f$ noise appear in a crystal in which the charge carriers interact only with phonons or with immobile scattering centers? This problem is worth attention because some authors attempted to explain $1/f$ noise by fluctuations of the phonon distribution. The phonons as excitations are not long-lived, especially at room temperature, due to anharmonic interactions between various modes of lattice vibrations. At temperatures T higher than the Debye temperature, the estimate for the inverse free time of a vibration mode due to anharmonic processes is of the order of $\tau^{-1} \sim k_B T/Mv_g a$, where M is the atomic mass, a is the interatomic distance in the crystal, and v_g is the group velocity of the vibration wave (see, for instance, Gurevich, 1986). According to this estimate, the inverse free time is several orders greater than the frequencies at which $1/f$ noise is usually measured. Thus, the phonon gas has no long relaxation times required to explain $1/f$ noise. Moreover, at low frequencies the phonon fluctuations are reduced to the fluctuations of lattice temperature. The latter have been shown not to be the source of $1/f$ noise (Sec. 8.7). One can add also that, according to well proved and generally accepted theory of electron conduction in solids with immobile scattering centers, no relaxation times much longer than the momentum and energy relaxation ones exist in this system.

According to some theories, $1/f$ noise is predicted to exist in some specific systems governed by dynamic equations. To the author's knowledge, no proof was presented that $1/f$ spectrum in these systems really survives down to the very low frequencies at which $1/f$ noise is measured or over long times. No experimental substantiation of these theories ever appeared.

Despite the great progress of $1/f$ noise physics in the last two decades, for the major part of systems with $1/f$ noise and/or random telegraph noise the concrete sources of these fluctuations remain unknown: this is the main unsolved problem. This problem is especially important in the case of semiconductors because of their technological importance and lack of any widely accepted mechanism of bulk $1/f$ noise.

9

Noise in superconductors

One more, the final record, and my annals
Are ended, and fulfilled the duty laid
By God on me, a sinner.
Alexander Pushkin,   *'Boris Godunov'.*

Many metals and alloys become superconductive at low temperatures owing to mutual attractive interaction between electrons and to the sharpness of the Fermi distribution edge at the Fermi energy E_F. Qualitatively, the electron system in the superconductive state may be viewed as being composed of electrons bound in electron pairs (Cooper pairs) each of which contains two electrons with opposite momenta and spins. The dimensions of the pairs are greater than the mean interelectron distance, i.e., the pairs strongly overlap. This state of paired electrons is called 'condensate'. It is superconductive, that is, current flows without any resistance. At temperatures T lower than the critical temperature T_c, the free energy of the condensate is lower than that of an unpaired electron gas of the same density. Therefore, at $T = T_c$ the metal undergoes a phase transition into the superconductive state.

The properties of a superconductor are determined by the binding energy of electron pairs in the condensate, 2Δ. This quantity is called also the superconductive energy gap, because just this energy is required to break an electron pair in the condensate and create two quasi-particles that are able, like electrons in a normal metal, to dissipate the current. The energy Δ decreases with increasing temperature T and becomes zero at $T = T_c$.

The condensate flows without resistance. At absolute temperatures $T > 0$ some of the electrons become unpaired. Their flow is dissipative and is possible, as the electron current in a normal metal, only at a nonzero electric field F. The non-random (mean) normal, dissipative, current density equals $j_n = \sigma_n(T)F$, where $\sigma_n(T)$ is the normal conductivity. At temperatures below T_c but close to T_c the normal conductivity is close to the conductivity of the metal in its normal, nonsuperconductive, state at $T > T_c$. However, at $T < T_c$ the normal current is completely shunted by the superconductive one: in the bulk of the superconductor the electric field $F = 0$. Consequently, $j_n = 0$ and no dissipation of current is measured.

The motion of quasi-particles (unpaired electrons) is random, like the motion of electrons in a normal conductor. It generates random currents. However, in the bulk of a superconductor these currents are completely shunted by the superconductive condensate, and no voltage fluctuations can be observed.

Under some conditions, in the so called resistive states, superconductivity and normal conduction under non-zero electric fields coexist, resulting in dissipation of dc current. The same kinetic processes that lead to current dissipation manifest themselves in electrical noise. Two systems that can be in resistive states are the most important in this field: (1) Josephson junctions and (2) type II superconductors. The noise in these resistive superconductive systems is discussed below in Secs. 9.1 and 9.2, respectively. The reader can find the list of recommended books and reviews on superconductivity in the list of references.

9.1 Noise in Josephson junctions

9.1.1 Introduction

In 1962 Brian Josephson predicted a fundamental phenomenon, called after him the Josephson effect (Josephson, 1962). He considered two bulk superconductors separated by a thin insulator layer, i.e., a tunnel junction (Sec. 5.5) in which both metals are in a superconductive state. It was well known that individual electrons can pass through the insulator layer owing to quantum-mechanical tunneling. Josephson has shown that the superconductive condensate can also flow through the tunnel barrier without any resistance (no voltage across the junction appears) if the current density j_s does not exceed a definite critical value j_c, which is the maximum superconductive current in this system. He has also shown that if a voltage U is applied across the junction, the superconductive current oscillates with an angular frequency ω given by the Josephson relationship: $\hbar\omega = 2eU$. These two effects are called the stationary and nonstationary Josephson effect, respectively.

A scheme of the Josephson tunnel junction is presented in Fig. 9.1. The right and left superconductors are separated by a thin dielectric layer. Their Fermi levels are μ_R and μ_L, respectively. The voltage across the junction $U = (\mu_R - \mu_L)/e$. In each superconductor the energies $\hbar\epsilon$ of the quasiparticles may be only beyond the superconductive gap of width 2Δ. The energies $\hbar\epsilon > \Delta$ and $\hbar\epsilon < -\Delta$ correspond to electron-like and hole-like quasi-particles, respectively.

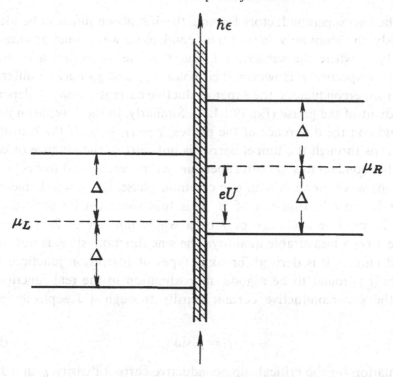

Fig. 9.1 Schematic diagram of the Josephson tunnel junction. 2Δ: superconductive energy gap, U: the voltage across the junction, μ_L and μ_R are the electron electro-chemical potentials of the left and right electrode, respectively. $\hbar\epsilon$ is the energy of the quasi-particle.

The Josephson effect has been found not only in tunnel junctions but also in a number of structures with weak coupling between two bulk supercon-ductors. The role of the thin insulator layer is played in these structures by a tiny microbridge, i.e., a narrow constriction ($S - c - S$ structure), a layer of a normal metal, and so on.

A very important notion in the theory of superconductivity is the conden-sate wave function $\psi(\mathbf{r}) = |\psi(\mathbf{r})|e^{i\chi(\mathbf{r})}$. Its absolute value squared is related, within the phenomenological Ginzburg–Landau theory of superconductivity, to the density n_s of 'superconductive electrons': $|\psi(\mathbf{r})|^2 = n_s/2$. The density of the superconductive current is given by the known quantum-mechanical expression for particles with charge $2e$:

$$\mathbf{j}_s = \frac{2e\hbar}{2im}[\psi^*(\mathbf{r})\vec{\nabla}\psi(\mathbf{r}) - (\vec{\nabla}\psi^*(\mathbf{r}))\psi(\mathbf{r})] = \frac{2e\hbar}{m}|\psi|^2\vec{\nabla}\chi(\mathbf{r}). \qquad (9.1.1.1)$$

The superconductive current is proportional to the gradient of the wave function's phase $\chi(\mathbf{r})$.

Let the two superconductors forming the Josephson junction be identical. Obviously, the absolute values of their condensate wave functions are equal: $|\psi_L| = |\psi_R|$, where the subscripts L and R relate to the left and the right electrode, respectively. However, their phases, χ_L and χ_R may be different. In uniform superconductors, the superconductive current density j_s depends on the gradient of the phase (Eq. (9.1.1.1)). Similarly, in the Josephson junction it depends on the difference of the phases, $\chi = \chi_L - \chi_R$. If the transmission of electrons through the tunnel barrier is not zero, in the absence of current, $j_s = 0$, the condensates in both superconductors correspond to one coherent (common) wave function with one common phase, i.e., $\chi = 0$. It means that j_s must be an odd function of χ. This function must be periodical with period 2π because a change of χ by a whole number of 2π cannot have any effect on a measurable quantity. The sine function, $\sin \chi$, is not only the simplest choice: it is derived for some types of Josephson junctions and in all cases it is found to be a good approximation to the real function $j_s(\chi)$. Thus, the superconductive current density through a Josephson junction equals:

$$j_s = j_c \sin \chi. \tag{9.1.1.2}$$

The equation for the critical superconductive current density j_c in a Josephson tunnel junction at any temperature $T < T_c$ was derived by Ambegaokar & Baratoff (1963):

$$j_c = \frac{\pi \Delta(T)}{2eAR_N} \tanh\left(\frac{\Delta(T)}{2k_B T}\right). \tag{9.1.1.3}$$

Here R_N is the resistance of the tunnel junction in the normal state ($T > T_c$), A is the area of the junction. One can see that the current decreases with R_N, i.e., as the tunnel barrier becomes less transparent for electron tunneling. It also decreases as the temperature approaches T_c because the superconductive energy gap tends to zero at $T \to T_c$.

If the current j, which in a circuit is determined by an external current source, exceeds j_c, the stationary Josephson effect gives way to the nonstationary one, which corresponds to a nonzero voltage across the junction $U \neq 0$. An exact relation between the difference of phases χ of the condensates' wave functions in the left and right electrode, on one hand, and the voltage across the junction, on the other hand, has been derived:

$$\frac{\partial \chi}{\partial t} = \frac{2e}{\hbar} U. \tag{9.1.1.4}$$

This is the Josephson equation. Equations (9.1.1.2) and (9.1.1.4) determine the dynamics of the superconductive current (supercurrent, for brevity) in a Josephson junction.

If the voltage U is time-independent, the phase $\chi(t) = \omega_J t$, where $\omega_J = 2|eU|/\hbar$ is called the Josephson frequency. Consequently, the supercurrent oscillates with angular frequency ω_J (Eq. (9.1.1.2)). The Josephson frequency has a simple meaning. When an electron pair of the superconductive condensate, having a total charge $2e$, passes from one superconductor to another one it gains an energy $2|eU|$. It generates oscillations with a quantum $\hbar\omega_J = 2|eU|$.

Aiming to outline the relations between the noise in Josephson junctions and the mean current, we present below a brief account of the microscopic theory of the Josephson current (see Larkin & Ovchinnikov, 1966). The tunnel current is obtained by averaging the current operator in Eq. (5.5.5). The terms which have been omitted for normal metals must be retained in the case of superconductors:

$$I(t) = e \int_{-\infty}^{t} dt' \{\langle [\hat{A}(t), \hat{A}^+(t')] \rangle - \langle [\hat{A}^+(t), \hat{A}(t')] \rangle + \langle [\hat{A}(t), \hat{A}(t')] \rangle$$

$$- \langle [\hat{A}^+(t), \hat{A}^+(t')] \rangle = 2e\operatorname{Re} \int_{-\infty}^{t} dt' \{\langle [\hat{A}(t), \hat{A}^+(t')] \rangle + \langle [\hat{A}(t), \hat{A}(t')] \rangle\}.$$

$$(9.1.1.5)$$

Equation (5.5.4) for $\hat{A}(t)$ has to be substituted into the last equation. We can decompose the average values of four operators of electron creation and annihilation into products of averages of pairs of operators, each related to a definite electrode. The main contribution to the current comes from electron energies in the vicinity of the Fermi energy. In particular, the contribution to the superconductive current is confined to energies $|E(\mathbf{p})| \sim 2\Delta$ (the electron energy is measured from the Fermi level). It allows us to express the sum over electron momenta in the right electrode ($\mathbf{k}$) and left electrode ($\mathbf{p}$) in terms of the resistance of the tunnel junction in the normal state (Eq. (5.5.12)):

$$\hbar^{-2} \sum_{\mathbf{pk}\sigma} |T(\mathbf{pk})|^2 \ldots = \sum_{\mathbf{pk}\sigma} |T(\mathbf{pk})|^2 \delta(E(\mathbf{p})) \delta(E(\mathbf{k})) \int \frac{dE(\mathbf{p})}{\hbar} \int \frac{dE(\mathbf{k})}{\hbar} \ldots$$

$$= \frac{\hbar}{4\pi e^2 R_N} \sum_{\sigma} \int \frac{dE(\mathbf{p})}{\hbar} \frac{dE(\mathbf{k})}{\hbar} \ldots$$

$$(9.1.1.6)$$

Here σ is the spin quantum number.

From Eqs. (9.1.1.5) and (9.1.1.6) one obtains the equation for the tunnel current between two superconductors:

$$I(t) = I_n(t) + I_s(t),$$

$$I_n(t) = \frac{\hbar}{2\pi e R_N} \sum_\sigma \mathrm{Re} \int_{-\infty}^t dt' \int \frac{dE(\mathbf{k})}{\hbar} \int \frac{dE(\mathbf{p})}{\hbar}$$

$$\times \{\langle a_R^+(\mathbf{k}\sigma t) a_R(\mathbf{k}\sigma t')\rangle \langle a_L(\mathbf{p}\sigma t) a_L^+(\mathbf{p}\sigma t')\rangle$$

$$- \langle a_R(\mathbf{k}\sigma t') a_R^+(\mathbf{k}\sigma t)\rangle \langle a_L^+(\mathbf{p}\sigma t') a_L(\mathbf{p}\sigma t)\rangle\}, \qquad (9.1.1.7)$$

$$I_s(t) = \frac{\hbar}{2\pi e R_N} \mathrm{Re} \int_{-\infty}^t dt' \int \frac{dE(\mathbf{k})}{\hbar} \int \frac{dE(\mathbf{p})}{\hbar}$$

$$\times \{-\langle a_R^+(\mathbf{k}\uparrow t) a_R^+(-\mathbf{k}\downarrow t')\rangle \langle a_L(\mathbf{p}\uparrow t) a_L(-\mathbf{p}\downarrow t')\rangle$$

$$- \langle a_R^+(-\mathbf{k}\downarrow t) a_R^+(\mathbf{k}\uparrow t')\rangle \langle a_L(-\mathbf{p}\downarrow t) a_L(\mathbf{p}\uparrow t')\rangle - (t \leftrightarrow t')\}.$$

It follows from Eqs. (9.1.1.5) and (9.1.1.7) that I_n is the current of quasiparticles, i.e., the normal (resistive) current, I_s is the current of electron pairs, i.e., the superconductive current. The up and down arrows in the equation for I_s correspond to spin quantum numbers $\sigma = \pm(1/2)$, respectively. The average matrix elements in the equation for I_n are independent of the spin quantum numbers; they can be calculated for any direction of spin and the result doubled.

The quantities $\langle a(\mathbf{p}\sigma t) a(-\mathbf{p}, -\sigma t')\rangle$ and $\langle a^+(\mathbf{p}\sigma t) a^+(-\mathbf{p}, -\sigma t')\rangle$ are nonzero only in superconductors, owing to the specific correlation between the electrons with opposite momenta and spin quantum numbers. They are matrix elements between such states of a given electrode (left or right), the number of electrons in which differ by two. Each of these average quantities corresponds to a definite electrode. Each electrode, taken separately, is in an equilibrium state. Nevertheless, in general, the mean values of products of operators in Eq. (9.1.1.7), in contrast with nonsuperconductive systems, do not depend on the difference of time instants $t - t'$ only.

Indeed, suppose all electron energies in an isolated metal are changed by the same amount $|e\phi|$. Of course, it does not result in any observable physical effect. However the matrix element of, say, a creation operator $a^+(t)$ between two states of the system, $|m_{N+1}\rangle$ with $N+1$ electrons and $|n_N\rangle$ with N electrons, is changed by a time-dependent phase factor:

$$\langle m_{N+1} | a^+(\mathbf{p}\sigma t) | n_N \rangle$$

$$= \exp\{(i/\hbar)[E_m + (N+1)|e\phi| - E_n - N|e\phi|]t\} \langle m_{N+1} | a^+(\mathbf{p}\sigma 0) | n_N \rangle$$

$$= e^{i|e\phi|t/\hbar} \langle m_{N+1} | a^+(\mathbf{p}\sigma t) | n_N \rangle_0.$$

$$(9.1.1.8)$$

Here the subscript 0 denotes the initial value of the matrix element before the electron energies changed, $\exp(i|e\phi|t/\hbar)$ is a time-dependent phase factor. In a more general case, when ϕ is time-dependent, the phase factor can be written as $\exp[-i\chi(t)/2]$. Obviously, the phase factor of the annihilation operator matrix element is $\exp[i\chi(t)/2]$.

The problem now is to separate out explicitly the phase factors $\exp[\pm i\chi(t)/2]$ from the mean products of electron creation and annihilation operators that enter into Eq. (9.1.1.7) for the current. In other words, we have to represent these mean quantities as products of time-dependent phase factors and functions (they are called Green functions) which depend on $t - t'$ only. It is accomplished in the theory of superconductivity by using a definite representation of operators a and a^+ (Larkin & Ovchinnikov, 1975; Shelankov, 1980):

$$a(\mathbf{k} \uparrow t) = a_1(\mathbf{k}t)\exp[i\chi(t)/2], \quad a(\mathbf{k} \downarrow t) = -a_2^+(\mathbf{k}t)\exp[i\chi(t)/2],$$
$$a^+(\mathbf{k} \uparrow t) = a_1^+(\mathbf{k}t)\exp[-i\chi(t)/2], \quad a^+(\mathbf{k} \downarrow t) = a_2(\mathbf{k}t)\exp[-i\chi(t)/2].$$
$$(9.1.1.9)$$

The sought Green functions which depend on the time-difference $t - t'$ only are defined in terms of the operators $a_i(\mathbf{k}t)$ and $a_i^+(\mathbf{k}t)$ $(i = 1, 2)$:

$$g_{ik}^{21}(t - t') = \frac{1}{\pi\hbar}\int dE(\mathbf{k})\langle a_i(\mathbf{k}t)a_k^+(\pm\mathbf{k}t')\rangle,$$
$$g_{ik}^{12}(t' - t) = -\frac{1}{\pi\hbar}\int dE(\mathbf{k})\langle a_i^+(\mathbf{k}t)a_k(\pm\mathbf{k}t')\rangle, \qquad (9.1.1.10)$$
$$g_{ik}^{R}(t - t') = \frac{1}{\pi\hbar}\int dE(\mathbf{k})\Theta(t - t')\langle\{a_i^+(\mathbf{k}t), a_k(\pm\mathbf{k}t')\}\rangle.$$

Here $\Theta(t)$ is the Heaviside step function (Eq. (1.6.22)), and $g_{ik}^{R}(t-t')$ is called the retarded Green function. The Green functions $g_{ik}^{21}(t - t')$, $g_{ik}^{12}(t - t')$, and $g_{ik}^{R}(t - t')$ are matrices. Their nondiagonal elements are specific for superconductors: they are directly related to the superconductive energy gap 2Δ. The plus and minus signs in the arguments correspond to diagonal and nondiagonal elements of the matrices g_{ik}, respectively. The nondiagonal elements, with opposite spins in the operators a and a^+, are antisymmetric: $g_{ik} = -g_{ki}$.

Let us now substitute Eqs. (9.1.1.9) and (9.1.1.10) in Eq. (9.1.1.7). In each term of Eq. (9.1.1.7) the phase exponents corresponding to different electrodes have opposite signs. Therefore the currents are expressed in terms of the difference $\chi(t) = \chi_L - \chi_R$ of phases in the left and right electrodes. The currents of quasi-particles and of Cooper pairs can be represented in the form:

$$I_n(t) = \frac{\pi\hbar}{eR_N}\mathrm{Re}\int_{-\infty}^{t} dt' [g_{11}^{12}(t-t')g_{11}^{21}(t'-t) - g_{11}^{21}(t-t')g_{11}^{12}(t'-t)]$$
$$\times \exp[i(\chi(t)-\chi(t'))/2]$$

$$I_s(t) = \frac{\pi\hbar}{eR_N}\mathrm{Re}\int_{-\infty}^{t} dt' [g_{12}^{12}(t-t')g_{12}^{21}(t'-t) - g_{12}^{21}(t-t')g_{12}^{12}(t'-t)]$$
$$\times \exp[i(\chi(t)+\chi(t'))/2].$$

$$(9.1.1.11)$$

Because each electrode, taken separately, is in an equilibrium state, the Fourier transforms of the functions $g_{ik}^{12}(\epsilon)$ and $g_{ik}^{21}(\epsilon)$ are expressed in terms of $\mathrm{Re}\,g_{ik}^{R}$, i.e., the real part of the retarded Green function:

$$g_{ik}^{12}(\epsilon) = -2f(\hbar\epsilon)\mathrm{Re}\,g_{ik}^{R}(\epsilon), \quad g_{ik}^{21}(\epsilon) = 2[1 - f(\hbar\epsilon)]\mathrm{Re}\,g_{ik}^{R}(\epsilon), \quad (9.1.1.12)$$

where $f(\hbar\epsilon)$ is the Fermi distribution function in which the energy $\hbar\epsilon$ is reckoned from the Fermi level. One can see that the current in a Josephson junction is expressed in terms of two quantities only: the functions $\mathrm{Re}\,g_{11}^{R}(\epsilon)$ and $\mathrm{Re}\,g_{12}^{R}(\epsilon)$.

These functions are found by solving the equations of superconductivity (see, for instance, Shelankov, 1980). Within the Bardeen–Cooper–Schrieffer theory of superconductivity, they are given by the following equations:

$$\mathrm{Re}\,g_{11}^{R} = \mathrm{sign}(\epsilon)\Theta(\hbar|\epsilon| - \Delta)\frac{\hbar\epsilon}{\xi(\epsilon)}, \quad \mathrm{Re}\,g_{12}^{R} = -\mathrm{sign}(\epsilon)\Theta(\hbar|\epsilon| - \Delta)\frac{\Delta}{\xi(\epsilon)},$$

$$(9.1.1.13)$$

where $\xi(\epsilon) = \sqrt{(\hbar\epsilon)^2 - \Delta^2}$, and $\mathrm{sign}(\epsilon) = \epsilon/|\epsilon|$. The currents can now be represented in the form:

$$I_n(t) = \frac{\hbar}{\pi e R_N}\mathrm{Re}\int_{-\infty}^{t} dt' \int d\epsilon d\epsilon' \exp[i(\epsilon'-\epsilon)(t-t') + i(\chi(t)-\chi(t'))/2]$$
$$\times [f(\hbar\epsilon') - f(\hbar\epsilon)]\mathrm{Re}\,g_{11}^{R}(\epsilon)\mathrm{Re}\,g_{11}^{R}(\epsilon'),$$

$$I_s(t) = \frac{\hbar}{e\pi R_N}\mathrm{Re}\int_{-\infty}^{t} dt' \int d\epsilon d\epsilon' \exp[i(\epsilon'-\epsilon)(t-t') + i(\chi(t)+\chi(t'))/2]$$
$$\times [f(\hbar\epsilon) - f(\hbar\epsilon')]\mathrm{Re}\,g_{12}^{R}(\epsilon)\mathrm{Re}\,g_{12}^{R}(\epsilon').$$

$$(9.1.1.14)$$

The mean voltage $\bar{U}$ can be separated out from the total voltage: $U(t) = \bar{U} + \tilde{U}(t)$. The last part is the variable part of the voltage. The function $\exp[i\chi(t)/2]$ can be represented as a Fourier integral (see Harris, 1975):

$$e^{i\chi(t)/2} = e^{i\Theta/2}\int_{-\infty}^{+\infty}\frac{d\omega}{2\pi}e^{-i\omega t}W(\omega). \quad (9.1.1.15)$$

Here $\Theta = \omega_J t + \Theta_0$, $\omega_J = 2e\bar{U}/\hbar$ is the Josephson frequency, Θ_0 is a constant phase. The Fourier amplitude $W(\omega)$, like the difference of phases $\chi(t)$, can be found by solving Eq. (9.1.1.4) of the Josephson junction's dynamics. The normal and superconductive currents can be represented in terms of the amplitudes $W(\omega)$:

$$
\begin{aligned}
I_n(t) &= \text{Im} \int \frac{d\omega}{2\pi} e^{-i\omega t} \int \frac{d\omega'}{2\pi} I_q(\omega' - \omega/2 - \omega_J/2) \\
&\quad \times W(\omega' + \omega/2) W^*(\omega' - \omega/2), \\
I_s(t) &= \text{Im}\, e^{i\omega_J t} \int \frac{d\omega}{2\pi} e^{-i\omega t} \int \frac{d\omega'}{2\pi} I_p(\omega' - \omega/2 + \omega_J/2) \\
&\quad \times W(\omega' + \omega/2) W(-\omega' + \omega/2).
\end{aligned}
$$

(9.1.1.16)

Two functions have been introduced:

$$
\begin{aligned}
I_q(\omega) &= \frac{\hbar}{\pi e R_N} \int d\epsilon d\epsilon' [f(\hbar\epsilon') - f(\hbar\epsilon)] \text{Re}\, g_{11}^R(\epsilon) \text{Re}\, g_{11}^R(\epsilon') \\
&\quad \times (\epsilon - \epsilon' + \omega - i\eta)^{-1}, \\
I_p(\omega) &= \frac{\hbar}{\pi e R_N} \int d\epsilon d\epsilon' [f(\hbar\epsilon) - f(\hbar\epsilon')] \text{Re}\, g_{12}^R(\epsilon) \text{Re}\, g_{12}^R(\epsilon') \\
&\quad \times (\epsilon - \epsilon' + \omega - i\eta)^{-1}.
\end{aligned}
$$

(9.1.1.17)

Here $\eta \to +0$.

If $U = 0$, the phase difference $\chi = \text{const}$. Therefore $\omega_J = 0$ and $W(\omega) = 2\pi\delta(\omega)e^{i\chi/2}$. Since the imaginary parts of I_q and I_p are odd in ω, the normal current $I_n = \text{Im}\, I_q(0) = 0$, $I_s = \text{Im}\, (I_p(0)e^{i\chi})$. I_s is given by Eq. (9.1.1.2) with $j_c = \text{Re}\, I_p(0)/A$, where A is the junction's area. The small-signal complex conductance (inverse impedance) is expressed in terms of the same functions $I_q(\omega)$ and $I_p(\omega)$ (see, for instance, the book by Likharev, 1986). Its real part equals:

$$
\text{Re}\, Z^{-1}(f) = \frac{e}{hf}[\text{Im}\, I_q(\omega) + \text{Im}\, I_p(\omega)\cos\chi].
$$

(9.1.1.18)

Apart from normal and superconductive currents, one has to take into account also the displacement current in the junction. It gives a purely reactive contribution $-i\omega C$ to the junction's complex conductance $Z^{-1}(f)$, where C is the junction's capacitance.

If the voltage across the junction is constant, $U = \bar{U} \neq 0$, the amplitude $W(\omega) = 2\pi\delta(\omega)$, $\omega_J \neq 0$. Then

$$
\begin{aligned}
I_s(t) &= \text{Im}\left(e^{i\omega_J t} I_p(\omega_J/2)\right) = \text{Re}\, I_p(\omega_J/2)\sin\chi + \text{Im}\, I_p(\omega_J/2)\cos\chi, \\
I_n &= \text{Im}\, I_q(-\omega_J/2).
\end{aligned}
$$

(9.1.1.19)

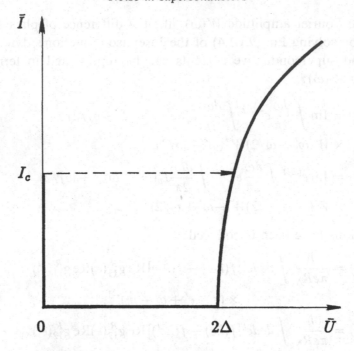

Fig. 9.2 The current–voltage characteristic of the Josephson tunnel junction at low temperatures. I_c is the critical current. The branch with zero dc current at voltages up to 2Δ corresponds to oscillatory superconductive current and absence of normal current at low temperatures.

The first equation means that, when the phase χ is not constant, the current contains a term linear in $\cos \chi$, apart from a $\sin \chi$ term as in Eq. (9.1.1.2) (Harris, 1974).

The current–voltage characteristic (CVC) of a tunnel Josephson junction, i.e., the dc current vs dc voltage, is shown in Fig. 9.2. It consists of two branches. The vertical part at $U = 0$ is the superconductive branch. The resistive branch corresponds to nonzero voltage U. If U is nonzero but constant, the phase, according to Eq. (9.1.1.4), $\chi(t) = (2eU/\hbar)t$. The superconductive current oscillates with the Josephson frequency (Eq. (9.1.1.19)), its dc (mean) component is zero, i.e., it does not contribute to the CVC. The normal current is constant in time and grows with U:

$$I_n = \operatorname{Im} I_q(-\omega_J/2) = \frac{\hbar}{eR_N} \int d\epsilon \, [f(\hbar\epsilon - eU/2) - f(\hbar\epsilon + eU/2)]$$
$$\times \operatorname{Re} g_{11}^R(\epsilon - eU/2\hbar) \operatorname{Re} g_{11}^R(\epsilon + eU/2\hbar). \tag{9.1.1.20}$$

At nonzero Δ, low temperatures $k_B T \ll 2\Delta$, and low voltages $e\bar{U} < 2\Delta$, the normal current is suppressed by the existence of the superconductive gap, formally, by the condition $|\hbar\epsilon| > \Delta$ in Eq. (9.1.1.13). In the limit $\Delta \to 0$, Eq. (9.1.1.20) reduces to Eq. (5.5.9) for the tunnel current between normal metals. When the voltage across the junction $\bar{U} > 2\Delta$, the normal current grows steeply with $\bar{U}$.

The dynamics of Josephson junctions is in many respects determined by the value of the dimensionless parameter (McCumber–Stewart parameter) $\beta = 2eI_cR^2C/\hbar$. Here R is the effective resistance of the junction, i.e., the tunnel resistance R_N for an unshunted junction or the shunting resistance R_e if $R_e \ll R_N$. At high C and R (unshunted tunnel junctions) the parameter $\beta \gg 1$. In this case, the response of the voltage $\delta U(t)$ to a small variation of the current $\delta I(t)$ contains only low-frequency components: the high-frequency ones are shunted by the capacitance. Therefore the inverse complex impedance equals:

$$Z^{-1}(f) = R_d^{-1} - i\omega C, \qquad (9.1.1.21)$$

where the small-signal conductance $R_d^{-1} = dI_n/d\bar{U}$ can be found from the CVC, that is, from Eq. (9.1.1.20).

The problem of the junction's impedance at $\beta \simeq 1$ or $\beta \ll 1$ requires a more general approach, which is presented in Sec. 9.1.3.

The equations presented above will be used in the next sections, in which the noise in Josephson junctions is discussed.

9.1.2 Noise in tunnel Josephson junctions

One of the sources of current fluctuations in the tunnel Josephson junctions is the random transitions of electrons between the electrodes through the tunnel barrier. These transitions cause current noise and, consequently, voltage noise. The main goal of this section is to present the equation for the current noise in tunnel Josephson junctions in terms of the properties of the superconductive electrodes, each of which, taken separately, is in an equilibrium state. This equation is a generalization of Eq. (5.5.16) for normal tunnel junctions. The probability of transmission of Cooper pairs and quasi-particles through the tunnel barrier is completely specified, as in Eqs. (5.5.16) and (9.1.1.6), by the tunnel resistance R_N in the normal state of the junction.

If the inelastic scattering processes in the tunnel layer can be neglected, the electron motion between the electrodes corresponds to a coherent wave function. It can be described, to a definite extent, in the same way as the motion of electrons in normal quantum ballistic contacts and normal tunnel junctions (Secs. 5.2 and 5.5). The correlation function of current fluctuations follows from the general quantum-mechanical Eq. (1.2.11) and Eq. (5.5.3) for the tunnel current operator $\hat{I}(t)$ (compare with Eq. (5.5.13)):

$$\psi_I(t_1, t_2) = \frac{1}{2}\langle\{\delta\hat{I}(t_1), \delta\hat{I}(t_2)\}\rangle = \frac{1}{2}\langle\{\hat{I}(t_1), \hat{I}(t_2)\}\rangle - \langle\hat{I}(t_1)\rangle\langle\hat{I}(t_2)\rangle$$

$$= -\frac{e^2}{2}\langle\{(\hat{A}(t_1) - \hat{A}^+(t_1)), (\hat{A}(t_2) - \hat{A}^+(t_2))\}\rangle. \tag{9.1.2.1}$$

Upon substitution of Eq. (5.5.4) for $\hat{A}(t)$ and decomposition of the mean product of four operators of electron creation and annihilation into products of two mean products of two operators (each of which is related to a definite electrode), one obtains:

$$\psi_I(t_1, t_2) = \frac{e^2}{2}\sum_{pk\sigma}|T(\mathbf{pk})|^2\{\langle a_L^+(\mathbf{p}\sigma t_1)a_L(\mathbf{p}\sigma t_2)\rangle\langle a_R(\mathbf{k}\sigma t_1)a_R^+(\mathbf{k}\sigma t_2)\rangle$$

$$+ \langle a_L(\mathbf{p}\sigma t_1)a_L^+(\mathbf{p}\sigma t_2)\rangle\langle a_R^+(\mathbf{k}\sigma t_1)a_R(\mathbf{k}\sigma t_2)\rangle \tag{9.1.2.2}$$

$$+ \langle a_L^+(\mathbf{p}\sigma t_1)a_L^+(-\mathbf{p}, -\sigma t_2)\rangle\langle a_R(\mathbf{k}\sigma t_1)a_R(-\mathbf{k}, -\sigma t_2)\rangle$$

$$+ \langle a_L(\mathbf{p}\sigma t_1)a_L(-\mathbf{p}, -\sigma t_2)\rangle\langle a_R^+(\mathbf{k}\sigma t_1)a_R^+(-\mathbf{k}, -\sigma t_2)\rangle + (t_1 \leftrightarrow t_2)\}.$$

The essential difference between Eq. (9.1.2.2) and the corresponding Eq. (5.5.14) for the tunnel junction between normal metals is the presence of mean products of two operators of the electrons' creation and two operators of the electrons' annihilation. As was mentioned in Sec. 9.1.1, these quantities in superconductors are nonzero and are proportional to the order parameter of superconductivity Δ, i.e., to the superconductor's energy gap.

Using Eqs. (9.1.1.6) and (5.5.12) one obtains:

$$\psi_I(t_1, t_2) = \frac{e^2}{2}\frac{\hbar}{4\pi e^2 R_N}\int\frac{dE(\mathbf{k})}{\hbar}\int\frac{dE(\mathbf{p})}{\hbar}$$

$$\times \left\{2\langle a_L^+(\mathbf{p}\uparrow t_1)a_L(\mathbf{p}\uparrow t_2)\rangle\langle a_R(\mathbf{k}\uparrow t_1)a_R^+(\mathbf{k}\uparrow t_2)\rangle\right.$$

$$+ 2\langle a_L(\mathbf{p}\uparrow t_1)a_L^+(\mathbf{p}\uparrow t_2)\rangle\langle a_R^+(\mathbf{k}\uparrow t_1)a_R(\mathbf{k}\uparrow t_2)\rangle$$

$$+ \langle a_L^+(\mathbf{p}\uparrow t_1)a_L^+(-\mathbf{p}\downarrow t_2)\rangle\langle a_R(\mathbf{k}\uparrow t_1)a_R(-\mathbf{k}\downarrow t_2)\rangle$$

$$+ \langle a_L(\mathbf{p}\uparrow t_1)a_L(-\mathbf{p}\downarrow t_2)\rangle\langle a_R^+(\mathbf{k}\uparrow t_1)a_R^+(-\mathbf{k}\downarrow t_2)\rangle$$

$$+ \langle a_L^+(\mathbf{p}\downarrow t_1)a_L^+(-\mathbf{p}\uparrow t_2)\rangle\langle a_R(\mathbf{k}\downarrow t_1)a_R(-\mathbf{k}\uparrow t_2)\rangle$$

$$+ \left.\langle a_L(\mathbf{p}\downarrow t_1)a_L(-\mathbf{p}\uparrow t_2)\rangle\langle a_R^+(\mathbf{k}\downarrow t_1)a_R^+(-\mathbf{k}\uparrow t_2)\rangle + (t_1 \leftrightarrow t_2)\right\}.$$

$$\tag{9.1.2.3}$$

The mean products of operators in Eq. (9.1.2.3) can be represented with phase factors separated out (see Eqs. (9.1.1.9) and (9.1.1.10)):

$$\psi_I(t_1, t_2) = -\frac{\pi\hbar}{2R_N} \Big\{ [g_{11}^{21}(t_1 - t_2)g_{11}^{12}(t_2 - t_1) + g_{11}^{12}(t_1 - t_2)g_{11}^{21}(t_2 - t_1)]$$
$$\times \cos[(\chi(t_1) - \chi(t_2)/2]$$
$$+ [g_{12}^{21}(t_1 - t_2)g_{12}^{12}(t_2 - t_1) + g_{12}^{12}(t_1 - t_2)g_{12}^{21}(t_2 - t_1)]$$
$$\times \cos[(\chi(t_1) + \chi(t_2))/2] \Big\}.$$

$$(9.1.2.4)$$

The first and second parts of ψ_I are the correlation functions of normal and superconductive currents, respectively.

Using Eqs. (9.1.1.12) for the Fourier components of the Green functions, one obtains:

$$\psi_I(t_1, t_2) = \frac{\hbar}{2\pi R_N} \int d\epsilon \int d\epsilon' \cos[(\epsilon - \epsilon')(t_1 - t_2)]$$
$$\times \{f(\hbar\epsilon)[1 - f(\hbar\epsilon')] + f(\hbar\epsilon')[1 - f(\hbar\epsilon)]\}$$
$$\times \{\mathrm{Re}\, g_{11}^R(\epsilon)\mathrm{Re}\, g_{11}^R(\epsilon') \cos[(\chi(t_1) - \chi(t_2))/2]$$
$$+ \mathrm{Re}\, g_{12}^R(\epsilon)\mathrm{Re}\, g_{12}^R(\epsilon') \cos[(\chi(t_1) + \chi(t_2))/2]\}.$$

$$(9.1.2.5)$$

Thus, the correlation function is expressed in terms of the same two functions $\mathrm{Re}\, g_{11}^R$ and $\mathrm{Re}\, g_{12}^R$ as the mean current (Eq. (9.1.1.14)).

Because the Josephson junction is, in general, not a stationary system, the correlation function depends not only on the time difference $t_1 - t_2$ but also on the 'mean' time $\bar{t} = (t_1 + t_2)/2$. We can substitute the Fourier expansion for $e^{i\chi(t)/2}$ (Eq. (9.1.1.15)) into Eq. (9.1.2.5) and represent $\psi_I(t_1, t_2)$ more explicitly as a function of these two time variables. The Fourier transform with respect to $t_1 - t_2$, multiplied by 2, is analogous to spectral density. However, under nonstationary conditions this quantity depends on $\bar{t}$:

$$S_I(f, \bar{t}) = 2 \int d(t_1 - t_2) \exp[i\omega(t_1 - t_2)]\psi_I(t_1, t_2)$$
$$= e \int \frac{d\omega_1 d\omega_2}{2\pi^2} \Big\{ \coth\Big(\frac{\omega_J/2 + \omega - \omega_1}{2k_B T}\Big) \mathrm{Im}\, I_q(\omega_J/2 - \omega_1 + \omega)$$
$$\times W(\omega_1 + \omega_2/2)W^*(\omega_1 - \omega_2/2)e^{-i\omega_2\bar{t}} + \coth\Big(\frac{\omega - \omega_1}{2k_B T}\Big)\mathrm{Im}\, I_p(\omega - \omega_1)$$
$$\times \frac{1}{2}\Big[W(\omega_1 + \omega_2/2)W(-\omega_1 + \omega_2/2)e^{i(\omega_J - \omega_2)\bar{t}}$$
$$+ W^*(\omega_1 + \omega_2/2)W^*(-\omega_1 + \omega_2/2)e^{-i(\omega_J - \omega_2)\bar{t}}\Big]\Big\}.$$

$$(9.1.2.6)$$

The quantities I_q and I_p have been introduced in Eq. (9.1.1.17).

In the absence of voltage ($U = 0$), i.e., on the vertical superconductive branch of the CVC, the Josephson junction is an equilibrium system. The phase difference $\chi = $ const., and the Fourier amplitudes $W(\omega) = 2\pi\delta(\omega)e^{i\chi/2}$. Then the spectral density, as in any stationary system, is independent of $\bar{t}$. It equals:

$$S_I(f) = 2e \coth\left(\frac{hf}{2k_B T}\right)[\operatorname{Im} I_q(\omega) + \operatorname{Im} I_p(\omega)\cos\chi]. \qquad (9.1.2.7)$$

Since the expression in rectangular brackets is proportional to the small-signal conductance, i.e., inverse impedance (Eq. (9.1.1.18)), one can see that Eq. (9.1.2.7) is in accordance with the Nyquist equation for the spectral density of current noise in equilibrium systems (Eq. (2.2.4)).

The frequencies f at which the noise is measured are usually so small that $hf \ll \Delta$. Therefore only quasiparticles with energies $\hbar|\epsilon| > \Delta$ contribute to the noise. The number of such quasiparticles at low temperatures, $T \ll \Delta/k_B$, is small, and the spectral density $S_I \propto \exp(-\Delta/k_B T)$.

At nonzero but constant voltage $U = \bar{U}$ across the junction, when $\omega_J \neq 0$, that part of $S_I(f, \bar{t})$ which is proportional to $\operatorname{Im} I_q$ and is created by random transitions of quasi-particles is stationary (independent of $\bar{t}$). It can be represented in a form similar to Eq. (5.5.16) (Dahm, Denenstein, Langenberg et al., 1969; Rogovin & Scalapino, 1974):

$$\begin{aligned} S_{I_n}(f) = e\Big\{ &I_n(U + hf/e)\coth\left(\frac{eU + hf}{2k_B T}\right) \\ &+ I_n(U - hf/e)\coth\left(\frac{eU - hf}{2k_B T}\right)\Big\}. \end{aligned} \qquad (9.1.2.8)$$

Here $I_n(U)$ is the tunnel quasi-particle (normal) current as a function of the voltage across the junction. At $hf \ll eU$ this equation can be simplified:

$$S_{I_n}(f) = 2e\coth\left(\frac{eU}{2k_B T}\right)I_n(U). \qquad (9.1.2.9)$$

At small voltages, $U \ll 2k_B T/e$, the spectral density corresponds to the Nyquist one in which the conductance is replaced by $I_n(U)/U$. The last differs from R_N^{-1} at low temperatures. At high voltages, $U \gg 2k_B T/e$, the current noise $S_I = 2eI_n$. It can be viewed as a shot noise (Sec. 1.5).

At low temperatures and voltages, when $k_B T, eU \ll 2\Delta$, the normal current through the junction is small (Sec. 9.1.1) and so is the normal current noise. At temperatures close to the critical one, T_c, when $\Delta \ll k_B T$, the current noise is close to the noise of a tunnel junction between two normal metals, $S_I \to 4k_B T/R_N$.

Under constant voltage, that part of $S_I(f, \bar{t})$ which corresponds to the tunneling of Cooper pairs oscillates with the Josephson frequency ω_J:

$$S_{I_s}(f, \bar{t}) = e \coth\left(\frac{hf}{2k_B T}\right) \operatorname{Im} I_p(\omega) \cos(\omega_J \bar{t}). \qquad (9.1.2.10)$$

Its low-frequency component ($f \ll \omega_J/2\pi$) is zero.

The random currents discussed above arise due to random electron transitions through the junction. They have to be considered as Langevin (intrinsic) current sources and, strictly speaking, their spectral density has to be denoted as S_I^{int} (for the Langevin approach see Sec. 1.10). These current fluctuations give rise to fluctuations of voltage $\delta U(t)$ across the junction. The ultimate sensitivity of devices using the Josephson effect, as well as the width of the oscillations' spectrum lines, are determined by the spectral density of voltage noise. The current and voltage fluctuations are related by the conditions imposed by the circuit, that is, by conditions on the total current. If it is fixed,

$$S_U(f) = |Z(f)|^2 S_I(f). \qquad (9.1.2.11)$$

One can see (Fig. 9.2) that, at a nonzero given total current I smaller than the critical current I_c, a Josephson junction with not too small parameter β can exist in two different states: (1) a purely superconductive state on the vertical branch of the CVC and (2) a resistive state with $U \neq 0$. The first state is separated from the second one by an energy barrier, i.e., the energy of the junction versus the phase difference χ has a barrier. A junction placed initially in the superconductive state by increasing the current from zero value can escape and switch to the resistive state. Ultimately, this transition is due to fluctuations and, of course, is a random process. The transition can take place over the barrier (at high enough temperatures) or by quantum tunneling. The latter process is called macroscopic tunneling because a macroscopic quantity, the phase difference χ, is changed. The lifetime of the superconductive state depends on the current I (it determines the height of the barrier) and temperature T. A comprehensive analysis of these transitions is presented in the book by Likharev (1986).

Noise in Josephson junctions with a small value of the parameter β is discussed in the next section.

9.1.3 Noise in Josephson junctions with direct conduction and in shunted tunnel junctions

One of the important types of junction exhibiting both the dc and ac Josephson effects is the microbridge, that is a narrow constriction between two bulk superconductors (the $S - c - S$ system). Its properties in many respects differ from those of the tunnel Josephson junctions. One of the differences is a much smaller capacitance, owing to which the effects of the displacement current on the dynamics of the junction may be neglected. This makes the $S - c - S$ junction similar to the tunnel junctions shunted by an external resistance R_e much smaller than the normal tunnel resistance R_N. Both types of Josephson junctions can be approximated by the so called resistively shunted junction (RSJ) model (McCumber, 1968; Stewart, 1968). According to this model, the total current in the absence of fluctuations is a sum of the normal and superconductive currents:

$$I = R^{-1}U + I_c \sin \chi. \tag{9.1.3.1}$$

Here R is the resistance R_N of the junction in the normal state, in the case of a junction with direct conduction, or R_e, the shunt's resistance, in the case of a shunted tunnel junction.

Using the Josephson equation (9.1.1.4), the voltage U can be expressed in terms of the time-derivative of the phase difference χ. Therefore the equation for χ reads:

$$\tilde{I} = \omega_c^{-1}\dot{\chi} + \sin \chi. \qquad \tilde{I} = I/I_c, \quad \omega_c = 2eI_cR/\hbar. \tag{9.1.3.2}$$

The solution to this equation at a fixed current I was found by Aslamazov & Larkin (1968):

$$\chi(t) = 2 \arctan\left[\frac{\sqrt{\tilde{I}^2 - 1}\tan(\Theta/2) + 1}{\tilde{I}}\right]. \qquad \Theta = \omega_J t + \Theta_0. \tag{9.1.3.3}$$

The time-derivative $\dot{\chi}$, which determines the voltage across the junction, equals:

$$\dot{\chi}(t) = \frac{(\tilde{I}^2 - 1)\omega_c}{\tilde{I} + \sin \Theta(t)}, \tag{9.1.3.4}$$

The dependence of the dc voltage $\bar{U}$ on the current, i.e., the current–voltage characteristic, and the Josephson frequency $\omega_J = 2e\bar{U}/\hbar$, are derived by averaging Eq. (9.1.3.4) over a period of oscillations, $2\pi/\omega_J$:

$$\bar{U} = I_cR\sqrt{\tilde{I}^2 - 1}. \tag{9.1.3.5}$$

In contrast to the unshunted tunnel junction, the dc voltage in junctions covered by the RSJ model does not switch discontinuously when the current exceeds the critical value I_c, but smoothly increases with the dc current $\tilde{I}$ (Eq. (9.1.3.5) and Fig. 9.3).

It is usually assumed that the sources of current and voltage noise in the

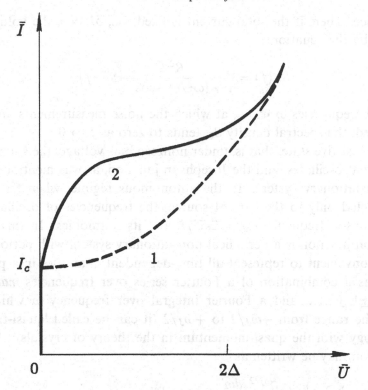

Fig. 9.3 The CVC of a resistively shunted Josephson junction and a Josephson junction with direct conduction. 1: the CVC of a resistively shunted junction in the limit of very small capacitance C: $R(2eI_cC/\hbar)^{1/2} \ll 1$. 2: the CVC of a S–c–S junction in which self-stimulation of superconductivity takes place.

Josephson junctions covered by the RSJ model are the current fluctuations $\delta I^{\mathrm{int}}(t)$ generated by the resistance R (Likharev & Semenov, 1972). The correlation function of this Langevin source is, consequently, assumed to be given by the Nyquist equation:

$$S_I^{\mathrm{int}}(f) = 2R^{-1}hf \coth\left(\frac{hf}{2k_BT}\right). \qquad (9.1.3.6)$$

The total current fluctuation consists of the normal current of quasi-particles δI_n, the superconductive current δI_s, the displacement current $C(d\delta U/dt)$, and the Langevin current source δI^{int}:

$$\delta I = \delta I_n + \delta I_s + C(d\delta U/dt) + \delta I^{\mathrm{int}}. \qquad (9.1.3.7)$$

On the superconductive (vertical) branch of the CVC the junction is a stationary, and even equilibrium, system. The spectral density has its common meaning. By linearization of Eq. (9.1.3.2) one can find the junction's complex impedance. Then, if the total current is fixed, i.e., $\delta I = 0$, the voltage noise is given by the equation

$$S_U(f) = \frac{R^2}{1 + (\omega_c^2/\omega^2)\cos^2\chi} S_I^{\text{int}}(f). \qquad (9.1.3.8)$$

At small frequencies $\omega \ll \omega_c$, at which the noise measurements are usually performed, the spectral density S_U tends to zero as $f \to 0$.

In the resistive state, that is, under nonzero bias voltage, the superconductive current oscillates, and the Josephson junction, as was mentioned above, is a nonstationary system. In the autonomous regime, when the junction is connected only to the current source, the frequencies of oscillations are the Josephson frequency $\omega_J = 2eU/\hbar$ and its harmonics. In this case the Josephson junction is a periodical nonstationary system with period $2\pi/\omega_J$.

It is convenient to represent all time-dependent quantities in a periodical system as a combination of a Fourier series over frequencies $m\omega_J$, where $m = 0, \pm1, \pm2, \ldots$, and a Fourier integral over frequency ω which varies within the range from $-\omega_J/2$ to $+\omega_J/2$ (it can be called 'quasi-frequency' by analogy with the quasi-momentum in the theory of crystals). A current fluctuation may be written as

$$\delta I(t) = \sum_m e^{-i\omega_m t} \int_{-\omega_J/2}^{\omega_J/2} \frac{d\omega}{2\pi} \delta I_m(\omega)e^{-i\omega t}, \qquad \omega_m = m\omega_J. \qquad (9.1.3.9)$$

The voltage $U(t)$ across an autonomous junction is a sum of a constant voltage $\bar{U}$ and an ac voltage which oscillates with frequencies $\omega_m = m\omega_J$. In accordance with the Josephson equation, the difference $\chi(t) - \omega_J t$ is a periodical function of time t. It means that the function $\exp[i(\chi - \omega_J t)/2]$ is also a periodical function of time and can be represented by a Fourier series (compare with Eq. (9.1.1.15)):

$$\exp[i\chi(t)/2] = \exp[i\omega_J t/2] \sum_m W_m \exp[i\omega_m t]. \qquad (9.1.3.10)$$

The Fourier amplitudes W_m can be found by solving the problem of the Josephson junction's dynamics (Zorin, 1981).

Small fluctuations of current and voltage in a periodical system are connected by the impedance matrix $Z_{mm'}(f)$:

$$\delta I_m(\omega) = \sum_{m'} Z_{mm'}^{-1}(f)\delta U_{m'}(\omega). \qquad (9.1.3.11)$$

Here $Z_{mm'}^{-1}(f)$ is the reciprocal impedance matrix. The impedances $Z_{mm'}$ with $m \neq m'$ are associated with the parametric transformation of the fluctuations' frequencies as a result of mixing with the harmonics of the Josephson oscillations.

The impedance matrix $Z_{mm'}(f)$ of the biased resistively shunted Josephson junction or of the junction with direct conductance is found as follows (Likharev & Semenov, 1972). A small variation δI of the current causes a small variation of the phase difference $\delta \chi$. Linearization of the equation for χ around the solution given by Eq. (9.1.3.3) yields:

$$\delta U(t) = \frac{\hbar}{2e}\delta \dot{\chi}$$

$$= R\left\{\delta I(t) + \frac{d}{dt}\left(\frac{1}{\tilde{I} + \sin \Theta}\right)\int^t dt' \delta I(t')[\tilde{I} + \sin \Theta(t')]\right\}.$$

(9.1.3.12)

The equation for the matrix of the junction's impedance at $I > I_c$ follows from Eq. (9.1.3.12):

$$Z_{mm'}(f) = Z^*_{-m,-m'}(-f) = R\left\{\delta_{mm'} + \tilde{I}\frac{\omega_{m-m'}p_{m-m'}}{\omega_{m'} + \omega + i\eta}\right.$$

$$\left. + \frac{1}{2i}\left[\frac{\omega_{m-m'+1}p_{m-m'+1}}{\omega_{m'-1} + \omega + i\eta} - \frac{\omega_{m-m'-1}p_{m-m'-1}}{\omega_{m'+1} + \omega + i\eta}\right]\right\}.$$

(9.1.3.13)

Here $\eta \to +0$, p_m is the Fourier amplitude of the periodical function of time, $(\tilde{I} + \sin \Theta(t))^{-1}$:

$$p_m = \frac{\left(\tilde{I} - \sqrt{\tilde{I}^2 - 1}\right)^{|m|}}{\sqrt{\tilde{I}^2 - 1}}\begin{cases}(-1)^{|m|/2}, & \text{if } m \text{ even};\\ i\,\text{sign}\,m\,(-1)^{(|m|+1)/2}, & \text{if } m \text{ odd}.\end{cases}$$

(9.1.3.14)

In time-periodical systems the correlation function, like all characteristic functions of two time variables t_1 and t_2, is invariant under the shift of both variables by the same number of periods. Therefore it can be represented as

$$\psi_I(t_1, t_2) = \sum_{mm'} e^{-i\omega_m t_1 + i\omega_{m'} t_2}\int_{-\omega_J/2}^{\omega_J/2}\frac{d\omega}{2\pi}e^{-i\omega(t_1-t_2)}\frac{1}{2}S_{I,mm'}(f).$$

(9.1.3.15)

The matrix $S_{I,mm'}$ plays the role of the spectral density of noise. If the fluctuations are averaged over the period of oscillations, only terms with $m = m' = 0$ remain in the r.h.s. of Eq. (9.1.3.15). The spectral density of these fluctuations, averaged over a period, is $S_{I,00}(f)$. As follows from Eq. (9.1.3.11), at a fixed total current,

$$S_{U,m_1m_2}(f) = \sum_{m'_1 m'_2} Z_{m_1 m'_1}(f)Z_{m_2 m'_2}(-f)S^{int}_{m'_1 m'_2}(f).$$

(9.1.3.16)

The spectral density of Langevin currents, given by Eq. (9.1.3.6), can be represented in the matrix form:

$$S_{I,mm'}^{\text{int}}(f) = \delta_{mm'} 2R^{-1} h(\omega_m/2\pi + f) \coth\left(\frac{h(\omega_m/2\pi + f)}{2k_B T}\right). \qquad (9.1.3.17)$$

Thus, the spectral density of voltage fluctuations averaged over the period of oscillations $2\pi/\omega_J$ equals:

$$S_{U,00}(f) = \sum_m Z_{0m}(f) Z_{0m}^*(f) 2R^{-1} h(\omega_m/2\pi + f) \coth\left(\frac{h(\omega_m/2\pi + f)}{2k_B T}\right).$$
$$(9.1.3.18)$$

Due to the specific properties of the nonlinearity of the Josephson junction, the voltage fluctuations at a low frequency f are determined not only by the current fluctuations at the same frequency, but also by those at the Josephson frequency and its harmonics. The frequency f at which the noise is usually measured, and the frequencies that determine the linewidth Γ (Sec. 9.1.4) usually satisfy the quasi-classical condition $hf \ll k_B T$. However, the Josephson frequency ω_J may be high enough for the opposite, quantum, inequality to be satisfied. In this case a part of the spectral density $S_{U,00}(0)$, which is given by the terms with $m \neq 0$ in the r.h.s. of Eq. (9.1.3.18), is determined by quantum fluctuations of the current, and can be measured. The theory of this noise has been developed by Koch, Van Harlingen & Clarke (1980), who also performed the experiment (Koch *et al.*, 1982). Let us estimate the magnitude of this effect.

According to Eqs. (9.1.3.3),(9.1.3.13), and (9.1.3.14),

$$Z_{00}(0) = R_d = \frac{d\bar{U}}{dI} = R\frac{\tilde{I}}{\sqrt{\tilde{I}^2 - 1}}, \quad Z_{01}(0) = -Z_{0,-1}(0) = \frac{1}{i}\frac{R_d}{2\tilde{I}}. \quad (9.1.3.19)$$

Koch *et al.* (1980) have shown that at $\omega \ll \omega_J$ the main contribution to the sum over $m \neq 0$ in Eq. (9.1.3.18) comes from the terms with $m = \pm 1$, the contribution of higher harmonics is negligible. Therefore,

$$S_{U,00}(0) = R_d^2 \left[S_I(0) + \frac{1}{2\tilde{I}^2} S_I(\omega_J/2\pi) \right]. \qquad (9.1.3.20)$$

That part of the low-frequency voltage noise which contains the quantum effects equals:

$$S_{U,00}(0) - R_d^2 S_I(0) = \frac{2e\bar{U}I_c^2 R_d^2}{RI^2} \coth\left(\frac{e\bar{U}}{k_B T}\right). \qquad (9.1.3.21)$$

The measured difference $S_U(0) - 4k_B T R_d^2/R$ is in satisfactory agreement with the r.h.s. of Eq. (9.1.3.21) (Fig. 9.4).

Despite the similarity of some properties of shunted tunnel junctions and microbridges ($S - c - S$ systems) they obviously strongly differ by their

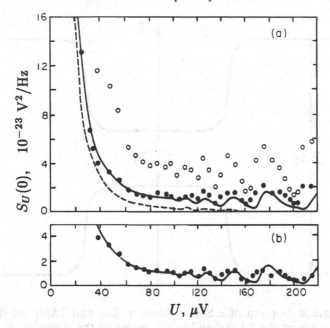

Fig. 9.4 Spectral density of voltage noise across a resistively shunted Josephson junction as a function of mean voltage $\bar{U}$. (a) Open circles: measured voltage noise at frequency $f = 100$ kHz, $T = 1.4$ K. Solid circles: mixed-down noise ($1/f$ noise subtracted). Solid and dashed lines are the results of computer simulations with and without the zero-point term, respectively. (b) Solid circles are data after the heating correction has been made, solid line is identical to that in (a). From Koch *et al.* (1982).

mechanisms of conduction. The microscopic theory of the Josephson effect in junctions with direct, nontunnel, conduction was developed by Aslamazov & Larkin (1976).

When a current I passes through the microbridge, the superconductive order parameter Δ is suppressed in the constriction, where the current density is high, and becomes smaller than its value in the electrodes ('banks') (see Fig. 9.5). The variation of $\Delta(x)$ in the constriction is smooth enough when the length L of the constriction is greater than the length $\eta = (\hbar D/\Delta)^{1/2}$ of the quasi-particles' diffusion in the characteristic time $\hbar/\Delta$. In this case the quasi-particles can be, at any point, only in states with energies $\hbar|\epsilon| > \Delta$. Since within the constriction $\Delta < \Delta_0$, the quasi-particles become trapped in the energy well created by the decrease of Δ. These quasi-particles can escape into the electrodes only after acquiring an energy $\hbar|\epsilon| > \Delta_0$. At currents $I > I_c$, where I_c is the critical current, a nonzero voltage, $U \neq 0$, appears across the junction. In accordance with the Josephson's equations, the superconducting current I_s and the energy gap 2Δ oscillate. The quasi-particles trapped in

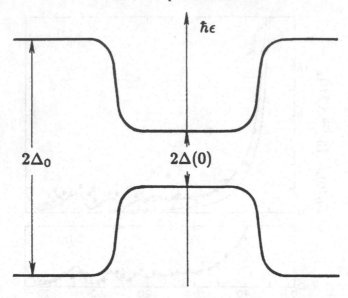

Fig. 9.5 Schematic diagram of a S–c–S junction. $2\Delta_0$ and $2\Delta(0)$ are the superconductive energy gap in the 'banks' and in the center of the constriction, respectively.
The superconductive energy gap is suppressed in the constriction due to the high
current density.

the energy well collide with its oscillating ('jittering') walls. As was shown
by Aslamazov & Larkin (1976), this interaction produces a diffusion flow
of quasi-particles toward higher energies. Their number in the vicinity of
the energy gap edge ($\hbar|\epsilon| \simeq \Delta$) decreases. The order parameter Δ and the
quasi-particles' distribution function $f(\hbar\epsilon)$ in superconductors are related by
the fundamental equation:

$$\Delta = \frac{\lambda}{4} \int d\hbar\epsilon [1 - 2f(\hbar\epsilon)] \operatorname{Re} g_{12}^R(\epsilon), \qquad (9.1.3.22)$$

where λ is the dimensionless constant of electron–phonon interaction, which
is the source of superconductivity, the function $\operatorname{Re} g_{12}^R(\epsilon)$ is given by Eq.
(9.1.1.13). It follows from Eq. (9.1.3.22) that the reduction in the density
of quasi-particles results in an increase of Δ, and, consequently, to an
increase of the superconducting current, that is, to self-stimulation of the
superconductivity. This phenomenon exhibits itself in a steep growth of the
current with dc voltage $\bar{U}$ even at small $\bar{U} \ll \Delta_0/e$ (Fig. 9.3). The region of
rapid growth of the current is followed by a saturation of the current, i.e., by
a plateau in the current–voltage characteristic, in the same region $\bar{U} \ll \Delta_0/e$.

   As was pointed out above, within the RSJ model the only source of noise
is the equilibrium Nyquist normal current noise (Likharev & Semenov, 1972).

The nonequilibrium processes in long ($L \gg \eta$) junctions result in different mechanisms of noise which have been shown to be important (Kogan & Nagaev, 1988; 1990).

Since the main contribution to the normal current fluctuations comes from the thermal quasi-particles with $\hbar|\epsilon| \sim k_B T$, which are not trapped ($k_B T \gg \Delta_0$) and remain in an equilibrium state, the spectral density of these fluctuations is the same as in the RSJ model. However, one more source of fluctuations appears, specifically, the fluctuations of the superconducting current, $\delta I_s(t)$, caused by the fluctuations of the trapped quasi-particles' distribution function (energies $\hbar|\epsilon| \sim \Delta < \Delta_0$). By virtue of the relation between the superconductive order parameter and the distribution function of quasi-particles (Eq. (9.1.3.22)), these fluctuations produce fluctuations of the order parameter, $\delta\Delta(t)$. At a nonzero mean superconducting current $\bar{I}_s$ these fluctuations give rise to fluctuations $\delta I_s(t)$.

Two mechanisms of the quasi-particles' distribution function fluctuations may be important. The first is related to the fluctuations of the flow of quasi-particles toward higher energies, which determines the nonequilibrium distribution function in the energy well. This flow is due to the scattering of quasi-particles by impurities and the oscillating walls of the well, and is a random process. Its spectral density is proportional to the coefficient of diffusion toward higher energies. The second mechanism of the distribution function's fluctuations is the random trapping of quasi-particles into and their emission from the well, which are accompanied by emission and absorption of phonons, respectively. Both mechanisms may produce a significant noise at low temperatures when the energy relaxation time is large.

Estimates (Kogan & Nagaev, 1988) have shown that the first mechanism of the superconductive current fluctuations, $\delta I_s(t)$, is the most important. Their low-frequency spectral density in the region of the CVC plateau (i.e., in the region of saturation of the diffusion flow towards higher energies) equals:

$$S_I^{(s)}(0) = \frac{\Delta_0^3}{e^2\overline{U^2}R} = \frac{4k_B T}{R}\frac{\Delta_0^3}{4e^2\overline{U^2}k_B T}, \tag{9.1.3.23}$$

where $\overline{U^2}$ is the voltage squared and averaged over the period of oscillations. Due to this quantity in the denominator, the spectral density decreases with dc voltage $\bar{U}$ above the lower edge of the plateau. In the range of the 'shoulder' (at the lower edge of the plateau) the second factor in the r.h.s. of Eq. (9.1.3.23) may be very large. Comparing $S_I^{(s)}$ with Eq. (9.1.3.6) for the noise in the resistive model one can conclude that the noise caused by the fluctuations of the diffusive flow of quasi-particles towards higher energies may be many times greater than the noise in the resistive model.

As long as the fluctuations $\delta I_s(t)$ are slower than the Josephson oscillations, the voltage response to these fluctuations is simply $\delta U = -R_d \delta I_s$. The corresponding contribution to the voltage noise $S_U(0)$ is

$$S_U^{(s)}(0) = R_d^2 S_I^{(s)}(0). \tag{9.1.3.24}$$

The noise $S_U(0)$ of $S - c - S$ junctions at $I > I_c$ was measured in several experiments (Decker & Mercereau, 1975; Pei & Lukens, 1975; Schwartz & Lukens, 1981; Schwartz, Mankievich, Jain & Lukens, 1981; Carroll & Paik, 1989). Usually it is higher than the noise calculated within the simple RSJ model by $1 - 2$ orders of magnitude. The possible extra noise is the noise $S_U^{(s)}$ produced by the fluctuations of the isotropic part of the quasi-particles' distribution function in the energy range $|\epsilon| \sim \Delta/\hbar$.

Since $S_U^{(s)}(0)$ is a product of R_d^2 times a decreasing function of $\bar{U}$, the maximum of $S_U^{(s)}(0)$ is expected to be shifted relative to the maximum of R_d toward smaller $\bar{U}$ by a voltage of the order of the width of the CVC 'shoulder'. Because the dc current increases monotonically with $\bar{U}$, the same shift is expected to be observed if S_U and R_d are considered as functions of $\bar{I}$, not $\bar{U}$. Such an effect was observed by Decker & Mercereau (1975). As the temperature approaches T_c and the critical current I_c drops, both the spectral density $S_U(0)$ (Decker *et al.*, 1975) and the spectral width of the Josephson oscillations (Schwartz & Lukens, 1981) decrease. These experimental findings are in agreement with Eqs. (9.1.3.23) and (9.1.3.24): the noise is proportional to Δ_0^3, which rapidly drops as $T \to T_c$.

9.1.4 Effect of noise on the spectral width of Josephson oscillations

The effects of fluctuations on the Josephson junctions are observed in several ways:

1. Direct measurements of noise spectral density,
2. Measurements of the Josephson high-frequency radiation line-width,
3. Measurements of the rounding of the current–voltage $(I - U)$ characteristic and its various singularities,
4. Measurements of the life-time of the superconductive state on the vertical branch of the CVC, i.e., the time of staying in this state before making a transition to the resistive state (end of Sec. 9.1.2).

Below, we discuss the effect of the noise on the linewidth of the Josephson oscillations (radiation).

As was shown in Sec. 9.1.1, the Josephson junction connected only to the dc voltage source $\bar{U}$ oscillates with the Josephson frequency, $\omega_J = 2e\bar{U}/\hbar$, and its harmonics. In the absence of voltage fluctuations, $\delta U(t)$, each of these oscillations is monochromatic, and the spectrum consists of infinitely narrow lines at frequencies $m\omega_J$, where $m = 1, 2, \ldots$. Voltage fluctuations broaden these lines.

The spectral density of emission of a system with an ac current $I(t)$ is known to be proportional to the Fourier transform of the correlation function:

$$G(\omega) = \int_{-\infty}^{+\infty} d\Delta t e^{i\omega\Delta t} \overline{I(t - \Delta t/2)I(t + \Delta t/2)}. \qquad (9.1.4.1)$$

The overline denotes averaging over a long enough time t_m of measurement.

The voltage fluctuations result in the fluctuations of the difference of the condensate wave functions' phases in the electrodes, $\delta\chi(t)$. The latter are found from the Josephson equation (9.1.1.4):

$$\delta\dot{\chi} = (2e/\hbar)\delta U(t). \qquad (9.1.4.2)$$

The total phase difference may be represented by $\chi(t) = \omega_0 t + \delta\chi(t)$, where ω_0 is the frequency of oscillations in the absence of fluctuations. Let us substitute the Josephson equation (9.1.1.2) for the superconductive current into Eq. (9.1.4.1) for $G(\omega)$. The product of two sine functions is equal to the sum of four exponents. Two of them oscillate with a high frequency $2\omega_0$ and drop out after averaging over time t_m. Therefore,

$$G(\omega) = \frac{1}{2}I_c^2 \int_{-\infty}^{+\infty} d\Delta t e^{i\omega\Delta t} \overline{\cos[\omega_0\Delta t + \delta\chi(t + \Delta t/2) - \delta\chi(t - \Delta t/2)]}.$$

That part of the integrand which, as a function of Δt, oscillates with a high frequency $\omega_0 + \omega$, may also be neglected. Hence,

$$G(\omega) = \frac{1}{4}I_c^2 \text{Re} \int_{-\infty}^{+\infty} d\Delta t e^{i(\omega-\omega_0)\Delta t} g(\Delta t), \qquad (9.1.4.3)$$

where

$$g(\Delta t) = \overline{\exp\{i[\delta\chi(t - \Delta t/2) - \delta\chi(t + \Delta t/2)]\}}. \qquad (9.1.4.4)$$

In the absence of fluctuations $g(\Delta t) = 1$, and the spectrum of oscillations is proportional to $\delta(\omega - \omega_0)$. Because the function $G(\omega)$ in the vicinity of the frequency ω_0 is a narrow peak with linewidth Γ, the function $g(\Delta t)$ in Eq. (9.1.4.3), according to the theory of Fourier integrals, decreases significantly at a time $|\Delta t| \simeq \Gamma^{-1}$. Only such comparatively long times $|\Delta t|$ are essential in Eq. (9.1.4.3). Hence, the fluctuation $\delta\chi(t - \Delta t/2) - \delta\chi(t + \Delta t/2)$ is determined by a great number of random events and may be considered as a Gaussian random variable (Sec. 1.1) at each given value of Δt. It means that

$$g(\Delta t) = \exp[-\frac{1}{2}\sigma^2(\Delta t)], \quad \sigma^2(\Delta t) = \overline{[\delta\chi(t - \Delta t/2) - \delta\chi(t + \Delta t/2)]^2},$$

(9.1.4.5)

where $\sigma^2(\Delta t)$ is the variance of the random quantity (Eq. (1.1.6)).

Integrating Eq. (9.1.4.2), one is able to express σ^2 in terms of the correlation function of voltage fluctuations:

$$\sigma^2(\Delta t) = \frac{4e^2}{\hbar^2} \int_{-\Delta t/2}^{\Delta t/2} dt'\, dt'' \overline{\delta U(t + t')\delta U(t + t'')}.$$

(9.1.4.6)

Upon averaging over a long time t_m, the correlation function of such a nonstationary system, as is the biased Josephson junction, depends on the difference $t' - t''$ only, as in a stationary system. One can define the spectral density of the voltage noise $S_U(f)$ as the doubled Fourier transform of the averaged correlation function. Then,

$$\sigma^2(\Delta t) = \frac{8e^2}{\hbar^2} \int_{-\infty}^{+\infty} \frac{d\omega}{2\pi} \frac{\sin^2(\omega\Delta t/2)}{\omega^2} S_U(f).$$

(9.1.4.7)

If at all frequencies at which $\omega\Delta t \leq 1$ the spectral density $S_U(f)$ is almost constant (and equal to its low-frequency value), it can be removed from within the integral in Eq. (9.1.4.7). In this case:

$$\frac{1}{2}\sigma^2(\Delta t) = \frac{e^2}{\hbar^2} S_U(0)|\Delta t|,$$

(9.1.4.8)

and, according to Eqs. (9.1.4.3) and (9.1.4.5), the emission line has a Lorentzian form (Larkin & Ovchinnikov, 1967):

$$G(\omega) \propto \frac{\Gamma/\pi}{(\omega - \omega_0)^2 + \Gamma^2},$$

(9.1.4.9)

with a half-width at half-maximum

$$\Gamma = \frac{e^2}{\hbar^2} S_U(0).$$

(9.1.4.10)

This case takes place when the half-width Γ is smaller than the cut-off frequency of the spectral density $S_U(f)$. In the opposite case the line has a Gaussian form (Larkin & Ovchinnikov, 1967).

9.2 Noise in type II superconductors

9.2.1 Introduction

In type II superconductors ('dirty' low-T_c alloys, all high-T_c superconductors) the process of magnetic field penetration into the bulk starts at a critical magnetic field H_{c1}. At fields exceeding H_{c1} the magnetic flux penetrates in the form of individual quantized vortices (Abrikosov, 1957). Each vortex (fluxon, flux line) is carrying a magnetic flux quantum $\Phi_0 = hc/2e = 2.07 \cdot 10^{-7}$ Gauss·cm^2. The superconductor is threaded by these vortices.

In the central part (core) of each vortex, the absolute value of the superconductive condensate wave function squared, $|\psi(\rho)|^2$, or the superconductive energy gap 2Δ, varies from zero at the center $\rho = 0$ up to the bulk value. This variation takes place in a region $\rho <\sim \xi$, where ξ is the coherence length of the superconductive state. This length determines the 'rigidity' of the superconductive condensate wave function. In type II superconductors the coherence length is smaller than the magnetic field penetration depth λ. In the outer region of the vortex, persistent superconductive currents are flowing around the core. The density of these currents and the magnetic flux associated with the vortex fall off at a distance $\sim \lambda$. It means that the vortex's dimensions in the plane normal to the magnetic field are of the order of λ. Thus, in type II superconductors the nondissipative current and magnetic induction in the bulk coexist.

The density of vortices increases with magnetic field. The superconductive state is destroyed at the upper critical field H_{c2}. At fields $H > H_{c2}$ the magnetic induction in the bulk is uniform, as in a normal metal. The upper critical field corresponds to such a density of the flux-lines at which the distance between the cores of adjacent vortices is of the order of the core width, i.e., coherence length ξ.

An external current creates a Lorentz force acting on the vortices. Its magnitude per unit volume equals:

$$\mathbf{f} = \frac{1}{c}\mathbf{j} \times \mathbf{B}, \qquad (9.2.1.1)$$

where $\mathbf{j}$ is the external current density, $\mathbf{B}$ is the bulk magnetic induction, c is the speed of light (we are using the Gaussian system of units). As the number of vortices per unit area is B/Φ_0, the force acting on a single vortex per unit length equals:

$$\mathbf{f}_1 = \frac{\Phi_0}{c}\mathbf{j} \times \mathbf{b}, \qquad (9.2.1.2)$$

where $\mathbf{b} = \mathbf{B}/|\mathbf{B}|$. Owing to this force, the vortices tend to move transverse to the current. If they are moving with drift velocity $\mathbf{v} \parallel \mathbf{f}_1$ (its value may be determined by the viscous resistance to the vortices' drift) an electric field is induced:

$$\mathbf{F} = \frac{1}{c} \mathbf{B} \times \mathbf{v}. \tag{9.2.1.3}$$

It follows from Eq. (9.2.1.2) that this electric field is parallel to $\mathbf{j}$. The density of power dissipation $\mathbf{j} \cdot \mathbf{F}$ is therefore nonzero. The power is not dissipated if the flux lines are pinned by various defects and immobilized ($\mathbf{v} = 0$).

The main dissipation phenomena and, consequently, the main noise phenomena in type II superconductors, are caused by various motions of vortices. These phenomena comprise flux creep, flux flow, and slow nonexponential decay of magnetization. In general, the noise phenomena in type II superconductors constitute an important part of their low-frequency kinetic phenomena.

9.2.2 Flux-creep and flux-flow noise

As was mentioned above, at low temperatures and small currents the vortices are pinned by defects. If all vortices are rigidly pinned and immobile, the superconductive current is not dissipated and the resistance of the superconductor is zero. However, the vortices can become depinned and start moving by hops. If the density of vortices is not too small, their interaction results in simultaneous (correlated) hops of bundles of vortices, each of which consists of a great number of vortices (Anderson, 1962). The rate of hops depends exponentially on the height E of the free-energy activation barrier that separates two stable states of the vortices' system. The rate of transitions over the barrier can be represented by:

$$r = \tau_0^{-1} \exp(-E/k_B T), \tag{9.2.2.1}$$

where τ_0^{-1} is the frequency of attempts to surmount the barrier. It is known to be $\sim 10^5 - 10^{11}$ s^{-1} in conventional, and $\sim 10^{11}$ s^{-1} in high-T_c superconductors, respectively. In the absence of any directed force acting on the vortices and of any gradient of their density, the hops of vortices are equally probable in the positive and negative directions along any line, and do not result in any mean directed magnetic flux motion. However, they result in a random flux motion which is a source of noise.

In the presence of a current, each vortex experiences a Lorentz force which is normal both to the magnetic induction $\mathbf{B}$ in the superconductor and to the current density $\mathbf{j}$ (Eq. 9.2.1.2)). This force adds a term $-\mathbf{f}_1 \cdot \mathbf{r}$ to the position-dependent energy of any vortex, where $\mathbf{r}$ is its radius-vector. The barriers become distorted, and the activation energy for a hop in the direction of the force is favored, as compared with the hops against the direction of force. This effect can be accounted for, in the linear approximation, by the following equation for the activation free energies for downstream and upstream hops, respectively:

$$E^{\pm} = E \mp |\mathbf{f}_1 \cdot \Delta\mathbf{r}|, \qquad (9.2.2.2)$$

where $\Delta\mathbf{r}$ is the vortex bundle displacement vector.

Owing to the decrease of the activation energies for downstream hops and their increase for hops in the opposite direction, a directed flow of vortices, i.e., of magnetic flux, appears. The magnetic flux motion with a velocity $\mathbf{v}$ induces an electric field $\mathbf{F}$ (Eq. (9.2.1.3)) parallel to the current, and dissipation of the electric power $\mathbf{j} \cdot \mathbf{F} \neq 0$. Thus the depinning of vortices results in a nonzero resistance of the superconductor. Until the pinning forces are, on average, dominant over the Lorentz force and the temperature is low, the motion of the vortices has the form of rare individual hops. This regime of flux motion is called flux-creep. At small currents I the resistance induced by flux-creep is immeasurably small. The current I_c, at which the resistance exceeds some experimental lower limit, is called the critical current. In the opposite case when, on average, the dominant force is the Lorentz one (high currents, high temperatures, small pinning forces) the flow of vortices becomes almost continuous and is called flux-flow. Under stationary conditions the Lorentz force is counterbalanced by viscosity due, in particular, to normal excitations in the cores of the vortices.

Since an electric current induces a nonzero voltage across a type II superconductor threaded by magnetic flux at $H_{c1} < H < H_{c2}$, making the superconductor's resistivity $\rho \neq 0$, an electric noise should appear in the form of random voltage fluctuations across the contacts of the superconductor. This kind of noise has been observed first by van Ooijen & van Gurp (1965, 1966) on samples of V foil at 4.2 K and subsequently studied by many authors in both low- and high-T_c superconductors. A review of the status of the problem in conventional low-T_c superconductors up to 1981 has been published by Clem (1981).

The problem of vortex kinetics in the presence of pinning centers is, in principle, extremely complex. First of all, the vortices interact with randomly distributed pinning centers, i.e., this system is disordered. Moreover, due to the interaction between vortices it is a many-body problem. The problem of vortex structure kinetics is not yet solved completely. Below we present some basic experimental results and some basic concepts for their interpretation.

The voltage fluctuations in type II superconductors are usually observed

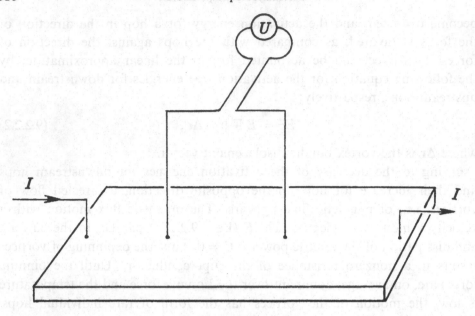

Fig. 9.6 Schematic diagram of voltage noise measurement in type II superconductors.

using samples having the form of thin films or foils. The magnetic field is perpendicular to the plane of the film. The voltage is measured between contacts soldered to the film: the line connecting these contacts is parallel to the current I, as shown in Fig. 9.6.

In the case of flux-creep, the voltage noise can be viewed as a result of random magnetic flux hops similar to random pulses (Sec. 1.5). This voltage noise across a type II superconductor placed in a normal magnetic field $H_{c1} < H < H_{c2}$ is, in principle, nonzero even at zero current, $\bar{I} = 0$, and at currents smaller than the critical one, I_c. It is then given by the Nyquist theorem: $S_U(f) = 4k_B T \operatorname{Re} Z(f)$. Here $\operatorname{Re} Z(f) = R$ is the resistance of the conductor in the limit of low currents and voltages ($I \to 0$, $U \to 0$). However, the spectral density of this noise is expected to be extremely small.

First of all, the resistance R at currents lower than I_c is immeasurably small, due to pinning of the vortices. Both in low- and high-T_c superconductors $S_U(f)$ becomes immeasurably small as the current is decreased. For example, in the experiments by van Gurp (1968) the resistance $R \approx 1$ mOhm. At $T = 4.2$ K the Nyquist noise $S_U(0) \approx 10^{-25}$ V^2/Hz. At the same time the minimal measurable spectral density was $\approx 10^{-18}$ V^2/Hz. Therefore the noise at $\bar{I} = 0$ is measured usually as magnetic noise (Sec. 9.2.3).

The simplest model of flux-flow noise assumes that the source of randomness of the vortex motion is their creation at one edge of the film. After entering the film the vortices move in bundles, m_ϕ vortices each, with a drift velocity **v** and disappear at the opposite edge of the film (van Ooijen & van Gurp, 1965; van Gurp, 1968). The noise is like a shot noise (Sec. 1.5) produced by identical pulses of equal duration $\tau_d = w/v$, where w is the width of the film. According to Eq. (1.5.10), the spectral density of the flux of bundles across the film equals

$$S_J = 2\bar{J}\left[\frac{\sin(\pi f\tau)}{\pi f\tau}\right]^2, \tag{9.2.2.3}$$

where $\bar{J} = LBv/m_\phi\Phi_0$ is the mean flux of bundles, L is the distance between contacts. The equation for voltage fluctuation follows from Eq. (9.2.1.3): $\delta U = (m_\phi\Phi_0/c)\delta J$. The spectral density of voltage noise in this model of flux-flow equals:

$$S_U = 2\frac{m_\phi\Phi_0 LBv}{c^2}\left[\frac{\sin(\pi f\tau)}{\pi f\tau}\right]^2, \tag{9.2.2.4}$$

The stationary voltage across the same contacts is $\bar{U} = BvL/c$. Denoting by $\Phi = m_\phi\Phi_0$ the magnetic flux through one bundle, one obtains:

$$S_U(f) = \frac{2\Phi\bar{U}}{c}\left[\frac{\sin(\pi f\tau)}{\pi f\tau}\right]^2. \tag{9.2.2.5}$$

This simple model completely neglects the pinning of vortices. Of course, by assuming a large enough number m_ϕ of vortices in a bundle, the motion of which is fully correlated, one is able to explain the rather high experimental values of $S_U(0)$. However, the frequency dependence of $S_U(f)$ does not agree with the measured one. The experimental noise $S_U(f)$ falls off monotonically without any oscillations predicted by Eq. (9.2.2.5). Secondly, the characteristic frequency f_c at which the spectral density is, say, twice as small as $S_U(0)$ is much greater than τ^{-1}. This means that the model of uniform motion of vortices from one edge of the sample to the opposite edge is inadequate. Therefore, Thompson & Joiner (1975, 1979) and Habbal & Joiner (1977) assumed that the vortices are pinned and depinned many times on their way, i.e., their motion is randomly interrupted. The free drift time τ_f is smaller than $\tau = w/v$ if the velocity v is calculated from the measured mean voltage $\bar{U}$. Since the time τ_f is a random quantity, the oscillations of $S_U(f)$ are smeared out and it becomes a monotonic function of f. Moreover, since $\tau_f^{-1} > \tau^{-1}$, the characteristic frequency of the spectrum f_c is higher than τ^{-1}. This simple model explains also that the measured f_c is proportional to the ratio U/B. The velocity of unpinned vortices increases with this ratio and the free drift time becomes smaller.

There are even more direct observations of the influence of defects on the vortex motion noise. Heiden & Kohake (1974), Heiden, Kohake *et al.* (1977) shifted the potential leads to a Nb foil along the line perpendicular to the current. The noise strongly increased when the leads were close to grain boundaries (see also Kohake & Heiden, 1980). Habbal & Joiner (1977a, 1977b) found that the value of $S_U(0)$ drops as the time of annealing is increased, that is, as the number of the pinning centers is reduced. Simultaneously, the shape of the spectrum changes: $S_U(f)$ falls off with f more steeply, i.e., the spectrum is confined to lower frequencies. It is interesting that in this experiment $S_U(0)$ dropped upon annealing by two orders, while the critical current decreased by $\approx 16\%$ only, indicating much greater sensitivity of noise to the defect structure of the superconductor.

According to Eq. (9.2.2.5), $S_U(0) = 2\Phi\bar{U}/c$. By measuring $\bar{U}$ and $S_U(0)$ one is able to find the typical number of vortices in a bundle, $m_\phi = \Phi/\Phi_0$. Van Gurp (1968) has found that m_ϕ decreases with the current density. At small j the number $m_\phi \simeq 10^5$, at large currents $m_\phi \simeq 10^3$. The decrease of m_ϕ with j was observed also by Jarvis & Park (1975). However, the measured values of Φ at large currents turned to be less than Φ_0, in contradiction with the initial model.

The noise can be measured not only in rectangular films but also in samples having the form of a Corbino disk (Jarvis & Park, 1975): one of the contacts is in the center of the disk, another is on its circumference (Fig. 9.7). Both the current and the induced electric field are directed along radii, perpendicular to the Lorentz force which, in a normal magnetic field, is at each point directed along the tangent, setting the vortices to circular rotation. In type II superconductors the noise in Corbino disks is only $\sim 1\%$ of the noise measured in rectangular samples (in the intermediate state of type I superconductors the noise is of the same order of magnitude in both samples). The Corbino disks differ from the rectangular films by the absence of edges: one may conclude that the disorder of the vortex structure arises just at the sample's edges. In the case of Corbino disks, the time of the drift between the opposite edges becomes meaningless.

Apart from Lorentz type noise spectra, $1/f$-type spectra have been also recorded in low-T_c superconductors. In V foil $1/f$ noise has been recorded at large j and/or high magnetic fields H (van Gurp, 1968). Such noise spectra are especially typical for high-T_c superconductors (Sec. 9.2.4).

9.2.3 Magnetic noise

Besides electrical noise magnetic flux noise is also observed. Yeh & Kao (1984; see also Yeh & Kao, 1991) were the first to measure the fluctuations

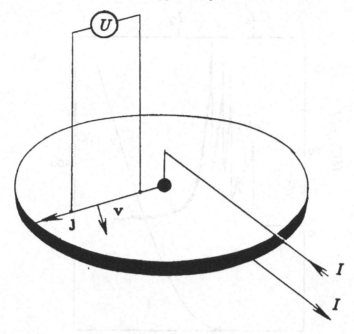

Fig. 9.7 Schematic diagram of voltage noise measurement in superconductors using specimens having the form of Corbino disks. The current I flows between the central electrode and the peripheral electrode. **j** is the current density, **v** is the velocity of vortices moving along circles.

of magnetic flux $\delta\Phi(t)$ through the loop area of a sensitive magnetometer (Superconductive QUantum Interference Device – SQUID) placed above a film of type II superconductor (low-T_c alloy $In_{0.9}Pb_{0.1}$). Since the ac voltage output $\delta U(t)$ of the SQUID is proportional to $\delta\Phi(t)$, the spectral density $S_{\Phi}(f)$ of magnetic flux fluctuations can be measured. Magnetic noise was measured as a function of the mean current $\bar{I}$ in the film at various magnetic fields. The random signal appeared almost abruptly at a critical current I_c at which the vortices started to move. This method yields more distinct values of I_c than the common method, in which a measurable dc voltage $\bar{U}$ has to be detected (Fig. 9.8). Several discontinuous augmentations of the noise signal are observed as $\bar{I}$ is increased. Possibly, they are caused by some outbursts of vortex depinning. The fluctuation signal falls off and ultimately becomes immeasurably small at large currents at which the mean voltage $\bar{U}$ is quasilinear in the mean current $\bar{I}$.

The magnetic fields in these experiments were rather high ($5-30$ mT as compared with $H_{c2} \simeq 120$ mT). Therefore the densities of vortices were not small, and they can not be considered as noninteracting. The magnetic flux fluctuations generated by the drift of the vortex structure (lattice or liquid)

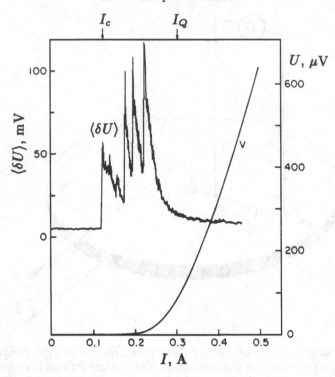

Fig. 9.8 Magnetic noise in $In_{0.9}Pb_{0.1}$ foils as a function of dc current I. $\langle \delta U \rangle$ is the broadband average (root-mean-square) ac voltage output from a SQUID magnetometer. I_c is the critical current, I_Q separates the near-onset region from the region at which the dc voltage grows quasi-linearly with dc current. Shown is the dependence of dc voltage U on current. Magnetic field $H = 227.5$ Oe. $T = 3.672$ K. From Yeh & Kao (1984).

are produced partly by the transfer of the static disorder of this structure, and partly by the random dynamic changes of the structure itself.

Ferrari, Johnson, Wellstood *et al.* (1988, 1989, 1990) and Johnson, Ferrari, Wellstood *et al.* (1990) studied magnetic flux fluctuations in high-T_c superconductors under different conditions: (1) no mean current passed through the film, $\bar{I} = 0$, and (2) the magnetic field was less than 1 μT (see the review by Ferrari, Johnson, Wellstood *et al.*, 1994). The mean distance between vortices is much greater than the magnetic field penetration depth, hence the interaction between vortices is weak and their motion is uncorrelated. The random fluctuations of the magnetic flux reveal the dynamics, e.g., *spontaneous* hops, of individual vortices, and provide information on their interaction with pinning centers. This noise technique differs from such experiments in which the system of vortices is set in motion by a strong enough driving force (flux-creep, flux-flow, thermally activated resistance).

The measurements have been performed in the temperature range from 4.2 K up to the transition region around T_c. The observed magnetic flux noise can be roughly divided in two types: random telegraph noise (RTN, Sec. 8.8) and $1/f$ noise (Ch. 8). In general, the improvement of crystalline quality of the films strongly reduced the noise.

Typically, a RTN is observed as random switching of the magnetic flux, measured by the SQUID, between two discrete values with an amplitude $\Delta\Phi \sim 10^{-3}\,\Phi_0$ which remains constant in time and is temperature-independent. However, the rate of the random switchings increases with temperature. An obvious explanation of this noise is that a vortex, or possibly a bundle of a few vortices, hops between two pinning sites overcoming an activation barrier.

As usual, the times τ_1 and τ_2 of staying in the two states depend exponentially on $1/k_B T$ (Eq. (8.8.1.12)). Very often the ratio of the two times is ≈ 1 and almost constant in the temperature range of measurements. It means that the difference of activation energies is small and the energies of the two vortex's states are almost equal (Eq. (8.8.1.13)). The spectral density of RTN is close to Lorentzian (Eqs. (1.7.10) and (8.8.1.9)). Its characteristic frequency τ^{-1}, being the sum of probabilities of transitions per unit time back and forth (Eq. (8.8.1.10)), steeply grows with temperature (for instance, from 20 Hz to 137 Hz when T is increased from 88.5 K to 88.6 K).

The length l of a vortex hop can be estimated from the measured amplitude of magnetic flux switchings. If w is the dimension of the SQUID's loop and the hop takes place at an angle θ from the radial direction, the hop results in the following change of the magnetic flux (Ferrari *et al.*, 1994):

$$\Delta\Phi/\Phi_0 = \alpha_s(l/w)\cos\theta, \qquad (9.2.3.1)$$

where $\alpha_s \approx 0.7$ is a geometrical factor. Since the maximum value of $\cos\theta$ is unity, the length of the hop $l \geq \alpha_s^{-1}w\Delta\Phi/\Phi_0$. This estimate yields, for hops contributing to RTN, hopping lengths from $\simeq 0.1\,\mu$m up to more than 30 μm.

In some samples the RTN was noisy, i.e., the value of the flux in one of the two states or in both states was not a constant. Such phenomena are commonly interpreted as the effect of interaction with another fluctuator (see Sec. 8.8). However, in the case of vortices which do not interact with one another, it is more plausible to assume that the vortex in one of the two states (or in both) moves freely in a random potential, and this motion produces higher frequency noise.

Usually the measured magnetic flux noise is not RTN but $1/f$ noise. Its spectral density increases with temperature, first slowly, then it rises steeply just below T_c. Above T_c the noise drops, hence it peaks sharply at T_c.

According to the common model (Sec. 8.3), $1/f$ noise is produced by a great number of individual vortices with various activation energies and relaxation times. When applied to pinned vortices in high-T_c superconductors, this model must be generalized to include the temperature dependence of the activation energies $E(T)$. The relaxation time of an elementary process with a temperature-dependent activation energy $E(T)$ is taken to be (Ferrari, Johnson, Wellstood *et al.*, 1990):

$$\tau(T) = \tau_0 \exp[E(T)/k_B T] = \tau_0 \exp[E_0 \beta(T)]. \tag{9.2.3.2}$$

Here τ_0^{-1} is the attempt frequency of the vortex to surmount the barrier, E_0 is the activation energy at $T = 0$, $\beta(T) = E(T)/E_0 k_B T$. If $F(E_0)$ is the density distribution function of the activation energies E_0, the noise spectral density equals:

$$S_\Phi(f, T) \propto \int_{E_{0min}}^{E_{0max}} dE_0 F(E_0) \frac{\tau_0 e^{\beta E_0}}{1 + \omega^2 \tau_0^2 e^{2\beta E_0}}. \tag{9.2.3.3}$$

Measuring the spectral density S_Φ vs frequency f and temperature T, the authors were able to find the distributions of activation energies $F(E_0)$ in the samples.

Nonzero mean supercurrent suppresses the noise (Ferrari, Wellstood, Kingston & Clarke, 1991). This effect is explained roughly as follows. The driving force makes the energies of the two states and, consequently, the two activation energies of each vortex unequal (Eq. (9.2.2.2)). As the difference of energies is increased, the noise spectral density decreases (Eq. (8.8.1.14)). It implies that in the absence of current most double wells between which the vortices are hopping are symmetric in energy (this symmetry was found by direct measurements for those hops that manifest themselves in RTN, see above).

9.2.4 Resistance noise in the transition region near T_c and in the normal state of high-T_c superconductors

Usually the transition into the superconductive zero-resistance state is not abrupt, especially in high-T_c superconductors. The resistance falls off to zero gradually in a rather broad temperature transition region, where superconductivity and normal conduction coexist. It is reasonable to consider all kinds of noise in high-T_c superconductors, in the transition region around

T_c and even in the normal state above T_c, as one problem. Measurements of resistance noise in the transition region and in the normal state have revealed some interesting phenomena.

In high-T_c materials the resistance noise of $1/f$ type in normal state and in the transition region is very intensive. The dimensionless parameter $\alpha = nVfS_R(f)/R^2$ (Sec. 8.2.6), which is used to compare $1/f$ noise in different sources, is several $(5 - 8)$ orders of magnitude higher than in conventional metals (Testa, Song, Chen *et al.*, 1988; Lee, Lee & Khim, 1989; Maeda, Nakayama, Takebayashi & Uchinokura, 1989; Maeda, Watanabe, Tsukada & Uchinokura, 1991; Song, Misra, Crooker & Gaines, 1991). However, this extremely high noise level seems not to be an inherent property of high-T_c materials. The noise is generally lower in samples with better crystalline quality (with less disorder). In some thin films of $Tl_2Ba_2CaCu_2O_8$ the resistance $1/f$ noise was found to be only by 1–2 orders higher than in conventional metals: in three films the parameter α was found to be 0.56, 2.1, and 1.2, respectively (Misra, Song, Crooker *et al.*, 1991). This reduction of noise depends on the degree of film alignment along the c-axis.

One more interesting feature of resistance noise in high-T_c superconductors is its temperature dependence in the transition region, specifically, an extremely high peak of resistance $1/f$ noise at temperatures in the transition resistive region near T_c (Testa, Song, Chen *et al.*, 1988 on $R_1Ba_2Cu_3O_{7-\delta}$, (R=Y, Eu); Maeda, Nakayama *et al.*, 1989, and many others). The noise as a function of temperature steeply increases in a narrow (several K) temperature interval immediately above the onset of full superconductivity, and drops at the high-temperature side of the transition region (Fig. 9.9).

In bulk $Tl_2Ba_2Ca_{n-1}Cu_nO_{4+n}$ ($n = 2$ and 3) the noise in the normal state decreases with decreasing temperature. Close to the transition to superconductivity, the noise is steeply growing when the temperature approaches the temperature of full superconductivity. However, this enhancement of $1/f$ noise is reduced or even disappears in some high quality samples (Song, Misra, Cao *et al.*, 1990).

The divergence of the noise spectral density S_U/U^2 as the state with $R = 0$ is approached was found in granular superconducting films of Y-Ba-Cu-O, Bi-Sr-Ca-Cu-O, and NbN, irrespective of the way the transition to $R = 0$ state was obtained: by lowering the temperature at fixed current or by lowering the current at fixed temperature (Aponte, Bellorin, Oentrich *et al.*, 1993).

Current $1/f$ noise in a high-T_c superconductor $YBa_2Cu_3O_{7-\delta}$, brought to a resistive state by application of a normal magnetic field, shows a hysteretical behavior (Celasco, Masoero, Mazetti & Stepanescu, 1991). The

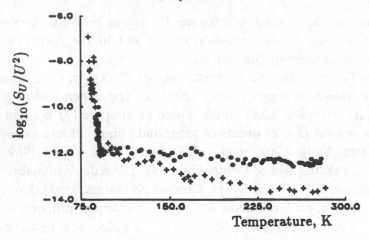

Fig. 9.9 Temperature dependence of the fractional spectral density S_U/U^2 of voltage noise in $YBa_2Cu_3O_{7-\delta}$. Frequency $f = 6$ Hz. From Testa *et al.* (1988).

noise is lower and the resistance is higher when the magnetic field is increased than when it is decreased. The current noise as a function of resistance is double-valued: it is different, up to an order of magnitude, when the same resistance is obtained by either increasing or decreasing the magnetic field (at increasing field it is lower).

The abnormally high resistance $1/f$ noise level and its dependence on the degree of disorder in the samples indicate that strong inhomogeneity of the material and, consequently, strongly nonuniform conductivity and current density are responsible for both the high noise level and its temperature dependence. Therefore, one has to apply the ideas of percolation theory (Ch. 7).

In the transition region the superconductor is, actually, a mixture of normal metallic and superconducting phases. The portion of the superconductive phase increases as the temperature is lowered. The simplest model of such a mixture is a lattice of normal resistors, a definite (non-fluctuating) but variable portion p of which are short-circuited (the latter simulate the superconductive links). Noise is generated by fluctuations $\delta r(t)$ of normal resistors. The relative number of short-circuited links p is assumed to increase as the temperature is decreased. In this percolation model, the spectral density of relative resistance fluctuations grows according to a power law as the portion of superconductor increases and approaches the critical value at which an infinite superconductive cluster appears, i.e., as the total resistance R decreases:

$$S_R(f)/R^2 \propto R^{-l_{rs}}. \tag{9.2.4.1}$$

According to Kiss, Larsson, Svedlindh *et al.* (1993); Kiss & Svedlindh (1993), who used data from the literature on percolation theory, in a two-dimensional system (film thickness less than the percolation length) $l_{rs} = 0.86 \pm 0.02$, in three dimensions $l_{rs} = 0.9 \pm 0.32$. Since this critical index is positive, the $1/f$ noise parameter $\alpha \propto S_R/R^2$ grows as the fully superconductive state is approached from $T > T_c$. It drops when full superconductivity is reached. Qualitatively, this prediction of the simplest percolation model agrees with experiments. However, the power l_{rs} measured in experiments differs from the values presented above.

According to one of the models of high-T_c superconductors, they consist of grains separated by thin high-resistance layers (see the review by Tinkham & Lobb, 1989). At the first, high-temperature, stage of the transition, the grains become superconductive at some temperature T_{c0}. However, the total resistance remains high due to the intergrain resistances which, at the same high temperatures, are Josephson-like junctions in the normal state (Sec. 9.1). At the second, low-temperature, stage the Josephson junctions become superconductive, one after another. Due to disorder, each Josephson junction (each pair of grains) has its own critical temperature of transition into superconductive state. The material is effectively a random resistor network (Ch. 7). Perturbations, which are qualitatively similar to those that modulate the normal resistances and generate resistance noise, may modulate the magnitudes of the critical currents of the Josephson junctions. At a given current the number of Josephson junctions in the superconductive state fluctuates. In other words, the short-circuits randomly switch 'on' and 'off'. In the percolation model of a normal-metal–superconductor mixture it is equivalent to fluctuations $\delta p(t)$ of the portion p of short-circuited resistors (Kiss, Larsson, Svedlindh *et al.*, 1993, and Kiss & Svedlindh, 1993; see also the review by Kiss & Svedlindh, 1994). This percolation model, in which p is also fluctuating, differs from the 'classical' one in which the noise is generated only by fluctuations of the non-short-circuited normal resistors. Its properties depend on the statistical properties of the switchings. The authors estimated the critical index l_{rs} of this percolation directly from experiments on noise in high-T_c superconductors. They found for two and three dimensions $l_{rs} \approx 1.54 \pm 0.09$ and $\approx 2.74 \pm 0.04$, respectively. These values are essentially higher than in the 'classical' percolation model.

References

General

Recommended books and reviews

A. Ambrozy. *Electronic Noise*. Akademiai Kiado, Budapest, 1982

D.A. Bell. *Noise and the Solid State*. Pentech Press, London, 1985

M.J. Buckingham. *Noise in Electronic Devices and Systems*. John Wiley & Sons, New York, 1983

R.E. Burgess, ed. *Fluctuation Phenomena in Solids*. Academic Press, New York, London, 1965

T. Chandrasekhar. Stochastic problems in physics and astronomy. *Rev. Mod. Phys.*, 1943, **15**, no. 1, 1

P. Gaspard, X.-J. Wang. Noise, chaos, and (ϵ, τ)-entropy per unit time. *Phys. Rep.*, 1993, **235**, no. 6, 291

B.K. Jones. Electrical noise as a measure of quality and reliability in electronic devices. *Adv. Electron. Electron Phys.*, 1994, **87**, 201

N.G. van Kampen. *Stochastic Processes in Physics and Chemistry*. North Holland, 1992

M. Lax. Quantum Noise. I. *Phys. Rev.*, 1958, **109**, 1921; II. Formal theory of quantum fluctuations from a driven state. *ibid*, 1963, **129**, no. 5, 2342; IV. Quantum theory of noise sources. *ibid*, 1966, **145**, no. 1, 110

M. Lax. Fluctuations from the nonequilibrium steady state. *Rev. Mod. Phys.*, 1960, **32**, no. 1, 25

M. Lax. Classical noise. III. Nonlinear Markoff processes. *Rev. Mod. Phys.*, 1966, **38**, no. 2, 359; IV. Langevin methods. *ibid*, 1966, **38**, no. 3, 541

D.K.C. MacDonald. *Noise and Fluctuations*, John Wiley, New York, 1962

V. Palenskis, ed. Fluctuation Phenomena in Physical Systems. *Proc. of the 7-th Vilnius Conference, 1994*. Vilnius University Press, Vilnius, Lithuania, 1995

S.M. Rytov. *Introduction to Statistical Radiophysics*, vol. 1. Nauka, Moscow, 1976 (in Russian)

S.M. Rytov, Yu.A. Kravtsov, V.I. Tatarskii. *Principles of Statistical Radiophysics.* **1**: *Elements of Random Process Theory*. Springer-Verlag, Berlin, 1987. **2**: *Correlation Theory of Random Processes. ibid*, 1988

A. van der Ziel. *Fluctuation Phenomena in Semiconductors*. Butterworths, London, 1959

A. van der Ziel. *Noise; Sources, Characterization, Measurement.* Prentice-Hall, Englewood Cliffs, N.J., 1970

A. van der Ziel. *Noise in Measurements.* Wiley Interscience, New York, 1976

A. van der Ziel. History of noise research. *Adv. Electron. Electron Phys.*, 1980, **50**, 351

Noise in Physical Systems. *Proc. 5-th Int. Conf. on Noise in Physical Systems, Bad-Nauheim.* Ed. E.D. Wolf, Springer-Verlag, Berlin, Heidelberg, New York, 1978.

Noise in Physical Systems. *Proc. 6-th Int. Conf. on Noise in Physical Systems, National Bureau of Standards, Gaithesburg, MD, 1981.* Ed. P.H.E. Meijer, R.D. Mountain, R.J. Soulen. National Bureau of Standards, no. 614. Washington, DC, 1981

Noise in Physical Systems and $1/f$ noise. *Proc. 7-th Int. Conf. on Noise in Physical Systems and 3-rd Int. Conf. on 1/f Noise, Montpellier, May 1983.* Ed. M. Savelli, G. Lecoy, J.P. Nougier. North Holland, Amsterdam, 1983

Noise in Physical Systems and $1/f$ noise. *Proc. 8-th Int. Conf. on Noise in Physical Systems. Rome, 1985.* Ed. A. D'Amico, P. Mazetti. North Holland, Amsterdam, 1986

Noise in Physical Systems. *Proc. 9-th Int. Conf., Montreal, 1987.* Ed. C.M. Van Vliet. World Scientific, Singapore, 1987

Noise in Physical Systems and $1/f$ fluctuations. *Proc. 10-th Int. Conf., Budapest, 1989.* Ed. A. Ambrozy, Academiai Kiado, Budapest, 1990

Noise in Physical Systems and $1/f$ fluctuations. *Proc. 11-th Int. Conf., Kyoto, 1991.* Ed. T. Musha, S. Sato, M. Yamamoto, Ohmsha Ltd., Japan, 1991

Noise in Physical Systems and $1/f$ Fluctuations. *Proc. 12-th Int. Conf., St. Louis, Missouri, USA, 1993.* Ed. P. Handel, A. Chang. *AIP Conference Proceedings #* 285. New York, 1993

Noise in Physical Systems and $1/f$ Fluctuations. *Proc. 13-th Int. Conf., Palanga, Lithuania, 1995.* Ed. V. Bareikis, R. Katilius. World Scientific, Singapore, 1995

Chapter 1

Recommended books and reviews

M.S. Bartlett. *An Introduction to Stochastic Processes, with special Reference to Methods and Applications.* Cambridge University Press, 1978

A.T. Bharucha-Reid. *Probabilistic Methods in Applied Mathematics.* Academic Press, New York, 1968

S. Chandrasekhar. Stochastic problems in physics and astronomy. *Rev. Mod. Phys.*, 1943, **15**, no. 1, 1

R.F. Fox. Gaussian stochastic processes in physics. *Phys. Rep.*, 1978, **48**, no. 3, 179

N.G. van Kampen. *Stochastic Processes in Physics and Chemistry.* North Holland, 1992

S. Karlin. *A First Course in Stochastic Processes.* Academic Press., New York, 1975

M. Lax. Quantum noise. I. Generalized mobility theory. *Phys. Rev.*, 1958, **109**, no. 6, 1921; II. Formal theory of quantum fluctuations from a driven state. *ibid*, 1963, **129**, no. 5, 2342; IV. Quantum theory of noise sources. *Ibid*, 1966a, **145**, no. 1, 110

M. Lax. Fluctuations from the nonequilibrium steady state. *Rev. Mod. Phys.*, 1960, **32**, no. 1, 25

M. Lax. Classical noise. III. Nonlinear Markoff processes. *Rev. Mod. Phys.*, 1966b, **38**, no. 2, 359. IV. Langevin methods. *ibid*, 1966c, **38**, no. 3, 541

A. Papoulis. *Probability, Random Variables, and Stochastic Processes*. McGraw Hill, New York, 1984

S.M. Rytov. *Introduction to Statistical Radiophysics*, **1**. Nauka, Moscow, 1976 (in Russian)

S.M. Rytov, Yu.A. Kravtsov, V.I. Tatarskii, *Principles of Statistical Radiophysics*. **1**: *Elements of Random Process Theory*. Springer-Verlag, Berlin 1987. **2**: *Correlation Theory of Random Processes. ibid*, 1988

K.M. van Vliet, J.R. Fassett. Fluctuations due to electronic transitions and transport in solids. In: *Fluctuation phenomena in solids*. Ed. R.E. Burgess. Academic Press, New York, 1965. p. 267

M.C. Wang, G.E. Uhlenbeck. On the theory of the Brownian motion. II. *Rev. Mod. Phys.*, 1945, **17**, 323

M.B. Weissman. $1/f$ noise and other slow, nonexponential kinetics in condensed matter. *Rev. Mod. Phys.*, 1988, **60**, no. 2, 537

M.B. Weissman. What is spin glass? A glimpse via mesoscopic noise. *Rev. Mod. Phys.*, 1993, **65**, no. 3, 829

A.M. Yaglom. *Correlation Theory of Stationary and Related Random Functions. I. Basic Results. II. Supplementary Notes and References*. Springer Series in Statistics. Springer-Verlag. New York, 1987

References

R. Brown. A brief account of microscopical observations made in the months of June, July, and August, 1827 on the particles contained in the pollen of plants, and on the general existence of active molecules in organic and inorganic bodies. *Edinburgh New Phil. J.*, 1828, **5**, 358 (see also *Phil. Mag.*, 1828, **4**, 161)

S.G. Brush. *The Kind of Motion we call Heat*. Book 2, ch. 15. North Holland, 1976

N. Campbell. The study of discontinuous phenomena, *Proc. Cambridge Philos. Soc.*, 1909, **15**, 117; *ibid*, **15**, 310

J.R. Carson. The statistical energy-frequency spectrum of random disturbances. *Bell System Tech. J.*, 1931, **10**, July, 374

A. Einstein. Über die von der molekularkinetischen Theorie der Warme gefordete Bewegung von in ruhenden Flussigkeiten suspendierten Teilchen. *Ann. Phys.*, 1905, **17**, 549

A. Einstein. Zur Theorie der Brownschen Bewegung. *Ann. Phys.*, 1906, **19**, 371

A. Einstein. Theoretische Bemerkungen uber die Brownsche Bewegung. *Z. Elektrochem.*, 1907, **132**, 41

H. Ekstein, N. Rostoker. Quantum theory of fluctuations. *Phys. Rev.*, 1955, **100**, no. 4, 1023

A.D. Fokker. Die mittlere Energie rotierender electrischer Dipole im Strahlungsfeld. *Ann. Phys.*, 1914, **43**, no. 5, 810

F.R. Gantmacher. *The Theory of Matrices*, **1**. Chelsea Publ. Co., New York, 1977

A. Khintchine. Korrelationstheorie der stationaren stochastischen Prozessen. *Math. Ann.*, 1934, **109**, no. 4, 604

P. Langevin. Sur la theorie du mouvement brownien. *Comptes Rendus (Paris)*, 1908, **146**, no. 10, 530

S. Machlup. Noise in semiconductors: spectrum of a two-parameter random signal. *J. Appl. Phys.*, 1954, **25**, no. 3, 341

L. Onsager. Reciprocal relations in irreversible processes. I. *Phys. Rev.*, 1931, **37**, no 4, 405; II. *ibid*, 1931, **38**, no. 12, 2265

R.G. Palmer. Broken Ergodicity, *Adv. Phys.*, 1982, **31**, no. 6, 669

C.M. Pease. *Methods of Matrix Algebra.* Academic Press, New York, 1965

M. Planck. Über einen Satz der statistischen Dynamik und seine Erweiterung in der Quantentheorie. *Sitzungsber. Preuss. Akad. Wiss.*, 1917, 324

W. Schottky. Über spontane Stromschwankungen in verschiedenen Elektrizitatleitern. *Ann. Phys.*, 1918, **57**, 541

M. von Smoluchowski. Zur kinetischen Theorie der Brownschen Molekularbewegung und der Suspensionen. *Ann. Phys.*, 1906a, **21**, no. 14, 756

M. von Smoluchowski. The mean free path of molecules in gases and its connection with the theory of diffusion. *Bull. Inter. de l'Acad. de Cracovie*, 1906b, 202

M. von Smoluchowski. Brownian molecular motion under the action of external forces and its connection with the generalized diffusion equation. *Ann. Phys.*, 1915, **48**, 1103

N. Wiener. Generalized harmonic analysis, *Acta Mathematica*, 1930, **55**, 117

Chapter 2

Recommended books and reviews

L.D. Landau, E.M. Lifshitz. *Statistical Physics, Part 1*, 3rd edition, Pergamon Press, 1985

M.L. Levin, S.M. Rytov. *Theory of Equilibrium Thermal Fluctuations in Electrodynamics.* Nauka, Moscow, 1967 (in Russian)

E.M. Lifshitz, L.P. Pitaevskii. *Statistical Physics, part 2*, Pergamon Press, 1980

S.M. Rytov. *Theory of Electrical Fluctuations and Thermal Radiation.* Publishing House USSR Acad. Sci., Moscow, 1953 (in Russian)

References

H.B. Callen, R.F. Green. On a theorem of irreversible thermodynamics. I. *Phys. Rev*, 1952, **86**, no. 5, 702

H.B. Callen, T.A. Welton. Irreversibility and generalized noise. *Phys. Rev.*, 1951, **83**, no. 1, 34

H.B. Callen, M.L. Barasch, J.L. Jackson. Statistical mechanics of irreversibility. *Phys. Rev.*, 1952, **88**, no. 6, 1382

R.F. Green, H.B. Callen. On a theorem of irreversible thermodynamics. II. *Phys. Rev.*, 1952, **88**, no. 6, 1387

G.L. de Haas-Lorentz. Die Brownsche Bewegung und einige verwandte Erscheinungen. *Die Wissenschaft (Braunschweig)*, 1913, **52**, 86

J.B. Johnson. Thermal agitation of electricity in conductors. *Phys. Rev.*, 1928, **32**, no. 1, 97

R.H. Koch, D.J. Van Harlingen, J. Clarke. Measurements of quantum noise in resistively shunted Josephson junctions, *Phys. Rev. B*, 1982, **26**, no. 1, 74

L.D. Landau, E.M. Lifshitz. *Electrodynamics of Continuous Media*, State Publ. House Tech.-Theor. Lit., Moscow, 1957 (in Russian)

H. Nyquist. Thermal agitation of electric charge in conductors. *Phys. Rev.*, 1928, **32**, no. 1, 110

Chapter 3

Recommended books and reviews

V. Bareikis, R. Katilius, R. Miliušytė. *Fluctuation Phenomena in Semiconductors in Nonequilibrium State*. Vilnius, Mokslas, 1989

V. Bareikis, R. Katilius, J. Pozhela, S.V. Gantsevich, V.L. Gurevich. Fluctuation spectroscopy of hot electrons in semiconductors. In: *Spectroscopy of non-equilibrium electrons and phonons*. Ch. 8, p. 327. Ed. C.V. Shank, B.P. Zakharchenya. Elsevier Science Publishers B.V., 1992

V. Bareikis, J. Liberis, I. Matulionene, A. Matulionis, P. Sakalas. Experiments on hot electron noise in semiconductor materials for high-speed devices. *IEEE Trans. Electron. Devices*, 1994, **41**, no. 11, 2050

S. Chapman, T.G. Cowling. *The Mathematical Theory of Non-uniform Gases, An account of the Kinetic Theory of Viscosity, Thermal Conduction and Diffusion in Gases*, 3-rd edn., Cambridge University Press, Cambridge, 1970

M.H. Ernst, E.G.D. Cohen. Nonequilibrium fluctuations in μ-space. *J. Stat. Phys.*, 1981, **25**, no. 1, 153

V.F. Gantmakher, I.B. Levinson. *Carrier scattering in metals and semiconductors*. North Holland, Amsterdam, 1987

S.V. Gantsevich, V.L. Gurevich, R. Katilius. Theory of fluctuations in nonequilibrium electron gas. *Rivista del Nuovo Cimento*, 1979, **2**, no. 5, 1

C. Jacoboni, L. Reggiani. The Monte Carlo method for the solution of charge transport in semiconductors with applications to covalent materials. *Rev. Mod. Phys.*, 1983, **55**, no. 3, 645

N.G. van Kampen. *Stochastic Processes in Physics and Chemistry*. North Holland, 1992

T. Komiyama, T. Kurosawa, T. Masumi. Streaming motion of carriers in crossed electric and magnetic field. In: *Hot-electron transport in semiconductors*. Ed. L. Reggiani. *Topics in Applied Physics*, **58**, Springer-Verlag, Berlin, 1985. p. 177

M. Lax. Quantum noise. I. *Phys.Rev.*, 1958, **109**, 1921; II. Formal theory of quantum fluctuations from a driven state. *ibid*, 1963, **129**, no. 5, 2342; IV. Quantum theory of noise sources. *ibid*, 1966c, **145**, no. 1, 110

M. Lax. Fluctuations from the nonequilibrium steady state. *Rev. Mod. Phys.*, 1960, **32**, no. 1, 25

M. Lax. Classical noise. III. Nonlinear Markoff processes. *Rev. Mod. Phys.*, 1966a, **38**, no. 2, 359

M. Lax. Classical noise. IV. Langevin methods. *Rev. Mod. Phys.*, 1966b, **38**, no. 3, 541

Noise in Physical Systems. *Proc. 5-th Int. Conf. on Noise in Physical Systems, Bad-Nauheim*. Ed. E.D. Wolf, Springer-Verlag, Berlin, 1978

Noise in Physical Systems. *Proc. 6-th Int. Conf. on Noise in Physical Systems, National Bureau of Standards, Gaithesburg, MD, 1981*. Ed. P.H.E. Meijer, R.D. Mountain, R.J. Soulen. National Bureau of Standards, no. 614. Washington, DC, 1981

Noise in Physical Systems and $1/f$ noise. *Proc. 7-th Int. Conf. on Noise in Physical Systems and 3-rd Int. Conf. on 1/f noise, Montpellier, May 1983*. Ed. M. Savelli, G. Lecoy, J.P. Nougier. North Holland, Amsterdam, 1983

A.-M.S. Tremblay. Theories of fluctuations in nonequilibrium systems. In : *Recent developments in Nonequilibrium Thermodynamics*. Barcelona, 1983. *Lecture notes in Physics*, no. 199. Ed.: H. Araki, J. Ehlers, *et al.*, p. 267. Springer-Verlag, 1984

A.F. Volkov, Sh.M. Kogan. Physical effects in semiconductors with negative differential conductivity. *Usp. Fiz. Nauk*, 1968, **96**, no. 4, 633 [*Sov. Phys. – Usp.*, 1969, **11**, no 6, 881]

References

V. Bareikis, A. Galdikas, R. Miliušytė, V. Viktoravičius. Noise of hot holes in Ge due to predominant inelastic scattering. In: *Proc. 6-th Int. Conf. on Noise in Physical Systems, National Bureau of Standarts, Gaithesburg, MD, 1981*. Ed. P.H.E. Meijer, R.D. Mountain, R.J. Soulen. National Bureau of Standards, no. 614. Washington, DC, 1981. p. 406

V. Bareikis, Yu. Liberis, Ch. Machulaitis, P. Mishkinis. Current fluctuations in an epitaxially-diffusion $p − n$ junction with a current filament. *Fiz. Tekh. Poluprovodn.*, 1977, **11**, no. 1, 110 [*Sov. Phys. – Semicond.*, 1977, **11**, no. 1]

V. Bareikis, J. Pozhela, I. Matulionene. Noise and diffusion of hot carriers in *p*-Ge. In: *Proc. 9-th Int. Conf. on the Physics of Semiconductors*, vol. 2. Ed. S.M. Ryvkin. Nauka, Leningrad, 1969. p. 760

M. Bixon, R. Zwanzig. Boltzmann–Langevin equation and hydrodynamic fluctuations. *Phys. Rev.*, 1969, **187**, no. 1, 267

B.J. Davydov. Über die Geschwindigkeitsverteilung der sich im elektrischen Felde bewegenden Electronen. *Phys. Z. Sowjetunion*. 1935, **8**, no. 1, 59; Teil II. *ibid*, 1936, **9**, no. 5, 433

B.J. Davydov. On the theory of electron motion in gases and semiconductors. *Zh. Eksp. Teor. Fiz.*, 1937, **7**, no. 9–10, 1069

E. Erlbach, J.B. Gunn. Noise temperature of hot electrons in germanium. *Phys. Rev. Lett.*, 1962, **8**, no. 7, 280

R.F. Fox, G.E. Uhlenbeck. Contribution to nonequilibrium thermodynamics. II. Fluctuation theory for the Boltzmann equation. *Phys. Fluids*, 1970, **13**, no. 12, 2881

S.V. Gantsevich, V.L. Gurevich, R. Katilius. Fluctuations in semiconductors in a strong electric field and the scattering of light by 'hot' electrons. *Zh. Eksp. Teor. Fiz.*, 1969, **57**, no. 2, 503 [*Sov. Phys. – JETP*, 1970, **30**, no. 2, 276]

V.L. Gurevich, R. Katilius. On the theory of hot electrons in an anisotropic semiconductor. *Zh. Eksp. Teor. Fiz.*, 1965, **49**, no. 4, 1145 [*Sov. Phys. – JETP*, 1966, **22**, no. 4, 796]

L.G. Hart. High-field current fluctuations in *n*-type germanium. *Can. J. Phys.*, 1970, **48**, no. 5, 531

N.A. Hashitsume. Statistical theory of linear dissipative systems. II. *Progress Theor. Phys.*, 1956, **15**, no. 4, 369

B.B. Kadomtsev. On fluctuations in gases. *Zh. Eksp. Teor. Fiz.*, 1957, **32**, no. 4, 943 [*Sov. Phys. – JETP*, 1957, **5**, no. 4, 771]

V.A. Kochelap, N.A. Zakhleniuk. Hydrodynamic fluctuations of a hot electron gas. *Phys. Rev. B*, 1994, **50**, no. 12, 8325

Sh.M. Kogan. Current filament in a semiconductor with an *S*-type current–voltage characteristic. *Zh. Eksp. Teor. Fiz.*, 1968, **54**, no. 4, 1228 [*Sov. Phys. – JETP*, 1968, **27**, no. 4, 656]

Sh.M. Kogan. On the theory of fluctuations in a gas with pairwise collisions between particles. *Teor. Mat. Fiz.*, 1972, **10**, no. 1, 143 [*Sov. Phys. - Theor. Math. Phys.*, 1972, **10**, no. 1, 94]

Sh.M. Kogan. Equations for the correlation functions using a generalized Keldysh technique. *Phys. Rev. A*, 1991, **44**, no. 12, 8072

Sh.M. Kogan, V.D. Shadrin. Scattering of infrared radiation in a semiconductor in high electric fields. *Fiz. Tekhn. Poluprovodn.*, 1971a, **5**, no. 11, 2224 [*Sov. Phys. – Semiconductors*, 1971, **5**, no. 11, 2113]

Sh.M. Kogan, V.D. Shadrin. The spectrum of fluctuations in a semiconductor with a current filament. *Fiz. Tverd. Tela*, 1971 b, **13**, no. 4, 1015 [*Sov. Phys. - Solid State*, 1971, **13**, no. 4, 845]

Sh.M. Kogan, A.Ya. Shulman. Electrical fluctuations in solid state plasma in high electric field. *Fiz. Tverd. Tela*, 1967, **9**, no. 8, 2259 [*Sov. Phys. – Solid State*, 1968, **9**, no. 8, 1771]

Sh.M. Kogan, A.Ya. Shulman. On the theory of fluctuations in a nonequilibrium electron gas. *Zh. Eksp. Teor. Fiz.*, 1969, **56**, no. 3, 862 [*Sov. Phys. – JETP*, 1969, **29**, no. 3, 467]

Sh.M. Kogan, A.Ya. Shulman. Extraneous random forces and equations for correlation functions in the theory of nonequilibrium fluctuations. *Fiz. Tverd. Tela*, 1970, **12**, no. 4, 1119 [*Sov. Phys. – Solid State*, 1970, **12**, no. 4, 874]

M.A. Leontovich. Fundamental equations of the kinetic theory of gases from the point of view of the theory of random processes. *Zh. Eksp. Teor. Fiz. (JETP)*, 1935, **5**, no. 3–4, 211 (in Russian)

I.B. Levinson, A.Yu. Matulis. Current fluctuations in a semiconductor in a high electric field. *Zh. Eksp. Teor. Fiz.*, 1968, **54**, no. 5, 1466 [*Sov. Phys. – JETP*, 1968, **27**, no. 5, 786]

J. Logan, M. Kac. Fluctuations and the Boltzmann equation. I. *Phys. Rev. A*, 1976, **13**, no. 1, 458

J.P. Nougier, M. Rolland. Mobility, noise temperature, and diffusivity of hot holes in germanium. *Phys. Rev. B*, 1973, **8**, no. 12, 5728

F.B. Pidduck. The kinetic theory of the motion of ions in gases. *Proc. London Math. Soc.*, 1916, Ser. 2, **15**, part 2, 89

W.E. Pinson, R. Bray. Experimental determination of the energy distribution functions and analysis of the energy-loss mechanisms of hot carriers in *p*-type germanium. *Phys. Rev.*, 1964, **136**, no. 5A, 1449

P.J. Price. Intervalley noise. *J. Appl. Phys.*, 1960, **31**, no. 6, 949

P.J. Price. Fluctuations of hot electrons. In: *Fluctuation Phenomena in Solids*, p. 355. Ed. R.E. Burgess, 1965, Academic Press, New York

W. Shockley. Hot electrons in germanium and Ohm's law. *Bell Syst. Tech. J.*, 1951, **30**, no. 4, part I, 990

A.Ya. Shulman. Correlation functions of nonequilibrium fluctuations in semiconductors in the electron-temperature approximation. *Fiz. Tverd. Tela*, 1970, **12**, no. 4, 1181 [*Sov. Phys. – Solid State*, 1970, **12**, no. 4, 922]

A.Ya. Shulman, Sh.M. Kogan. On the theory of fluctuations in a nonequilibrium gas. II. Spatially inhomogeneous fluctuations. *Zh. Eksp. Teor. Fiz.*, 1969, **57**, no. 6, 2112 [*Sov. Phys. – JETP*, 1970, **30**, no. 6, 1146]

I.I. Vosilyus, I.B. Levinson. Optical phonon generation and galvanomagnetic effects for a large-anisotropy electron distribution. *Zh. Eksp. Teor. Fiz.*, 1966, **60**, no. 6, 1660 [*Sov. Phys. – JETP*, 1966, **23**, no. 6, 1104]

Chapter 4

Recommended books and reviews

V.N. Abakumov, V.I. Perel', I.N. Yassievich. *Nonradiative Recombination in Semiconductors.* North Holland, 1991

K.M. van Vliet, J.R. Fassett. Fluctuations due to electronic transitions and transport in solids. In: *Fluctuation Phenomena in Solids.* Ed. R.E. Burgess. Academic Press, New York and London, 1965. p. 265

References

V.N. Abakumov, V.I. Perel', I.N. Yassievich. Theory of electron capture by attracting centers in photoexcited semiconductors. *Zh. Eksp. Teor. Fiz.*, 1977, **72**, no. 2, 674 [*Sov. Phys. – JETP*, 1977, **45**, no. 2, 354]

M. Lax. Cascade capture of electrons in solids. *Phys. Rev.* 1960, **119**, no. 5, 1502

M. Lax, P. Mengert. Influence of trapping, diffusion and recombination on carrier concentration fluctuations, *J. Phys. Chem. Solids*, 1960, **14**, 248

R.A. Suris, B.I. Fuks. Noise in finite compensated semiconductors. I. Generation–recombination noise. *Fiz. Tekhn. Poluprovodn.*, 1980, **14**, no. 6, 1077 [*Sov. Phys. – Semiconductors*, 1980, **14**, no. 6, 641]; II. Johnson noise and contact noise. *ibid*, 1980, **14**, no. 6, 1088 [*Sov. Phys. - Semiconductors*, 1980, **14**, no. 6, 647]

W. van Roosbroeck. The transport of added current carriers in a homogeneous semiconductor. *Phys. Rev.*, 1953, **91**, no. 2, 282

Chapter 5

Recommended books and reviews

S. Datta. Quantum interference devices. Ch. 10 In: *Physics of Quantum Electron Devices.* Ed. F. Capasso, Springer Series in Electronics and Photonics, no. 28, 1990, p. 321

H. Grabert, M.H. Devoret, eds. *Single Charge Tunneling. Coulomb Blockade Phenomena in Nanostructures.* Plenum Press, New York, 1992

M. Heiblum, M.V. Fischetti. Ballistic electron transport in hot electron transistors. In: *Physics of Quantum Electron Devices.* Ed. F. Capasso, Springer Series in Electronics and Photonics, no. 28, 1990, p. 271

W.J. Skocpol. Quantum effects in quasi-one-dimensional MOSFETs. In: *Physics of Quantum Electron Devices.* Ed. F. Capasso, Springer Series in Electronics and Photonics, no. 28, 1990, Ch. 12, p. 367

H. Van Houten, C.W.J. Beenakker, B.J. Van Wees. *Semiconductors and Semimetals*, **35**, Ed. M. Reed, Academic Press, New York, 1990, p. 9

References

A.I. Akimenko, A.B. Verkin, I.K. Yanson. Point-contact noise spectroscopy of phonons in metals. *J. Low Temp. Phys.*, 1984, **54**, no. 3/4, 247

J. Bardeen. Tunneling from a many-particle point of view. *Phys. Rev. Lett.*, 1961, **6**, no. 2, 57

C.W.J. Beenakker, M. Büttiker. Suppression of shot noise in metallic diffusive conductors. *Phys. Rev. B*, 1992, **46**, no. 3, 1889

C.W.J. Beenakker, H. van Houten. Semiclassical theory of shot noise and its suppression in a conductor with deterministic scattering. *Phys.Rev. B*, 1991, **43**, no. 14, 12066

M. Büttiker. Scattering theory of thermal and excess noise in open conductors. *Phys. Rev. Lett.*, 1990, **65**, no. 23, 2901

M. Büttiker. Scattering theory of current and intensity noise correlations in conductors and wave guides. *Phys. Rev. B*, 1992, **46**, no. 19, 12485

D.H. Cobden, N.K. Patel, M. Pepper, D.A. Ritchie, J.E.F. Frost, G.A.C. Jones. Noise and reproducible structure in a $GaAs/Al_xGa_{1-x}As$ one-dimensional channel. *Phys. Rev. B*, 1991, **44**, no. 4, 1938

M.H. Cohen, L.M. Falikov, J.C. Philips. Superconductive tunneling. *Phys. Rev. Lett.*, 1962, **8**, no. 8, 316

A.J. Dahm, A. Denenstein, D.N. Langenberg, W.H. Parker, D. Rogovin, D.J. Scalapino. *Phys. Rev. Lett.*, 1969, **22**, no. 26, 1416

C. Dekker, A.J. Scholten, F. Liefrink, R. Eppenga, H. van Houten, C.T. Foxon. Spontaneous resistance switching and low-frequency noise in quantum point contacts. *Phys. Rev. Lett.*, 1991, **66**, no. 16, 2148

L.I. Glazman, G.B. Lesovik, D.E. Khmel'nitskii, R.I. Shekhter. Reflectionless quantum transport and fundamental ballistic-resistance steps in microscopic constrictions. *Pis'ma Zh. Eksp. Teor. Fiz.*, 1988, **48**, no. 4, 218 [*Sov. Phys. – JETP Lett.*, 1988, **48**, no. 4, 238]

S. Hershfield, J.H. Davies, P. Hyldgaard, C.J. Stanton, J.W. Wilkins. Zero-frequency current noise for the double-tunnel-junction Coulomb blockade. *Phys. Rev. B*, 1993, **47**, no. 4, 1967

A. Kawabata. Theory of ballistic transport through a constriction-quantization of conductance. *J. Phys. Soc. Japan*, 1989, **58**, no. 2, 372

V.A. Khlus. Current and voltage fluctuations in the microcontacts of normal and superconductive metals. *Zh. Eksp. Teor. Fiz.*, 1987, **93**, no. 6, 2179 [*Sov. Phys. – JETP*, 1987, **66**, 1243]

Sh.M. Kogan. Theoretical study of a quantum-point-contact model. *Phys. Rev. B*, 1994, **50**, no. 23, 17230

Sh.M. Kogan, A.Ya. Shulman. On the theory of fluctuations in a nonequilibrium electron gas. *Zh. Eksp. Teor. Fiz.*, 1969, **56**, no. 3, 862 [*Sov. Phys. – JETP*, 1969, **29**, no. 3, p. 467]

A.N. Korotkov. Intrinsic noise of the single-electron transistor. *Phys. Rev. B*, 1994, **49**, no. 15, 10381

I.O. Kulik, A.N. Omel'yanchuk. Nonequilibrium fluctuations in normal-metal point contacts. *Fiz. Nizk. Temp.*, 1984, **10**, no. 3, 305 [*Sov. J. Low-Temp. Phys.*, 1984, **10**, no. 3, 158]

L.D. Landau, E.M. Lifshitz. *Quantum Mechanics*. Pergamon Press, Oxford, 1991

R. Landauer. Spatial variation of currents and fields due to localized scatterers in metallic conduction. *IBM J. Res. Dev.*, 1957, **1**, no. 3, 223

R. Landauer. Electrical transport in closed systems. *Z. Phys. B*, 1987, **68**, no. 2/3, 217

R. Landauer, Th. Martin. Equilibrium and shot noise in mesoscopic systems. *Physica B*, 1991, **175**, 167

G.B. Lesovik. Quantum excess noise in two-dimensional ballistic microcontacts. *Pis'ma Zh. Eksp. Teor. Fiz.*, 1989, **49**, no. 9, 513 [*Sov. Phys. – JETP Lett.*, 1989, **49**, no. 9, 592]

L.S. Levitov, G.B. Lesovik. Charge distribution in quantum shot noise. *Pis'ma Zh. Eksp. Teor. Fiz.*, 1993, **58**, no. 3, 225 [*JETP Lett.*, 1993, **58**, no. 3, 230]

Y.P. Li, D.C. Tsui, J.J. Heremans, J.A. Simmons, G.W. Weimann. Low-frequency noise in transport through quantum point contacts. *Appl. Phys. Lett.*, 1990, **57**, no. 8, 774

F. Liefrink, R.W. Stok, J.I. Dijkhuis, M.J.M. de Jong, H. van Houten, C.T. Foxon. Reduced shot noise in a quasi-one-dimensional channel. *Proc. 12-th Int. Conf. on Noise in Physical Systems and 1/f Fluctuations. St. Louis, 1993.* AIP Conference Proceedings # 285. p. 272

F. Liefrink, J. I. Dijkhuis, M. J. M. de Jong, L.W. Molenkamp, H. van Houten. Experimental study of reduced shot noise in a diffusive mesoscopic conductor. *Phys. Rev. B*, 1994, **49**, no. 19, 14066

Th. Martin, R. Landauer. Wave-packet approach to noise in multichannel mesoscopic systems. *Phys. Rev. B*, 1992, **45**, no. 4, 1742

K.E. Nagaev. On the shot noise in dirty metal contacts. *Phys. Lett. A*, 1992, **169**, no. 1–2, 103

K.E. Nagaev. Influence of electron-electron scattering on shot noise in diffusive contacts. *Phys. Rev. B*, 1995, **52**, no. 7, 4740

M. Reznikov, M. Heiblum, H. Shtrikman, D. Mahalu. Temporal correlation of electrons: suppression of shot noise in a ballistic quantum point contact. *Phys. Rev. Lett.*, 1995, **75**, no. 18, 3340

Yu. V. Sharvin. A possible method for studying Fermi surfaces. *Zh. Eksp. Teor. Fiz.*, 1965, **48**, 984 [*Sov. Phys. – JETP*, 1965, **21**, no. 3, 655]

A. Shimizu, M. Ueda. Effects of dephasing and dissipation on quantum noise in conductors. *Phys. Rev. Lett.*, 1992, **69**, no. 9, 1403

G. Timp, R.E. Behringer, J.E. Cunningham. Suppresion of impurity scattering in a one-dimensional wire. *Phys. Rev. B*, 1990, **42**, no. 14, 9259

B.J. van Wees, H. van Houten, C.W.J. Beenakker, J.G. Williamson, L.P. Kowenhoven, D. van der Marel, C.T. Foxon. Quantized conductance of point contacts in a two-dimensional electron gas. *Phys. Rev. Lett.*, 1988, **60**, no. 9, 848

D.A. Wharam, T.J. Thornton, R. Newbury, M. Pepper, H. Ahmed, J.E.F. Frost, D.G. Hasko, D.C. Peacock, D.A. Ritchie, G.A.C. Jones. One-dimensional transport and the quantization of the ballistic resistance. *J. Phys., C : Solid State Phys.*, 1988, **21**, no. 8, L209

I.K. Yanson. Nonlinear effects in the electric conductivity of point junctions and electron-phonon interaction in normal metals. *Zh. Eksp. Teor. Fiz.*, 1974, **66**, no. 3, 1035 [*Sov. Phys. – JETP*, 1974, **39**, no. 3, 506]

Chapter 6

Recommended books and reviews

B.L. Altshuler, P.A. Lee, R.A. Webb, eds. *Mesoscopic Phenomena in Solids*. Elsevier Science Publishers B.V., 1991

S. Feng. Conductance fluctuations and $1/f$ noise magnitudes in small disordered structures: theory. In: *Mesoscopic Phenomena in Solids*. Ed. B.L. Altshuler, P.A. Lee, R.A. Webb, Elsevier Science Publishers B.V., 1991. Ch. 4, p. 107

N. Giordano. Conductance fluctuations and low-frequency noise in small disordered systems: experiment. In: *Mesoscopic Phenomena in Solids*. Ed. B.L. Altshuler, P.A. Lee, R.A. Webb, Elsevier Science Publishers B.V., 1991. Ch. 5, p. 131

R. Schmitz. Fluctuations in nonequilibrium fluids. *Phys. Rep.*, 1988, **171**, no. 1, 1

K.M. van Vliet, J.R. Fassett. Fluctuations due to electronic transitions and transport in solids. In: *Fluctuation Phenomena in Solids*. Ed. R.E. Burgess, Academic Press, New York, 1965. p. 265

S. Washburn, R.A. Webb. Quantum transport in small disordered samples from the diffusive to the ballistic regime. *Rep. Progr. Phys.*, 1992, **55**, 1311

References

G.B. Alers, M.B. Weissman, R.S. Averback, H. Shyu. Resistance noise in amorphous Ni−Zr: Hydrogen diffusion and universal conductance fluctuations. *Phys. Rev. B*, 1989, **40**, no. 2, 900

B.L. Altshuler. Fluctuations in the extrinsic conductivity of disordered conductors. *Pis'ma Zh. Eksp. Teor. Fiz.*, 1985, **41**, no. 12, 530 [*Sov. Phys. – JETP Lett.*, 1985, **41**, no. 12, 648]

B.L. Altshuler, B.Z. Spivak. Variations of the random potential and the conductivity of samples of small dimensions. *Pis'ma Zh. Eksp. Teor. Fiz.*, 1985, **42**, no. 9, 363 [*Sov. Phys. – JETP Lett.*, 1986, **42**, no. 9, 447]

N.O. Birge, B. Golding, W.H. Haemmerle. Electron quantum interference and $1/f$ noise in bismuth. *Phys. Rev. Lett.*, 1989, **62**, no. 2, 195

N.O. Birge, B. Golding, W.H. Haemmerle. Conductance fluctuations and $1/f$ noise in Bi. *Phys. Rev. B*, 1990, **42**, no. 5, 2735

Sh. Feng, P.A. Lee, A.D. Stone. Sensitivity of the conductance of a disordered metal to the motion of a single atom: implications for $1/f$ noise. *Phys. Rev. Lett.*, 1986, **56**, no. 18, 1960. See also comment by J. Pelz, J. Clarke, *Phys. Rev. Lett.*, 1987, **59**, no. 9, 1061

R.P. Feynman, A.R. Hibbs. *Quantum Mechanics and Path Integrals*. McGraw-Hill Book Company, 1965.

C.L. Kane, R.A. Serota, P.A. Lee. Long-range correlations in disordered metals. *Phys. Rev. B*, 1988, **37**, no. 12, 6701

J. Kilmer, E.K. Chenette, C.M. van Vliet, P.H. Handel. Absence of temperature fluctuations in $1/f$ noise correlation experiments in silicon. *Phys. Status Solidi A*, 1982, **70**, 287

Sh.M. Kogan, K.E. Nagaev. Low-frequency current noise in solids and the internal friction, *Fiz. Tverd. Tela*, 1982, **24**, no. 11, 3381 [*Sov. Phys. – Solid State*, 1982, **24**, 1921]

Sh.M. Kogan, K.E. Nagaev. On the low-frequency current noise in metals. *Solid State Commun.*, 1984, **49**, no. 4, 387

L.D. Landau, E.M. Lifshitz. *Statistical Physics, part 1*. 3rd edn., Pergamon Press, 1985

P.A. Lee, A.D. Stone. Universal conductance fluctuations in metals. *Phys. Rev. Lett.*, 1985, **55**, no. 15, 1622

E.M. Lifshitz, L.P. Pitaevskii. *Statistical Physics, part 2*. Pergamon Press, 1980

D. Mailly, M. Sanquer. Sensitivity of quantum conductance fluctuations and of $1/f$ noise to time reversal symmetry. *J. Phys. I, Paris*, 1992, **2**, 357

P. McConville, N.O. Birge. Weak localization, universal conductance fluctuations, and $1/f$ noise in Ag. *Phys. Rev. B*, 1993, **47**, no. 24, 16667

J. Pelz, J. Clarke. Quantitative 'local-interference' model for $1/f$ noise in metal films. *Phys. Rev. B*, 1987, **36**, no. 8, 4479

J.H. Scofield, W.W. Webb. Resistance fluctuations due to hydrogen diffusion in niobium. *Phys. Rev. Lett.*, 1985, **54**, no. 4, 353

A.D. Stone. Reduction of low-frequency noise in metals by a magnetic field: Observability of the transition between random-matrix ensembles. *Phys. Rev. B*, 1989, **39**, no. 15, 10736

V.M. Vinokur, S.P. Obukhov. $1/f$ noise due to the motion of dislocations and intergrain boundaries. *Zh. Eksp. Teor. Fiz.*, 1989, **95**, no. 1, 223 [*Soviet Phys. – JETP*, 1989, **68**, no. 1, 126]

R.F. Voss, J. Clarke. Flicker $(1/f)$ noise: equilibrium temperature and resistance fluctuations. *Phys. Rev. B*, 1976, **13**, no. 2, 556

M.B. Weissman. Reconcilliation between the thermodynamics and noise measurements on metal film resistors. *Phys. Rev. Lett.*, 1978, **41**, no. 1, 1

M.B. Weissman, G.D. Dollinger. Noise from equilibrium enthalpy fluctuations. *J. Appl. Phys.*, 1981, **52**, no. 5, 3095

N.M. Zimmerman, W.W. Webb. Microscopic scatterer displacements generate the $1/f$ resistance noise of H in Pd. *Phys. Rev. Lett.*, 1988, **61**, no. 7, 889

Chapter 7

Recommended books and reviews

D.J. Bergman, D. Stroud. Physical properties of macroscopically inhomogeneous media. *Solid State Phys.*, 1992, **46**, 147

P.A. Lee, T.V. Ramakrishnan. Disordered electronic systems. *Rev. Mod. Phys.*, 1985, **57**, no. 2, 287

M.B. Isichenko. Percolation, statistical topography, and transport in random media, *Rev. Mod. Phys.*, 1992, **64**, no. 4, 961

B.I. Shklovskii, A.L. Efros. *Electronic Properties of Doped Semiconductors*. Springer-Verlag, Berlin, 1984

D. Stauffer, A. Aharoni. *Introduction to Percolation Theory*. Taylor & Francis, London, 1992

References

J. Adler, Y. Meir, A. Aharony, A.B. Harris. Series study of percolation moments in general dimension. *Phys. Rev. B*, 1990, **41**, no. 13, 9183

P.W. Anderson. Absence of diffusion in certain random lattices. *Phys. Rev.*, 1958, **109**, no. 5, 1492

S.R. Broadbent, J.M. Hammersley. Percolation processes. I. Crystals and mazes. *Proc. Cambridge Philos. Soc.*, 1957, **53**, part 3, 629

M. Celasco, A. Masoero, P. Mazetti, A. Stepanescu. Electrical conduction and
current noise mechanism in discontinuous metal films. II. Experimental. *Phys.
Rev. B*, 1978, **17**, no. 6, 2564

M. Celasco, A. Masoero, A. Stepanescu. Electrical conduction and current noise in
discontinous platinum films. *Phys. Rev. B*, 1978, **17**, no. 8, 3040

C.C. Chen, Y.C. Chou. Electrical-conductivity fluctuations near the percolation
threshold. *Phys. Rev. Lett.*, 1985, **54**, no. 23, 2529

C.J. Christensen, G.L. Pearson. Spontaneous resistance fluctuations in carbon
microphones and other granular resistances. *Bell System Tech. J.*, 1936, **15**, no.
2, 197

O. Cohen, Z. Ovadyahu, M. Rokni. $1/f$ noise and incipient localization. *Phys. Rev.
Lett.*, 1992, **69**, no. 24, 3555

O. Cohen, Z. Ovadyahu. Resistance noise near the Anderson transition. *Phys. Rev.
B*, 1994, **50**, no. 15, 10442

G.A. Garfunkel, M.B. Weissman. Noise scaling in continuum percolating films.
Phys. Rev. Lett., 1985, **55**, no. 3, 296

D.S. Gaunt, M.F. Sykes. Series study of random percolation in three dimensions. *J.
Phys. A*, 1983, **16**, no. 4, 783

J. Kertesz. Extrapolation of transfer matrix data for percolation and lattice animals
by the Romberg–Beleznay algorithm. *J. Phys. A*, 1986, **19**, 599

L.B. Kiss, P. Svedlindh. New noise exponents in random conductor–superconductor
and conductor–insulator mixtures. *Phys. Rev. Lett.*, 1993, **71**, no. 17, 2817

R.H. Koch, R.B. Laibowitz, E.I. Alessandrini, J.M. Viggiano. Resistivity-noise
measurements in thin gold films near the percolation threshold. *Phys. Rev. B*,
1985, **32**, no. 10, 6932

Sh.M. Kogan, B.I. Shklovskii. Excess low-frequency noise in hopping conduction.
Fiz. Tekhn. Poluprovdn., 1981, **15**, no. 6, 1049 [*Sov. Phys. – Semiconductors*,
1981, **15**, no. 6, 605]

J.V. Mantese, W.W. Webb. $1/f$ noise of granular metal–insulator composites. *Phys.
Rev. Lett.*, 1985, **55**, no. 20, 2212

J.V. Mantese, W.A. Curtin, W.W. Webb. Two-component model for the resistivity
and noise of tunneling metal–insulator composites. *Phys. Rev. B*, 1986, **33**, no.
12, 7897

J.V. Mantese, W.I. Goldburg, D.H. Darling, H.G. Craighead, U.J. Gibson, R.A.
Buhrman, W.W. Webb. Excess low frequency conduction noise in a granular
composite, *Solid State Commun.*, 1981, **37**, no. 4, 353

M. Murat, S. Marianer, D.J. Bergman. A transfer matrix study of conductivity and
permeability exponents in continuum percolation. *J. Phys. A*, 1986, **19**, L275

B. Nienhuis. Exact critical point and critical exponents of $O(n)$ models in two
dimensions. *Phys. Rev. Lett.*, 1982, **49**, no. 15, 1062

C. Pierre, R. Deltour, J. Van Bentum, J.A.A.J. Perenboom, R. Rammal.
Electrical-conduction mechanisms in polymer–copper-particle composites. II.
$1/f$ noise measurements in the percolation limit. *Phys. Rev. B*, 1990, **42**, no. 6,
3386

R. Rammal. Flicker noise near the percolation threshold. *J. Phys. (Paris) – Lett.*,
1985, **46**, no. 4, L129

R. Rammal, C. Tannous, P. Breton, A.-M.S. Tremblay. Flicker $(1/f)$ noise in
percolation networks. A hierarchy of exponents. *Phys. Rev. Lett.*, 1985, **54**, no.
15, 1718

M.F. Sykes, J.W. Essam. Some exact critical percolation probabilities for bond and site problems in two dimensions. *Phys. Rev. Lett.*, 1963, **10**, no. 1, 3; *J. Math. Phys.*, 1964, **5**, no. 8, 1117

M.F. Sykes, D.S. Gaunt, M. Glen. Percolation processes in two dimensions. II. Critical concentration and mean size index. *J. Phys. A*, 1976, **9**, 97; III. High density series expansions. *ibid*, 715; IV. Percolation probability. *ibid*, 725

J.L. Williams, R.K. Burdett. Current noise in thin gold films. *J. Phys. C: Solid State Phys.*, 1969, **2**, 298

J.L. Williams, I.L. Stone. Current noise in thin discontinous films. *J. Phys. C: Solid State Phys.*, 1972, **5**, no. 16, 2105

Chapter 8

Recommended books and reviews

D.A. Bell. A survey of $1/f$ noise in electrical conductors. *J. Phys. C: Solid State Phys.*, 1980, **13**, no. 24, 4425

S.M. Bezrukov, I. Vodyanoy. Noise in biological membranes and relevant ionic systems. *Advances in Chemistry Series. 235. Biomembrane Electrochemistry*. M. Blank, I. Vodyanoy, eds. Amer. Chem. Soc., Washington, DC, 1994

K. Binder, A.P. Young. Spin glasses: experimental facts, theoretical concepts, and open questions. *Rev. Mod. Phys.*, 1986, **58**, no. 4, 801

J.L. Black. Low-energy excitations in metallic glasses. In: *Glassy Metals*. I. Eds. H.-J. Günterodt, H. Beck. Springer-Verlag, Berlin, 1981, 167

H. Bouchiat, M. Ocio. Experimental investigation of $1/f$ magnetic fluctuations in spin glasses. *Comments Cond. Matt. Phys.*, 1988, **14**, no. 3, 163

M.J. Buckingham. *Noise in Electronic Devices and Systems*. Ellis Horwood Ltd., New York, 1983

L.J. DeFelice. *Introduction to Membrane Noise*. Plenum, New York, 1981

V.S. Dotsenko, M.V. Feigelman, L.B. Ioffe. Spin glasses and related phenomena. *Soviet Scientific Reviews, Section A: Physics Reviews*, 1990, **15**, part 1, 1. Harwood, Glasgow, UK

P. Dutta, P.M. Horn. Low-frequency fluctuations in solids. *Rev. Mod. Phys.*, 1981, **53**, no. 3, 497

N.V. Dyakonova, M.E. Levinshtein, S.L. Rumyantsev. Nature of the bulk $1/f$ noise in GaAs and Si (review). *Fiz. Tekhn. Poluprovodn.*, 1991, **25**, no. 12, 2065 [*Sov. Phys. – Semiconductors*, 1991, **25**, no. 12, 1241]

K.H. Fisher, J.A. Hertz. *Spin Glasses*. Cambridge University Press, Cambridge, England, 1991

G. Grüner. The dynamics of charge density waves. *Rev. Mod. Phys.*, 1988, **60**, no. 4, 1129

F.N. Hooge, T.G.M. Kleinpenning, L.K.J. Vandamme. Experimental studies on $1/f$ noise. *Rep. Prog. Phys.*, 1981, **44**, no. 5, 479

B.K. Jones. Electrical noise as a measure of quality and reliability in electron devices. *Adv. Electron. Electron Phys.*, 1994, **87**, 201

M.S. Keshner. $1/f$ noise, *Proc. IEEE*, 1982, **70**, no. 3, 212

M.J. Kirton, M.J. Uren. Noise in solid-state microstructures: A new perspective on individual defects, interface states and low-frequency $(1/f)$ noise. *Adv. Phys.*, 1989, **38**, no. 4, 367

Sh.M. Kogan. Current noise with $1/f$ type spectra in solids. *Uspekhi Fiz. Nauk*, 1985, **145**, no. 2, 285 (*Sov. Phys. – Uspekhi*, 1985, **28**, no. 2, 170)

A.N. Malakhov. On the problem of flicker-noise spectrum. *Radiotekh. Elektron.*, 1959, **4**, no. 1, 54

W.H. Press. Flicker noises in astronomy and elsewhere. *Comments Astrophys. Space Phys.*, 1978, **7**, no. 4, 103

B.P. Smolyakov, E.P. Khaimovich. Dynamic processes in dielectric glasses at low temperatures. *Usp. Fiz. Nauk*, 1982, **136**, 317 [*Sov. Phys. – Usp.*, 1982, **25**, no. 2, 102]

A. van der Ziel. Flicker noise in electron devices. *Adv. Electron. Electron Phys.*, 1979, **49**, 225

A. van der Ziel. History of noise research. *Adv. Electron. Electron Phys.*, 1980, **50**, 351

M.B. Weissman. $1/f$ noise and other slow nonexponential kinetics in condensed matter. *Rev. Mod. Phys.*, 1988, **60**, no. 2, 537

M.B. Weissman. What is spin-glass? A glimpse via mesoscopic noise. *Rev. Mod. Phys.*, 1993, **65**, no. 3, 829

References

G. Abowitz, E. Arnold, E.A. Leventhal. Surface states and $1/f$ noise in MOS transistors. *IEEE Trans. Electron. Devices*, 1967, **ED-14**, no. 11, 775

C.J. Adkins, R.H. Koch. Noise in inversion layers near the metal–insulator transition. *J. Phys. C: Solid State Phys.*, 1982, **15**, 1829

M. Alba, J. Hammann, M. Ocio, P. Refregier, H. Bouchiat. Spin-glass-dynamics from magnetic noise, relaxation, and susceptibility measurements. *J. Appl. Phys.*, 1987, **61**, no. 8, part IIA, 3683

G.B. Alers, M.B. Weissman. Mechanical relaxations and $1/f$ noise in Bi, Nb, and Fe films. *Phys. Rev. B*, 1991, **44**, no. 14, 7192

G.B. Alers, M.B. Weissman, R.S. Averback, H. Shyu. Resistance noise in amorphous Ni–Zr: Hydrogen diffusion and universal conductance fluctuations. *Phys. Rev. B*, 1989, **40**, no. 2, 900

P.W. Anderson, B.I. Halperin, C. Varma. Anomalous low-temperature properties of glasses and spin glasses. *Philos. Mag.*, 1972, **25**, no. 1, 1

V.N. Andreev, B.P. Zakharchenya, Yu.S. Kapshin, V.A. Noskin, F.A. Chudnovskii. Low-frequency noise in vanadium dioxide undergoing a metal–semiconductor phase transition. *Zh. Eksp. Teor. Fiz.*, 1980, **79**, no. 4, 1353 [*Sov. Phys. – JETP*, 1980, **52**, no. 4, 684]

M.N. Baibich, J.M. Broto, A. Fert, F. Nguyen van Dau, F. Petroff, P. Eitenne, G. Creuzet, A. Friederich, J. Chazelas. Giant magnetoresistance of (001) Fe/(001) Cr magnetic superlattices. *Phys. Rev. Lett.*, 1988, **61**, no. 21, 2472

P. Bak, C. Tang, K. Wiesenfeld. Self-organized criticality: an explanation of $1/f$ noise. *Phys. Rev. Lett.*, 1987, **59**, no. 4, 381

H.G.E. Beck, W.P. Spruit. $1/f$ Noise in the variance of Johnson noise. *J. Appl. Phys.*, 1978, **49**, no. 6, 3384

S. Bhattacharya, J.P. Stokes, M.O. Robbins, R.A. Klemm. Origin of broadband noise in charge density wave conductors. *Phys. Rev. Lett.*, 1985, **54**, no. 22, 2453

G. Binach, P. Grunberg, F. Sauerenbach, W. Zinn. Enhanced magnetoresistance in layered magnetic structures with antiferromagnetic interlayer exchange. *Phys. Rev. B*, 1989, **39**, no. 7, 4828

N.O. Birge, B. Golding, W.H. Haemmerle. Electron quantum interference and $1/f$ noise in bismuth. *Phys. Rev. Lett.*, 1989, **62**, no. 2, 195

N.O. Birge, B. Golding, W.II. Hacmmcrlc. Conductance fluctuations and $1/f$ noise in Bi. *Phys. Rev. B*, 1990, **42**, no. 5, 2735

R.D. Black, P.J. Restle, M.B. Weissman. Nearly traceless $1/f$ noise in bismuth. *Phys. Rev. Lett.*, 1983, **51**, no. 16, 1476

R.D. Black, W.M. Snow, M.B. Weissman. Nonscalar $1/f$ conductivity fluctuations in carbon, gold and chrome films. *Phys. Rev. B*, 1982, **25**, no. 4, 2955

R.D. Black, M.B. Weissman, F.M. Fliegel. $1/f$ noise in metal films lacks spatial correlation. *Phys. Rev. B*, 1981, **24**, no. 12, 7454

I. Bloom, A.C. Marley, M.B. Weissman. Nonequilibrium dynamics of discrete fluctuators in charge-density waves in NbSe₃. *Phys. Rev. Lett.*, 1993, **71**, no. 26, 4385

I. Bloom, A.C. Marley, M.B. Weissman. Discrete fluctuators and broadband noise in the charge-density wave in NbSe₃. *Phys. Rev. B*, 1994, **50**, no. 8, 5081

J.J. Brophy. Excess noise in deformed germanium. *J. Appl. Phys.*, 1956, **27**, no. 11, 1383

J.J. Brophy. Statistics of $1/f$ noise. *Phys. Rev.*, 1968, **166**, no. 3, 827

J.J. Brophy. Variance fluctuations in flicker noise and current noise. *J. Appl. Phys.*, 1969, **40**, no. 9, 3551

J.J. Brophy. Low-frequency variance noise. *J. Appl. Phys.*, 1970, **41**, no. 7, 2913

M.A. Caloyannides. Microcycle spectral estimates of $1/f$ noise in semiconductors. *J. Appl. Phys.*, 1974, **45**, no. 1, 307

M.J. Campbell, B.W. Jones. Cyclic changes in insulin needs of an unstable diabetic. *Science*, 1972, **177**, no. 4052, 889

P.M. Campbell, E.S. Snow, W.J. Moore, O.J. Glembocki, S.W. Kirchoefer. Light-activated telegraph noise in AlGaAs tunnel barriers: Optical probing of a single defect. *Phys. Rev. Lett.*, 1991, **67**, no. 10, 1330

Z. Çelik-Butler, T.Y. Hsiang. Spatial correlation measurements of $1/f$ noise in semiconductors. *Solid-State Electron.*, 1988, **31**, no. 2, 241

C.J. Christensen, G.L. Pearson. Spontaneous resistance fluctuations in carbon microphones and other granular resistances. *Bell System Tech. J.*, 1936, **15**, no. 2, 197

J. Clarke, T.Y. IIsiang. Low-frequency noise in tin and lead films at the superconducting transition. *Phys. Rev. B*, 1976, **13**, no. 11, 4790

L.J. De Felice. $1/f$ resistor noise. *J. Appl. Phys.*, 1976, **47**, no. 1, 350

T. Dilmi, A. Chovet, P. Viktorovitch. Influence of a magnetic field on $1/f$ noise in ambipolar semiconductors: Evidence of its surface origin. *J. Appl. Phys.*, 1979, **50**, no. 8, 5348

F.K. du Pré. A suggestion regarding the spectral density of flicker noise. *Phys. Rev.*, 1950, **78**, no. 5, 615

G. Durin, P. Falferi, M. Cerdonio, G.A. Prodi, S. Vitale. Low-temperature properties of soft magnetic materials: Magnetic viscosity and $1/f$ thermal noise. *J. Appl. Phys.*, 1993, **73**, no. 10, 5363

P. Dutta, P. Dimon, P.M. Horn. Energy scales for noise processes in metals. *Phys. Rev. Lett.*, 1979, **43**, no. 9, 646

P. Dutta, J.W. Eberhard, P.M. Horn. $1/f$ noise in metal films: the role of substrate. *Solid State Commun.*, 1978, **27**, no. 12, 1389

N.V. Dyakonova, M.E. Levinshtein, S.L. Rumyantsev. Illumination induced modification of the $1/f$ noise in gallium arsenide. *Fiz. Tekhn. Poluprovodn.*,

1988, **22**, no. 6, 1049 [*Sov. Phys. – Semiconductors*, 1988, **22**, no. 6, 661]

N.V. Dyakonova, M.E. Levinshtein. Effect of temperature on optical tuning of the $1/f$ noise in GaAs. *Pis'ma Zh. Tekhn. Fiz.*, 1988, **14**, no. 21, 1978 [*Sov. Phys. – Tech. Phys. Lett.*, 1988, **14**, no. 11, 857]

J.W. Eberhard, P.M. Horn. Temperature dependence of $1/f$ noise in silver and copper. *Phys. Rev. Lett.*, 1977, **39**, no. 10, 643

J.W. Eberhard, P.M. Horn. Excess $1/f$ noise in metals, *Phys. Rev. B*, 1978, **18**, no. 12, 6681

M. Epstein. Current noise in evaporated films of InSb and InAs. *J. Appl. Phys.*, 1965, **36**, no. 8, 2590

G. Fantozzi, A. Vincent, eds. Internal friction and ultrasonic attenuation in solids. Fourth European Conf., 1983. *J. Phys. (Paris)*, 1983, **44**, C9, Suppl. no. 12

K.R. Farmer, C.T. Rogers, R.A. Buhrman. Localized-state interactions in metal–oxide–semiconductor tunnel diodes. *Phys. Rev. Lett.*, 1987, **58**, no. 21, 2255

K.R. Farmer, R. Saletti, R.A. Buhrman. Current fluctuations and silicon oxide wear-out in metal–oxide–semiconductor tunnel junctions. *Appl. Phys. Lett.*, 1988, **52**, no. 20, 1749

D.S. Fisher, D.A. Huse. Nonequilibrium dynamics of spin glasses. *Phys. Rev. B*, 1988a, **38**, no. 1, 373

D.S. Fisher, D.A. Huse. Equilibrium behavior of the spin-glass ordered phase. *Phys. Rev. B*, 1988b, **38**, no. 1, 386

D.M. Fleetwood, N. Giordano. Experimental study of excess low-frequency noise in tin. *Phys. Rev. B*, 1982, **25**, no. 2, 1427

D.M. Fleetwood, N. Giordano. Resistivity dependence of $1/f$ noise in metal films. *Phys. Rev. B*, 1983a, **27**, no. 2, 667

D.M. Fleetwood, N. Giordano. Effect of strain on the $1/f$ noise of metal films. *Phys. Rev. B*, 1983b, **28**, no. 6, 3625

D.M. Fleetwood, N. Giordano. Direct link between $1/f$ noise and defects in metal films. *Phys. Rev. B*, 1985, **31**, no. 2, 1157

D.M. Fleetwood, J.T. Masden, N. Giordano. $1/f$ noise in platinum films and ultrathin platinum wires: evidence for a common, bulk origin. *Phys. Rev. Lett.*, 1983, **50**, no. 6, 450

B.R. Frieden, R.J. Hughes. Spectral $1/f$ noise derived from extremized physical information. *Phys. Rev. E*, 1994, **49**, no. 4, 2644

J.J. Gagnepain, J. Uebersfeld. $1/f$ Noise in quartz resonators. In: *Proc. Symp. $1/f$ Fluctuations*. Ed. T. Musha, Tokyo, 1977, p. 173

G.A. Garfunkel, G.B. Alers, M.B. Weissman. Mesoscopic noise studies of atomic motions in cold amorphous conductors. *Phys. Rev. B*, 1990, **41**, no. 8, 4901

C.G.B. Garrett, W.H. Brattain. Physical theory of semiconductor surfaces. *Phys. Rev*, 1955, **99**, no. 2, 376

L.J. Greenstein, J.J. Brophy. Influence of lower cutoff frequency on the measured variance of $1/f$ noise. *J. Appl. Phys.*, 1969, **40**, no. 2, 682

E.G. Guk, M.E. Levinshtein, N.V. Dyakonova. Optical suppression of $1/f$ noise in silicon. *Fiz. Tekhn. Poluprovodn.*, 1988, **22**, no. 6, 1120 [*Sov. Phys. – Semiconductors*, 1988, **22**, no. 6, 707]

K. Gunnarson, P. Svedlindh, P. Nordblad, L. Lundgren, H. Aruga, A. Ito. Measurements of the equilibrium magnetic noise of a short-range Ising spin glass. *Phys. Rev. B*, 1989, **40**, no. 10, 7162

V.L. Gurevich. *Transport in Phonon Systems*. North-Holland, 1986

D. Halford. A general mechanical model for $|f|^\alpha$ spectral density random noise with special reference to flicker noise $1/|f|$. *Proc. IEEE*, 1968, **56**, no. 3, 251

J. Hammann, M. Ocio, E. Vincent. Attempt at a comprehensive description of the slow spin glass dynamics. In: *Relaxation in Complex Systems and Related Topics*. Ed. I.A. Campbell, C. Giovannella. Plenum Press, New York, 1990, p. 11

Z. Harangozo, F.J. Kedves. High-temperature background of internal friction in metals. In: *Internal Friction in Solids. Proc. Summer Sch., Cracow, 14–17 June 1984*, Ed. S. Gorczyca, L.B. Magalas. Wyd. AGH, Krakow, p. 209

H.T. Hardner, S.S.P. Parkin, M.B. Weissman, M.B. Salamon, E. Kita. $1/f$ noise in giant magnetoresistive materials. *J. Appl. Phys.*, 1994, **75**, no. 10, 6531

H.T. Hardner, M.B. Weissman, M.B. Salamon, S.S.P. Parkin. Fluctuation–dissipation relation for giant magnetoresistive $1/f$ noise. *Phys. Rev. B*, 1993, **48**, no. 21, 16156

F.N. Hooge. $1/f$ noise is no surface effect. *Phys. Lett.*, 1969, **29A**, no. 3, 139

F.N. Hooge, A. Hoppenbrouwers. Amplitude distribution of $1/f$ noise. *Physica*, 1969, **42**, no. 2, 331

N.E. Israeloff, G.B. Alers, M.B. Weissman. Spin-fluctuation statistics in CuMn. *Phys. Rev. B*, 1991, **44**, no. 22, 12613

N.E. Israeloff, M.B. Weissman, G.J. Nieuwenhuys, J. Kosiorowska. Electrical noise from spin fluctuations in CuMn. *Phys. Rev. Lett.*, 1989, **63**, no. 7, 794

H.J. Jensen. Lattice gas as a model of $1/f$ noise. *Phys. Rev. Lett.*, 1990, **64**, no. 26, 3103

J.B. Johnson. The Schottky effect in low frequency circuits. *Phys. Rev.*, 1925, **26**, no. 1, 71

K. Kandiah, F.B. Whiting. Low-frequency noise in junction field effect transistors. *Solid-State Electron.*, 1978, **21**, no. 8, 1079

V.G. Karpov, M.I. Klinger, F.N. Ignat'ev. Theory of low-temperature anomalies in the thermal properties of amorphous structures. *Zh. Eksp. Teor. Fiz.*, 1983, **84**, no. 2, 760 [*Sov. Phys. – JETP*, 1983, **57**, no. 2, 439]

C.D. Keener, M.B. Weissman. $1/f$ noise in bismuth consistent with defect motion. *Phys. Rev. B*, 1991, **44**, no. 17, 9178

J. Kilmer, E.K. Chenette, C.M. van Vliet, P.H. Handel. Absence of temperature fluctuations in $1/f$ noise correlation experiments in silicon. *Phys. Status Solidi A*, 1982, **70**, no. 1, 287

R.H. Koch, J.R. Lloyd, J. Cronin. $1/f$ noise and grain-boundary diffusion in aluminum and aluminum alloys. *Phys. Rev. Lett.*, 1985, **55**, no. 22, 2487

R.H. Koch, W. Reim, A.P. Malozemoff, M.B. Ketchen. Temperature dependence of the magnetic $1/f$ noise in $Eu_{0.4}Sr_{0.6}S$. *J. Appl. Phys.*, 1987, **61**, no. 8, part IIA, 3678

Sh.M. Kogan. $1/f$ Noise in spin-glasses and in the disordered kinetic Ising model. *Solid State Commun.*, 1981, **38**, no. 11, 1015

Sh.M. Kogan, K.E. Nagaev. Low-frequency current noise in solids and the internal friction. *Fiz. Tverd. Tela*, 1982, **24**, no. 11, 3381 [*Sov. Phys. – Solid State*, 1982, **24**, 1921]

Sh.M. Kogan, K.E. Nagaev. On the low-frequency current noise in metals. *Solid State Commun.*, 1984a, **49**, no. 4, 387

Sh.M. Kogan, K.E. Nagaev. Noise in tunnel junctions due to two-level systems in the dielectric layer. *Pis'ma Zh. Tekhn. Fiz.*, 1984b, **10**, no. 5, 313

L.D. Landau, E.M. Lifshitz. *Electrodynamics of Continuous Media*. Pergamon Press, Oxford, 1984

L.D. Landau, E.M. Lifshitz. *Statistical Physics. Part I*. 3rd edn. Pergamon Press, Oxford, 1985

B.M. Lebed', I.I. Marchik, V.A. Noskin, V.G. Shirko. Noise of $1/f$ type in ferromagnets. *Fiz. Tverd. Tela*, 1982, **24**, no. 7, 1997 [*Sov. Phys. – Solid State*, 1982, **24**, no. 7, 1140]

C. Leemann, M.J. Scove, E.P. Stillwell. Excess '$1/f$' noise in bismuth whiskers. *Solid State Commun.*, 1980, **35**, no. 1, 97

M.E. Levinshtein, S.L. Rumyantsev. $1/f$ noise under the conditions of strong geometric magnetoresistance. *Fiz. Tekhn. Poluprovodn.*, 1983, **17**, no. 10, 1830 [*Sov. Phys. – Semiconductors*, 1983, **17**, no. 10, 1167]

T.M. Lifshits, L.Ya. Pervova. Measurement of noise spectrum in planar germanium photodiodes in a broad frequency range. *Radiotekhn. Elektron.*, 1959, **4**, no. 9, 1543

S.B. Lowen, M.C. Teich. Fractal renewal processes generate $1/f$ noise. *Phys. Rev. E*, 1993, **47**, no. 2, 992

A. Ludviksson, R. Kree, A. Schmid. Low-frequency $1/f$ fluctuations of resistivity in disordered metals. *Phys. Rev. Lett.*, 1984, **52**, no. 11, 950

N.B. Lukyanchikova. New 'universal' relation concerning $1/f$ noise. *Phys. Lett. A*, 1993, **180**, no. 3, 285

J.V. Mantese, W.I. Goldburg, D.H. Darling, H.G. Craighead, U.J. Gibson, R.A. Buhrman, W.W. Webb. Excess low-frequency conduction noise in a granular composite. *Solid State Commun.*, 1981, **37**, no. 4, 353

T.G. Mapple, L. Bess, H.A. Gebbie. Variation of noise with ambient in germanium filaments. *J. Appl. Phys.*, 1955, **26**, no. 4, 490

E. Marinari, G. Parisi, D. Ruelle, P. Windey. Random walk in a random environment and $1/f$ noise. *Phys. Rev. Lett.*, 1983, **50**, no. 17, 1223

P. McConville, N.O. Birge. Weak localization, universal conductance fluctuations, and $1/f$ noise in Ag. *Phys. Rev. B*, 1993, **47**, no. 24, 16667

A.U. McRae, H. Levinshtein. Surface-dependent $1/f$ noise in germanium. *Phys. Rev.*, 1960, **119**, no. 1, 62

A.L. McWhorter. $1/f$ Noise and germanium surface properties. In: *Semiconductor Surface Physics*, ed. R.H. Kingston, Univ. of Philadelphia Press, Philadelphia, 1957, 207

A. Mircea, A. Roussel, A. Mittonneau. $1/f$ Noise: still a surface effect. *Phys. Lett.*, 1972, **41A**, no. 4, 345

H.C. Montgomery. Electrical noise in semiconductors. *Bell System Tech. J.*, 1952, **31**, no. 5, 950

W.J. Moore. Statistical studies of $1/f$ noise from carbon resistors. *J. Appl. Phys.*, 1974, **45**, no. 4, 1896

M. Nelkin, A.-M. Tremblay. Deviation of $1/f$ voltage fluctuations from scale-similar Gaussian behavior. *J. Stat. Phys.*, 1981, **25**, no. 2, 253

V.E. Noble, J.E. Thomas, Jr. Effects of gaseous ambients upon $1/f$ noise in germanium filaments. *J. Appl. Phys.*, 1961, **32**, no. 9, 1709

P. Nordblad, P. Svedlindh, L. Lundgren, L. Sandlund. Time decay of the remanent magnetization in a CuMn spin glass. *Phys. Rev. B*, 1986, **33**, no. 1, 645

A.S. Nowick, B.S. Berry. *Anelastic Relaxation in Crystalline Solids*. Academic Press, New York, 1972

M. Ocio, H. Bouchiat, P. Monod. Observation of $1/f$ magnetic fluctuations in a spin glass. *J. Phys. (Paris) Lett.*, 1985, **46**, L-647

M. Ocio, H. Bouchiat, P. Monod. Observation of $1/f$ magnetic fluctuations in spin glasses. *J. Magn. Magn. Mat.*, 1986, **54-57**, no. 1, 11

V. Palenskis. Transport phenomena and electric fluctuations in thin island Cr-based films. PhD dissertation, Vilnius University, Vilnius, 1976

C. Parman, J. Kakalios. Nonlinear $1/f$ noise in amorphous silicon. *Phys. Rev. Lett.*, 1991, **67**, no. 18, 2529

C.E. Parman, N.E. Israeloff, J. Kakalios. Conductance-noise power fluctuations in hydrogenated amorphous silicon. *Phys. Rev. Lett.*, 1992, **69**, no. 7, 1097

J. Pelz, J. Clarke. Dependence of $1/f$ noise on defects induced in copper films by electron irradiation. *Phys. Rev. Lett.*, 1985, **55**, no. 7, 738

J. Pelz, J. Clarke. Quantitative 'local-interference' model for $1/f$ noise in metal films. *Phys. Rev. B*, 1987, **36**, no. 8, 4479

J. Pelz, J. Clarke, W.E. King. Flicker $(1/f)$ noise in copper films due to radiation induced defects. *Phys. Rev. B*, 1988, **38**, no. 15, 10371

W.A. Phillips. Tunneling states in amorphous solids. *J. Low Temp. Phys.*, 1972, **7**, no. 3/4, 351

V.Ya. Pokrovskiĭ, S.V. Zaitsev-Zotov. Spontaneous resistance fluctuations and transition of the charge-density waves into disordered states in o-TaS nanosamples. *Europhys. Lett.*, 1990, **13**, no. 4, 361

G.A. Prodi, S. Vitale, M. Cerdonio, P. Falferi. Thermal magnetization noise as a function of frequency in amorphous ferromagnets. *J. Appl. Phys.*, 1989, **66**, no. 12, 5984

W.E. Purcell. Variance noise spectra of $1/f$ noise. *J. Appl. Phys.*, 1972, **43**, no. 6, 2890

K.S. Ralls, R.A. Buhrman. Defect interactions and noise in metallic nanoconstrictions. *Phys. Rev. Lett.*, 1988, **60**, no. 23, 2434

K.S. Ralls, R.A. Buhrman. Microscopic study of $1/f$ noise in metal nanobridges. *Phys. Rev. B*, 1991, **44**, no. 11, 5800

K.S. Ralls, W.J. Skocpol, L.D. Jackel, R.E. Howard, L.A. Fetter, R.W. Epworth, D.M. Tennant. Discrete resistance switching in submicrometer silicon inversion layers: individual interface traps and low-frequency $(1/f)$ noise. *Phys. Rev. Lett.*, 1984, **52**, no. 3, 228

P. Refregier, M. Alba, J. Hammann, M. Ocio. Dynamic behaviour of the insulating spin glass $CsNiFeF_6$. *J. Phys. C: Solid State Phys.*, 1987, **20**, 5545

P. Refregier, M. Ocio, H. Bouchiat. Equilibrium magnetic fluctuations in spin glasses: Temperature dependence and deviations from $1/f$ behaviour. *Europhys. Lett.*, 1987, **3**, no. 4, 503

P. Refregier, M. Ocio, J. Hammann, E. Vincent. Nonstationary spin glass dynamics from susceptibility and noise measurements. *J. Appl. Phys.*, 1988, **63**, no. 8, part IIB, 4343

W. Reim, R.H. Koch, A.P. Malozemoff, M.B. Ketchen, H. Maletta. Magnetic equilibrium noise in spin-glasses: $Eu_{0.4}Sr_{0.6}S$. *Phys. Rev. Lett.*, 1986, **57**, no. 7, 905

P.J. Restle, R.J. Hamilton, M.B. Weissman, M.S. Love. Non-Gaussian effects in $1/f$ noise in small silicon-on-sapphire resistors. *Phys. Rev. B*, 1985, **31**, no. 4, 2254

P.J. Restle, M.B. Weissman, R.D. Black. Tests of Gaussian statistical properties of $1/f$ noise. *J. Appl. Phys.*, 1983, **54**, no. 10, 5844

J. Richard, P. Monceau, M. Papoular, M. Renard. $f^{-\alpha}$ noise in $NbSe_3$. *J. Phys. C: Solid State Phys.*, 1982, **15**, 7157

C.T. Rogers, R.A. Buhrman. Composition of $1/f$ noise in metal–insulator–metal tunnel junctions. *Phys. Rev. Lett.*, 1984, **53**, no. 13, 1272

C.T. Rogers, R.A. Buhrman. Nature of single-localized-electron states derived from tunneling measurements. *Phys. Rev. Lett.*, 1985, **55**, no. 8, 859

B.V. Rollin, I.M. Templeton. Noise in semiconductors at very low frequencies. *Proc. Phys. Soc. London*, 1953, **B 66**, no. 2, 259

C.T. Sah, F.H. Hielscher. Evidence of the surface origin of the $1/f$ noise. *Phys. Rev. Lett.*, 1966, **17**, no. 18, 956

F.W. Schmidlin. Enhanced tunneling through dielectric films due to ionic defects. *J. Appl. Phys.*, 1966, **37**, no. 7, 2823

W. Schottky. Über spontane Stromschwankungen in verschiedenen Elektrizitatleitern. *Ann. Physik*, 1918, **57**, 541

W. Schottky. Small-shot effect and flicker effect. *Phys. Rev.*, 1926, **28**, 74

J.H. Scofield, D.H. Darling, W.W. Webb. Exclusion of temperature fluctuations as the source of $1/f$ noise in metal films. *Phys. Rev. B*, 1981, **24**, no. 12, 7450

J.H. Scofield, J.V. Mantese, W.W. Webb. $1/f$ noise of metals: A case of extrinsic origin. *Phys. Rev. B*, 1985, **32**, no. 2, 736

M.-Ho Song, H.S. Min. Influence of magnetic field on $1/f$ noise in GaAs Corbino disks. *J. Appl. Phys.*, 1985, **58**, no. 11, 4221

M.-Ho Song, A.N. Birbas, A. van der Ziel, A.D. van Rheenen. Influence of magnetic field on $1/f$ noise in GaAs resistors without surface effects. *J. Appl. Phys.*, 1988, **64**, no. 2, 727

M. Stoisiek, D. Wolf. Recent investigations on the stationarity of $1/f$ noise. *J. Appl. Phys.*, 1976, **47**, no. 1, 362

M. Stoisiek, D. Wolf, W. Werner. Bipolar transistors with low $1/f$ noise. *Electron. Lett.*, 1980, **16**, no. 10, 372

M. Surdin. Fluctuations de courant thermoionique et le 'flicker effect'. *J. Phys. Radium*, 1939, **10**, 188; *ibid*, 1951, **12**, 777

M. Tacano, H. Tanoue, Y. Sugiyama. Dependence of Hooge parameter of compound semiconductors on temperature. *Jpn. J. Appl. Phys.*, 1992, **31**, part 2, no. 3B, L316

J.L. Tandon, H.R. Bilger. $1/f$ Noise as a nonstationary process: Experimental evidence and some analytical conditions. *J. Appl. Phys.*, 1976, **47**, no. 4, 1697

A.-M.S. Tremblay, M. Nelkin. Equilibrium resistance fluctuations. *Phys. Rev. B*, 1981, **24**, no. 5, 2551

L.A. Vainstein. On the theory of flicker noise. *Zh. Eksp. Teor. Fiz.*, 1982, **83**, no. 5, 1841 [*Sov. Phys. – JETP*, 1982, **56**, no. 5, 1064]

S.N. Vainshtein, M.E. Levinshtein, S.L. Rumyantsev. Light-quenching of the $1/f$ noise in gallium arsenide. *Pis'ma Zh. Tekhn. Fiz.*, 1987, **13**, no. 11, 645 [*Sov. Phys. – Tech. Phys. Lett.*, 1987, **13**, 267]

L.K.J. Vandamme. $1/f$ noise in homogeneous single crystals of III-V compounds. *Phys. Lett.*, 1974, **A49**, no. 3, 233

A. van der Ziel. On the noise spectra of semiconductor noise and of flicker effect. *Physica*, 1950, **16**, no. 4, 359

P. Vande Voorde, C.K. Iddings, W.F. Love, D. Halford. Structure in the flicker-noise spectrum of n-InSb. *Phys. Rev. B*, 1979, **19**, no. 8, 4121

P. Vande Voorde, W.F. Love. Magnetic effects on $1/f$ noise in n-InSb. *Phys. Rev. B*, 1981a, **24**, no. 8, 4781

P. Vande Voorde, W.F. Love. Asymmetric current effects of $1/f$ noise in metal to n-InSb contacts with an applied magnetic field. *Appl. Phys. Lett.*, 1981b, **39**, no. 7, 575

K.M. van Vliet, A. van der Ziel, R.R. Schmidt. Temperature-fluctuation noise of thin films supported by a substrate. *J. Appl. Phys.*, 1980, **51**, no. 6, 2947

A.H. Verbruggen, R.H. Koch, C.P. Umbach. Correlation between $1/f$ noise and grain boundaries in thin gold films. *Phys. Rev. B*, 1987, **35**, no. 11, 5864

A.A. Verveen, H.E. Derksen. Fluctuation phenomena in nerve membrane. *Proc. IEEE*, 1968, **56**, no. 6, 906

R.F.C. Vessot. Lectures on frequency stability and clocks and on the gravitational red-shift experiment. In: *Experimental Gravitation, Proc. Int. School of Physics 'Enrico Fermi', Course 56*, Academic Press, New York, 1974, p. 111

S. Vitale, A. Cavalleri, M. Cerdonio, A. Maraner, G.A. Prodi. Thermal equilibrium noise with $1/f$ spectrum in a ferromagnetic alloy: anomalous temperature dependence. *J. Appl. Phys.*, 1994, **76**, no. 10, 6332

S. Vitale, R. Tommasini, M. Cerdonio, M. Bonaldi, A. Cavalleri, G. Durin. Magnetic viscosity, thermal relaxation, and thermal equilibrium noise in Co-based amorphous alloys at milliKelvin temperatures. *J. Appl. Phys.*, 1992, **72**, no. 10, 4820

R.F. Voss. Linearity of $1/f$ noise mechanisms. *Phys. Rev. Lett.*, 1978 a, **40**, no. 14, 913

R.F. Voss. $1/f$ Noise and percolation in impurity bands in inversion layers. *J. Phys. C: Solid State Phys.*, 1978b, **11**, L923

R.F. Voss, J. Clarke. Flicker ($1/f$) noise: Equilibrium temperature and resistance fluctuations. *Phys. Rev. B*, 1976, **13**, no. 2, 556

R.T. Wakai, D.J. Van Harlingen. Direct life-time measurements and interactions of charged defect states in submicron Josephson junctions. *Phys. Rev. Lett.*, 1987, **58**, no. 16, 1687

M.B. Weissman. Relations between geometrical factors for noise magnitudes in resistors, *J. Appl. Phys.*, 1980, **51**, no. 11, 5872

M.B. Weissman, R.D. Black, W.M. Snow. Calculations of experimental implications of tensor properties of resistance fluctuations. *J. Appl. Phys.*, 1982, **53**, no. 9, 6276

M.B. Weissman, N.E. Israeloff, G.B. Alers. Spin-glass fluctuation statistics: mesoscopic experiments in CuMn. *J. Magn. Magn. Mater.*, 1992, **114**, no. 1/2, 87

M. Weller, P. Moser. Internal friction of chromium after 20 K electron irradiation. *J. Phys. (Paris)*, 1981, **42**, Colloque C5, Suppl. no. 10, part 2, 741

J.L. Williams, I.L. Stone. Current noise in thin discontinuous films. *J. Phys. C: Solid State Phys.*, 1972, **5**, no. 16, 2105

S.V. Zaitsev-Zotov, V.Ya. Pokrovskii. Solitary two-level fluctuators in extremely small samples of quasi-one-dimensional TaS_3 conductor. *Pis'ma Zh. Eksp. Teor. Fiz.*, 1989, **49**, no. 8, 449 [*Sov. Phys. – JETP Lett.*, 1989, **49**, no. 8, 514]

S. Zhang, P.M. Levy, A. Fert. Conductivity and magnetoresistance of magnetic multilayered structures. *Phys. Rev. B*, 1991, **45**, no. 15, 8689

G.P. Zhigal'skii. Relationship between $1/f$ noise and nonlinearity effects in metal films. *Pis'ma Zh. Eksp. Teor. Fiz.*, 1991, **54**, no. 9, 510 [*Sov. Phys. – JETP Lett.*, 1991, **54**, no. 9, 513]

N.M. Zimmerman, J.H. Scofield, J.V. Mantese, W.W. Webb. Volume versus surface origin of $1/f$ noise in metals. *Phys. Rev. B*, 1986, **34**, no. 2, 773

N.M. Zimmerman, W.W. Webb. Microscopic scatterer displacements generate the $1/f$ resistance noise of H in Pd. *Phys. Rev. Lett.*, 1988, **61**, no. 7, 889

N.M. Zimmerman, W.W. Webb. $1/f$ Resistance noise complements anelasticity measurements of hydrogen motion in amorphous Pd – Si. *Phys. Rev. Lett.*, 1990, **65**, no. 8, 1040

Chapter 9

Recommended books and reviews

A.A. Abrikosov. *Fundamentals of the Theory of Metals*. North Holland, Amsterdam, 1988

A.A. Abrikosov, L.P. Gor'kov, I.E. Dzialoshinski. *Quantum Field Theoretical Methods in Statistical Physics*. Pergamon Press, Oxford, 1965

J. Bindslev Hansen. Noise in superconducting Josephson junctions. In: *Nonlinear Superconductive Electronics and Josephson Devices*, p. 161. Ed. G. Costabile *et al.* Plenum Press, New York, 1991

J.R. Clem. Flux-flow noise in superconductors, *Phys. Rep.*, 1981, **75**, no. 1, 1

M.J. Ferrari, M. Johnson, F.C. Wellstood, J.J. Kingston, T.J. Shaw, J. Clarke. Magnetic flux noise in copper oxide superconductors. *J. Low Temp. Phys.*, 1994, **94**, no. 1/2, 15

P.G. de Gennes. *Superconductivity of Metals and Alloys*. Addison-Wesley, Redwood City, California, 1989

L.B. Kiss, P. Svedlindh. Noise in high-T_c superconductors. *IEEE Trans. Electron. Devices*, 1994, **41**, no. 11, 2112

K.K. Likharev. Superconducting weak links, *Rev. Mod. Phys.*, 1979, **51**, no. 1, 101

K.K. Likharev. *Dynamics of Josephson Junctions and Circuits*. Gordon & Breach, New York, 1986

D. Rogovin, D.J. Scalapino. Fluctuation phenomena in tunnel junctions. *Ann. Phys. (NY)*, 1974, **86**, no. 1, 1

M. Tinkham. *Introduction to Superconductivity*. Krieger, New York, 1980

M. Tinkham, C.J. Lobb. Physical properties of the new superconductors. In: *Solid State Physics*, 1989, **42**, 91

References

A.A. Abrikosov. On magnetic properties of type II superconductors. *Zh. Eksp. Teor. Fiz.*, 1957, **32**, no. 6, 1442 [*Sov. Phys. – JETP*, 1957, **5**, no. 6, 1174]

V. Ambegaokar, A. Baratoff. Tunneling between superconductors. *Phys. Rev. Lett.*, 1963, **10**, no. 11, 486; Erratum, *ibid*, 1963, **11**, no. 2, 104

P.W. Anderson. Theory of flux creep in hard superconductors. *Phys. Rev. Lett.*, 1962, **9**, no. 7, 309

J.M. Aponte, A. Bellorin, R. Oentrich, J. van der Kuur, G. Guttierez, M. Octavio. Resistance noise in high-T_c and low-T_c granular superconducting films. *Phys. Rev. B*, 1993, **47**, no. 14, 8964

L.G. Aslamazov, A.I. Larkin. Josephson effect in superconducting point contacts. *Pis'ma Zh. Eksp. Teor. Fiz.*, 1968, **9**, no. 2, 150 [*Sov. Phys. – JETP Lett.*, 1968, **9**, no. 2, 87]

L.G. Aslamazov, A.I. Larkin. Superconducting contacts with a nonequilibrium electron distribution function. *Zh. Eksp. Teor. Fiz.*, 1976, **70**, no. 4, 1340 [*Sov. Phys. – JETP*, 1976, **43**, no. 4, 698]

K.R. Carroll, H.J. Paik. Noise in a point contact dc SQUID, *J. Low Temp. Phys.*, 1989, **75**, 187

M. Celasco, A. Masoero, P. Mazetti, A. Stepanescu. Evidence of current-noise hysteresis in superconducting $YBa_2Cu_3O_{7-\delta}$ specimens in a magnetic field. *Phys. Rev. B*, 1991, **44**, no. 10, 5366

A.J. Dahm, A. Denenstein, D.N. Langenberg, W.H. Parker, D. Rogovin, D.J. Scalapino. Linewidth of the radiation emitted by a Josephson junction. *Phys. Rev. Lett.*, 1969, **22**, no. 26, 1416

S.K. Decker, J.E. Mercereau. Noise measurements in superconducting proximity bridges. *Appl. Phys. Lett.*, 1975, **27**, no. 8, 466; *IEEE Trans. Magn.*, 1975, **Mag-11**, no. 2, 848

M.J. Ferrari, M. Johnson, F.C. Wellstood, J. Clarke, P.A. Rosenthal, R.H. Hammond, M.R. Beasley. Magnetic flux noise in thin-film rings of $YBa_2Cu_3O_{7-\delta}$. *Appl. Phys. Lett.*, 1988, **53**, no. 8, 695

M.J. Ferrari, M. Johnson, F.C. Wellstood, J. Clarke, P.A. Rosenthal, R.H. Hammond, M.R. Beasley. Flux noise and flux creep in YBCO thin films. *IEEE Trans. Magn.*, 1989, **25**, no. 2, 806

M.J. Ferrari, M. Johnson, F.C. Wellstood, J. Clarke, D. Mitzi, P.A. Rosenthal, C.B. Eom, T.H. Geballe, A. Kapitulnik, M.R. Beasley. Distribution of flux-pinning energies in $YBa_2Cu_3O_{7-\delta}$ and $Bi_2Sr_2CaCu_2O_{8+\delta}$ from flux noise. *Phys. Rev. Lett.*, 1990, **64**, no. 1, 72

M.J. Ferrari, F.C. Wellstood, J.J. Kingston, J. Clarke. Suppression of magnetic-flux noise in $YBa_2Cu_3O_{7-x}$ by a supercurrent. *Phys. Rev. Lett.*, 1991, **67**, no. 10, 1346

F. Habbal, W.C.H. Joiner. Correlation between changes in pinning strength and flux flow noise due to annealing in a type II superconductor. *Phys. Lett.*, 1977a, **60A**, no. 5, 434

F. Habbal, W.C.H. Joiner. Flux flow noise power spectra in the presence of local pinning interactions. *J. Low Temp. Phys.*, 1977b, **28**, no. 1/2, 83

R.E. Harris. Cosine and other terms in the Josephson tunneling current. *Phys. Rev. B*, 1974, **10**, no. 1, 84

R.E. Harris. Josephson tunneling current in the presence of a time-dependent voltage. *Phys. Rev. B*, 1975, **11**, no. 9, 3329

C. Heiden, D. Kohake. Local generation of flux flow noise by pinning due to grain boundaries. *Phys. Status Solidi B*, 1974, **64**, no. 2, K83

C. Heiden, D. Kohake, W. Krings, L. Ratke. Correlation studies on flux flow noise of polycrystalline niobium and vanadium foils. *J. Low Temp. Phys.*, 1977, **27**, no. 1/2, 1

P. Jarvis, J.G. Park. Noise voltages produced by flux motion in superconductors. *J. Phys. F: Metal Phys.*, 1975, **5**, no. 8, 1573

M. Johnson, M.J. Ferrari, F.C. Wellstood, J. Clarke, M.R. Beasley, A. Inam, X.D. Wu, L. Nazar, T. Venkatesan. Random telegraph signals in high-temperature superconductors. *Phys. Rev. B*, 1990, **42**, no. 16, 10792

B.D. Josephson. Possible new effects in superconducting tunneling. *Phys. Lett.*, 1962, **1**, no. 7, 251

L.B. Kiss, T. Larsson, P. Svedlindh, L. Lundgren, H. Ohlsen, M. Ottoson, J.

Hudner, L. Stolt. Conductance noise and percolation in $YBa_2Cu_3O_7$ thin films. *Physica C*, 1993, **207**, 318

L.B. Kiss, P. Svedlindh. New noise exponents in random conductor–superconductor and conductor–insulator mixtures. *Phys. Rev. Lett.*, 1993, **71**, no. 17, 2817

R.H. Koch, D.J. Van Harlingen, J. Clarke. Quantum-noise theory for the resistively shunted Josephson junction. *Phys. Rev. Lett.*, 1980, **45**, no. 26, 2132

R.H. Koch, D.J. Van Harlingen, J. Clarke. Measurements of quantum noise in resistively shunted Josephson junctions. *Phys. Rev. B*, 1982, **26**, no. 1, 74

Sh.M. Kogan, K.E. Nagaev. Fluctuations' kinetics in superconductors at frequencies low compared with the energy gap, *Zh. Eksp. Teor. Fiz.*, 1988, **94**, no. 3, 262 [*Sov. Phys. – JETP*, 1988, **67**, no. 3, 579]

Sh.M. Kogan, K.E. Nagaev. The kinetics of fluctuations and the spectral density of noise in resistive states of superconductors, *Proc. 10-th Int. Conf. on Noise in Physical Systems*, Ed. A. Ambrozy, Academiai Kiado, Budapest, 1990, p. 129

D. Kohake, C. Heiden. Local correlation between flux flow noise and isolated pinning structures. *J. Low Temp. Phys.*, 1980, **40**, no. 5/6, 531

A.I. Larkin, Yu.N. Ovchinnikov. Tunnel effect between superconductors in an alternating field. *Zh. Eksp. Teor. Fiz.*, 1966, **51**, no. 5, 1535 [*Sov. Phys. – JETP*, 1967, **24**, no. 5, 1035]

A.I. Larkin, Yu.N. Ovchinnikov. Radiation line width in the Josephson effect. *Zh. Eksp. Teor. Fiz.*, 1967, **53**, no. 6, 2159 [*Sov. Phys. – JETP*, 1968, **26**, no. 6, 1219]

A.I. Larkin, Yu.N. Ovchinnikov. Nonlinear conductivity of superconductors in the mixed state. *Zh. Eksp. Teor. Fiz.*, 1975, **68**, no. 5, 1915 [*Sov. Phys. – JETP*, 1975, **41**, no. 5, 960]

J.H. Lee, S.C. Lee, Z.G. Khim. Noise measurement near the transition region in $YBa_2Cu_3O_{7-\delta}$ thin film superconductors. *Phys. Rev. B*, 1989, **40**, no. 10, 6806

K.K. Likharev, V.K. Semenov. Fluctuation spectrum in superconducting point junctions. *Pis'ma Zh. Eksp. Teor. Fiz.*, 1972, **15**, no. 10, 625 [*Sov. Phys. – JETP Lett.*, 1972, **15**, no. 10, 442]

A. Maeda, Y. Nakayama, S. Takebayashi, K. Uchinokura. $1/f$ conduction noise in the high-temperature superconductor Bi-Sr-Ca-Cu-O system. *Physica C*, 1989, **160**, 443

A. Maeda, H. Watanabe, I. Tsukada, K. Uchinokura. The origin of the huge $1/f$ conduction noise in high-T_c cuprates. *Physica C*, 1991, **185-189**, 1301

D.E. McCumber. Effect of ac impedance on dc voltage–current characteristics of superconductor weak-link junctions. *J. Appl. Phys.*, 1968, **39**, no. 7, 3113

A. Misra, Y. Song, P.P. Crooker, J.R. Gaines, A.H. Cardona. $1/f$ noise in $Tl_2Ba_2CaCu_2O_8$. *Appl. Phys. Lett.*, 1991, **59**, no. 7, 863

S.S. Pei, J.E. Lukens. Observation of the intrinsic noise of a thin-film microbridge Josephson junction. *Appl. Phys. Lett.*, 1975, **26**, no. 8, 480

D.B. Schwartz, J.E. Lukens. Anomalies in the Josephson radiation linewidth. *Proc. LT-16, Physica B+C*, 1981, **108**, no. 1–3, part 2, 1297

D.B. Schwartz, P.M. Mankievich, A.K. Jain, J.E. Lukens. Microwave power spectra of variable thickness sub-micron bridges. *IEEE Trans. Magn.*, 1981, **Mag-17**, no. 1, 92

A.L. Shelankov. Dragging of normal component by the condensate in nonequilibrium superconductors. *Zh. Eksp. Teor. Fiz.*, 1980, **78**, no. 6, 2359 [*Sov. Phys. – JETP*, 1980, **56**, no. 6, 1186]

Y. Song, A. Misra, Y. Cao, A. Querubin, Jr., X.-D. Chen, P.P. Crooker, J.R. Gaines.

$1/f$ noise power measurements on $Tl_2Ba_2Ca_{n-1}Cu_nO_{4+2n}$ ($n = 2$ and 3). *Physica C*, 1990, **172**, no. 1, 1

Y. Song, A. Misra, P.P. Crooker, J.R. Gaines. $1/f$ Noise and morphology of $YBa_2Cu_3O_{7-\delta}$ single crystals. *Phys. Rev. Lett.*, 1991, **66**, no. 6, 825

W.C. Stewart. Current–voltage characteristics of Josephson junctions. *Appl. Phys. Lett.*, 1968, **12**, no. 8, 277

J.A. Testa, Yi Song, X.D. Chen, J. Golben, S.-I. Lee, B.R. Patton, J.R. Gaines. $1/f$-noise-power measurements of copper oxide superconductors in the normal and superconducting states. *Phys. Rev. B*, 1988, **38**, no. 4, 2922

J.D. Thompson, W.C.H. Joiner. Flux flow noise spectra of type II superconductors. *Solid State Commun.*, 1975, **16**, 849

J.D. Thompson, W.C.H. Joiner. Flux-flow noise in a $Pb_{0.8}In_{0.2}$ superconducting alloy. *Phys. Rev. B*, 1979, **20**, no. 1, 91

G.J. van Gurp. Flux-transport noise in type II superconductors. *Phys. Rev.*, 1968, **166**, no. 2, 436

D.J. van Ooijen, G.J. van Gurp. Measurement of noise in the resistive state of type II superconductors. *Phys. Lett.*, 1965, **17**, no. 3, 230

D.J. van Ooijen, G.J. van Gurp. Motion and pinning of flux in superconducting vanadium foils, studied by means of noise. *Philips Res. Rep.*, 1966, **21**, no. 5, 343

W.J. Yeh, Y.H. Kao. Measurements of flux-flow and $1/f$ noise in superconductors. *Phys. Rev. Lett.*, 1984, **53**, no. 16, 1590

W.J. Yeh, Y.H. Kao. Flux-flow noise in type II superconductors. *Phys. Rev. B*, 1991, **44**, no. 1, 360

A.B. Zorin. Fluctuations in finite capacitance Josephson tunnel junctions. *Fiz. Nizkikh Temp.*, 1981, **7**, no. 6, 709 [*Soviet J. Low Temp. Phys.*, 1981, **7**, no. 6, 346]

Subject index